Optical Fibre Technology: Advances and Applications

Optical Fibre Technology: Advances and Applications

Editor: Bob Tucker

MURPHY & MOORE
www.murphy-moorepublishing.com

Published by Murphy & Moore Publishing
1 Rockefeller Plaza,
New York City, NY 10020, USA
www.murphy-moorepublishing.com

Optical Fibre Technology: Advances and Applications
Edited by Bob Tucker

International Standard Book Number: 978-1-63987-412-5 (Hardback)

Cataloging-in-Publication Data

Optical fibre technology : advances and applications / edited by Bob Tucker.
p. cm.
Includes bibliographical references and index.
ISBN 978-1-63987-412-5
1. Optical fibers. 2. Fiber optics. 3. Optical measurements. 4. Optical communications. I. Tucker, Bob.
TK8306 .O68 2022
621.36--dc23

Contents

Preface

An optical fibre is a flexible, transparent fibre made of silica or plastic. Optical fibres are often used as a means to transmit light between the two ends of the fibre and find wide use in fibre-optic communications. They allow transmission over greater distances and at higher bandwidths than electrical cables. Fibres are used instead of metal wires because the signals travel through them with less loss, and they are immune to electromagnetic interference. Specially designed fibres are also used for many other applications, some of them being fibre optic sensors and fibre lasers. The ever growing need of advanced technology is the reason that has fueled the research in the field of optical fibre in recent times. This book strives to provide a fair idea about this discipline and to help develop a better understanding of the latest advances within this field. As optical fibre technology is evolving at a rapid pace, its contents will help the readers understand the modern concepts and applications of the subject.

The information contained in this book is the result of intensive hard work done by researchers in this field. All due efforts have been made to make this book serve as a complete guiding source for students and researchers. The topics in this book have been comprehensively explained to help readers understand the growing trends in the field.

I would like to thank the entire group of writers who made sincere efforts in this book and my family who supported me in my efforts of working on this book. I take this opportunity to thank all those who have been a guiding force throughout my life.

Editor

1

Fiber-Based Cylindrical Vector Beams and its Applications to Optical Manipulation

Renxian Li, Lixin Guo, Bing Wei, Chunying Ding and Zhensen Wu

1. Introduction

Radiation pressure force (RPF) indeced by a focused laser beam has bean widely utlized for the manipulation of small particles, and has found more and more applications in various fields including physics [1], biology [2], and optofludics [3, 4]. Accurate prediction of optical force exerted on particles enables better understanding of the physical mechanicsm, and is of great help for the design and improvement of optical tweezers.

Many researches have been devoted to the prediction of radiation pressure force (RPF), and different approaches have been developed for the theoretical calculation of RPF exerted on a homogeneous sphere. The geometrical optics [5-7] and Rayleigh theory [8] are respectively considered for the particles much larger and smaller than the wavelength of incident beam. Since geometrical optics and Rayleigh theory are both approximation theories, rigorous theories based on Maxwell's theory have been considered [9-13]. Generalized Lorenz-Mie Theory (GLMT) [14] has been used to investigate the RPFs exerted on some regular particles[10-13, 15, 16] induced by a Gaussian beam. GLMT can rigorously calculate RPF induced by any beam. To isolate the contribution of various scattering process to RPF, Debye series is introduced [17, 18].

Traditional optical tweezers use Gaussian beams as trapping light sources. This approach works well for the manipulation of microscopic spheres. However, the deveopment of science and technology brings new challenges to optical tweezers, and several approaches have been developed. Holographic methods have been used to increase the strength and dexterity of optical trap [19]. Another approch is the employment of non-Gaussian beam including Laguerre-Gaussian beams [20] and Bessel beams [28]. Laguerre-Gaussian beams have zero on-axis intensity, and can increase the strength of optical trap. Bessel beams consist of a series of concentric rings of decreasing intensity, and have characteristics of non-diffraction and self-reconstruction. A single Bessel beam can be used to simultaneously trap and manipulate, accelerate, rotate, or guide many particles. Bessel beams can trap and manipulate both high-index and low-index particles.

In addition to Laguerre-Gaussian and Bessel beams, there is a speical class of beams which have cylindrical symmetry in both amplitude and polarization, hence the name Cylindrical Vector Beams (CVBs) [29-32]. CVBs are solutions of vector wave equation in the paraxial limit. The special features of CVBs have attracted considerable interest for a variety of novel applications, including lithography, particle acceleration, material processing, high-resolution metrology, atom guiding, optical trapping and manipulation. The most interesting features for optical trapping arise from the focusing properties of CVBs. A radially polarized beam focused by a high numerical aperture objective has a peak at the focus, and can trap a high-index particle. On the contrary, an azimuthally polarized beam has null central intensity, and can trap low-index particle. These two kinds of beams can be experimentally switched.

CVBs can be generated by many methods, which are categorized as active or passive depending on whether amplifying media is used. The simplest mothod is to convert an incident Gaussian beam to a radially polarized beam using a radial polariser. However this method does not produce very high purity tansverse modes. Moer efficient methods use interferometry. Since a CVB can be expressed as the linear superpostion of two Hermite-Gaussian or Laguerre-Gaussian beams with different orientations of polarization. Another efficient method is based on optical fiber [33]. This technique takes advantage of the similarity between the poarization propeties of the modes that propagate inside a step-index optical fiber and CVBs. When TE_{01} or TM_{01} is excited in the fibre, it excites a CVB in free space. Fiber-generated CVB, taking Bessel-Gaussian as example, and its applicaions top optical manipulations will be discussed in this chapter.

2. Mathematical description of cylindrical vector beams

Cylindrical vector beams are solutions of vector wave equation

$$\nabla \times \nabla \times \vec{E} + k^2 \vec{E} = 0, \tag{1}$$

where $k=2\pi/\lambda$ is wavenumber with λ being the wavelength. In the paraxial approximation, the radially and azimuthally polarized vector Bessel-Gaussian beams, two kinds of typical CVBs, can be expressed as

$$\vec{E}_{rad} = E_0 \frac{\rho}{w_0} e^{-\frac{\rho^2}{w_0^2}} e^{i(\omega t - kz)} \hat{e}_\rho \tag{2}$$

$$\vec{E}_{azi} = E_0 \frac{\rho}{w_0} e^{-\frac{\rho^2}{w_0^2}} e^{i(\omega t - kz)} \hat{e}_\phi \tag{3}$$

where r and ϕ are respectively the radial and azimuthal coordinates, $\hat{e}_\rho$ and $\hat{e}_\phi$ are unit vectors in ρ and ϕ directions, and the subscripts *rad* and *azi* denote the polarization state. w_0 is the width of beam waist, and E_0 is a constant. Fig. 1(a) and (b) respectively give the intensity distribution of radially and azimuthally polarized Bessel-Gaussian beam in the plane *z=0*. Note that the longitudial component of CVB is negligible under the condition of paraxial approximation. A general CVB can be considered as a linear superposition of a radially polarized CVB and an azimuthally polarized one.

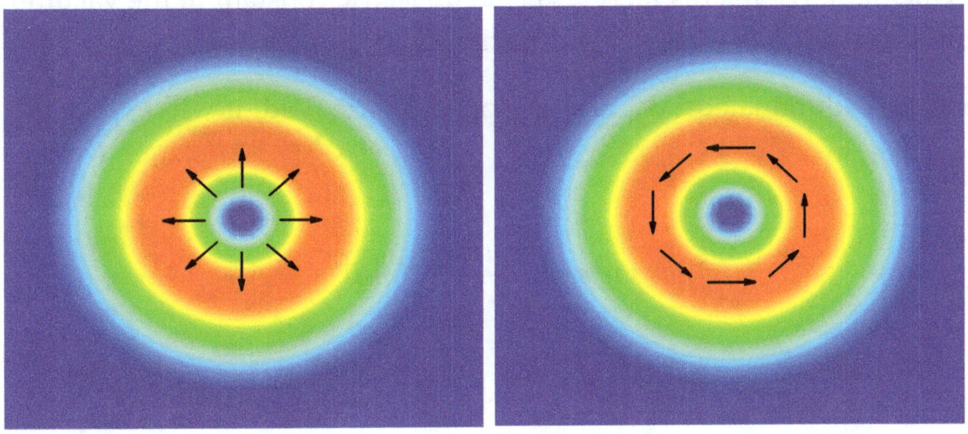

(a) Radially polarized CVB (b) Azimuthally polarized CVB

Figure 1. Intensity distribution of CVB. The arrows indicate the direction of polarization

3. Radiation force induced by CVB

3.1. Optical force on Rayleigh particles

In the Rayleigh regime, particles can be considered as infinitesimal induced dipoles which interact with incident beam. Here we assume that the particle is a microsphere. RPF will be decomposed into scattering force and gradient force.

The oscillating dipole, which is induced by time-harmonic fields, can be considered as an antenna. The antenna will radiate energy. The difference between energy removed from incident beam and energy radiated by the antenna accounts for the change of momentum flux, and hence rusults in a scattering force. The scattering force can be expressed as

$$\vec{F}_{scat} = \hat{e}_z C_{pr} n_1^2 \varepsilon_0 |E|^2 \tag{4}$$

with

$$C_{pr} = C_{scat} = \frac{8}{3}\pi(ka)^4 a^2 \left(\frac{m^2-1}{m^2+2}\right)^2 \tag{5}$$

and

$$m = n_2 / n_1 \tag{6}$$

where n_1 is the refractive index of surrounding media, and n_2 is the refractive index of the particle. a is the radius of microsphere. ε_0 is the dielectric constant in the vacumm. Note that the scattering force always points in the direction of incident beam.

When a particle is illuminated by a non-uniform electric field, it will experience a gradient force.

$$\vec{F}_{grad} = \pi n_1^2 \varepsilon_0 a^3 \left(\frac{m^2-1}{m^2+2}\right) \nabla |E|^2 \tag{7}$$

For a time-harmonic field, the gradient force can also be expressed in terms of the intensity I of incident beam:

$$\vec{F}_{grad} = \frac{2\pi n_1 a^3}{c} \left(\frac{m^2-1}{m^2+2}\right) \nabla I \tag{8}$$

where c is the speed of light in the vacumm. It is obvious that the gradient force depends on the gradient of the intensity. By sbustituting Eqs. (2) and (3) into Eqs. (4) and (7), we can obtain the scattering and gradient force of vector Bessel-Gaussian beams exerted on a microsphere. For radially polarized Bessel-Gaussian beam, they can be expressed as

$$\vec{F}_{scat} = \hat{e}_z C_{pr} n_1^2 \varepsilon_0 E_0^2 \frac{\rho^2}{w_0^2} \left(e^{-\frac{\rho^2}{w_0^2}}\right)^2 \tag{9}$$

$$\vec{F}_{grad} = \pi n_1^2 \varepsilon_0 a^3 \left(\frac{m^2-1}{m^2+2}\right) \left[\frac{2E_0^2 \rho}{w_0^2} \left(e^{-\frac{\rho^2}{w_0^2}}\right)^2 - \frac{4E_0^2 \rho^3}{w_0^4} \left(e^{-\frac{\rho^2}{w_0^2}}\right)^2\right] \hat{e}_\rho \tag{10}$$

From Eq. (10), we can find that the gradient force has only ϱ component. This is because $|E|^2$ is only dependent on ϱ. Here we give only the force for radially polarized beam incidence, and that for azimuthally polarized beam incidence can be derived in the same way.

3.2. Radiation force exerted on Mie particles

Many practical particles manipulated with optical tweezers, such as bioloical cells, are Mie particles, whose size is in the order of the wavelength of trapping beam. To calculate the radiation force exerted on such particles, a rigorous electromagnetic theory based on the Maxwell equations must be considered. Generalized Lorenz-Mie Thoery (GLMT) developed by Gouesbet et al. can solve the interaction between homogeneous spheres and focused beams with any shape, and has been entended to solve the scattering of shaped beam by multilayered spheres, homogeneous and multilayered cylinders, and homogeneous and multilayered spheroids. GLMT has been applied to the rigorous calculation of radiation pressure and optical torque. In GLMT, the incident beam is described by a set of beam shape coefficients(BSCs), which can be evaluated by integral localized approximation (ILA) [34].

This section is devoted to the GLMT for radiation force exerted on a sphere illuminated by a vetor Bessel-Gaussian beam. The general theory for radiation force based on electromagnetic scattering theory is followed by BSCs for CVB. To clarify the physical interpretation of various features of RPF that are implicit in the GLMT, Debye Series Expansion (DSE) is introduced.

3.2.1. Generization Lorenz-Mie theory

Consider a sphere with radius a and refractive index m_1 illuminated by a CVB of wavelength λ in the surrounding media. The center of the sphere is located at O_P, origin of the Cartesian coordinate system $O_{P\text{-}xyz}$. The beam center is at O_G, origin of coordinate system $O_{G\text{-}uvw}$, with u axis parallel to x and similarly for the others. The coordinates of O_G in the system $O_{P\text{-}xyz}$ is (x_0, y_0, z_0). The refractive index of surrounding media is m_2. The other parameters are defined in Fig. 2.

When a sphere is illuminated by focused beam, the RPF is proportional to the net momentum removed from the incident beam, and can be expressed in terms of the surface integration of Maxwell stress tensor

$$<\mathbf{F}>=<\oint_S \widehat{\mathbf{n}\cdot\vec{T}dS}> \tag{11}$$

where < > represents a time average, $\hat{n}$ the outward normal unit vector, and S a surface enclosing the particle. The Maxwell stress tensor $\overleftrightarrow{T}$ is given by

$$<\vec{T}>=\frac{1}{4\pi}\left(\varepsilon\mathbf{EE}+\mathbf{HH}-\frac{1}{2}\left(\varepsilon E^2+H^2\right)\vec{I}\right) \tag{12}$$

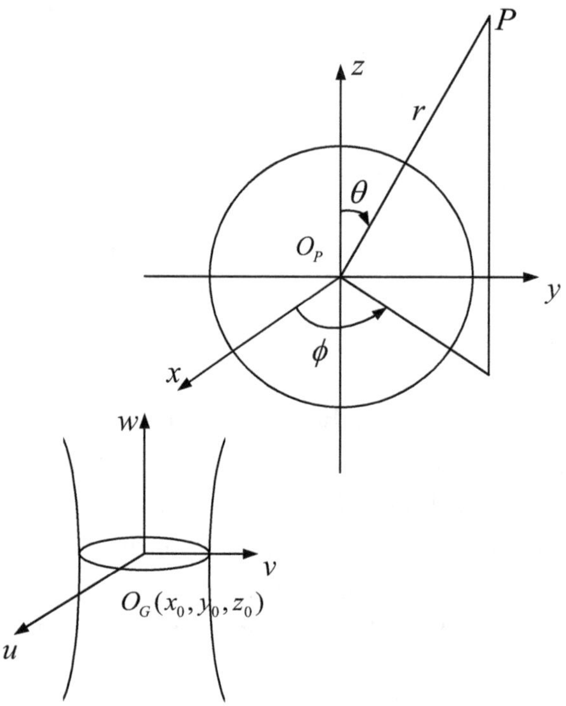

Figure 2. Coordinate systems in GLMT. $O_{P\text{-}xyz}$ is attached to the sphere, and $O_{G\text{-}uvw}$ to the incident beam.

Where the electromagnetic fields $\boldsymbol{E}$ and $\boldsymbol{H}$ are the total fields, namely the sum of the incident and scattered fields, given by

$$\mathbf{E} = \mathbf{E}_i + \mathbf{E}_s, \mathbf{H} = \mathbf{H}_i + \mathbf{H}_s \tag{13}$$

$\boldsymbol{E}_i$ and $\boldsymbol{H}_i$ are the incident electromagnetic wave, and can be expaned as:

$$E_r^i = \frac{E_0}{k_0^2 r^2} \sum_{n=1}^{\infty} \sum_{m=-n}^{+n} c_n^{pw} g_{n,TM}^m n(n+1) \psi_n(k_0 r) P_n^{|m|}(\cos\theta) e^{(im\varphi)} \tag{14}$$

$$E_\theta^i = \frac{E_0}{k_0 r} \sum_{n=1}^{\infty} \sum_{m=-n}^{+n} c_n^{pw} \left[g_{n,TM}^m \psi_n'(k_0 r) \tau_n^{|m|}(\cos\theta) + m g_{n,TE}^m \psi_n(k_0 r) \pi_n^{|m|}(\cos\theta) \right] e^{(im\varphi)} \tag{15}$$

$$E_\varphi^i = i \frac{E_0}{k_0 r} \sum_{n=1}^{\infty} \sum_{m=-n}^{+n} c_n^{pw} \left[m g_{n,TM}^m \psi_n'(k_0 r) \pi_n^{|m|}(\cos\theta) + g_{n,TE}^m \psi_n(k_0 r) \tau_n^{|m|}(\cos\theta) \right] e^{(im\varphi)} \tag{16}$$

$$H_r^i = \frac{H_0}{k_0^2 r^2} \sum_{n=1}^{\infty} \sum_{m=-n}^{+n} c_n^{pw} g_{n,TE}^m n(n+1) \psi_n(k_0 r) P_n^{|m|}(\cos\theta) e^{(im\varphi)} \tag{17}$$

$$H_\theta^i = -\frac{H_0}{k_0 r}\sum_{n=1}^{\infty}\sum_{m=-n}^{+n} c_n^{pw}\left[m g_{n,TM}^m \psi_n(k_0 r)\pi_n^{|m|}(\cos\theta) - g_{n,TE}^m \psi_n'(k_0 r)\tau_n^{|m|}(\cos\theta)\right] e^{(im\varphi)} \tag{18}$$

$$H_\varphi^i = -\frac{iH_0}{k_0 r}\sum_{n=1}^{\infty}\sum_{m=-n}^{+n} c_n^{pw}\left[g_{n,TM}^m \psi_n(k_0 r)\tau_n^{|m|}(\cos\theta) - m g_{n,TE}^m \psi_n'(k_0 r)\pi_n^{|m|}(\cos\theta)\right] e^{(im\varphi)} \tag{19}$$

with

$$\pi_n^m(\cos\theta) = \frac{dP_n^m(\cos\theta)}{d\theta} \tag{20}$$

$$\tau_n^m(\cos\theta) = m\frac{P_n^m(\cos\theta)}{\sin\theta} \tag{21}$$

$$c_n^{pw} = (-i)^{n+1}\frac{2n+1}{n(n+1)} \tag{22}$$

where $P_n^m(\cos\theta)$ represents the associated Legendre polynomials of degree n and order m, $\psi(\cdot)$ is the spherical Ricatti-Bessel functions of first kind, and the prime indicates the derivative of the function with respect to its argument. $g_{n,TM}^m$ and $g_{n,TE}^m$ are so-called BSCs and will be discussed in next subsection.

Similarly, the scattered fields $\boldsymbol{E}_s$ and $\boldsymbol{H}_s$ have the expression :

$$E_r^s = -\frac{E_0}{k_0^2 r^2}\sum_{n=1}^{\infty}\sum_{m=-n}^{+n} c_n^{pw} A_n^m n(n+1)\xi_n(k_0 r)P_n^{|m|}(\cos\theta)e^{(im\varphi)} \tag{23}$$

$$E_\theta^s = -\frac{E_0}{k_0 r}\sum_{n=1}^{\infty}\sum_{m=-n}^{+n} c_n^{pw}\left[A_n^m \xi_n'(k_0 r)\tau_n^{|m|}(\cos\theta) + m B_n^m \xi_n(k_0 r)\pi_n^{|m|}(\cos\theta)\right] e^{(im\varphi)} \tag{24}$$

$$E_\varphi^s = -i\frac{E_0}{k_0 r}\sum_{n=1}^{\infty}\sum_{m=-n}^{+n} c_n^{pw}\left[m A_n^m \xi_n'(k_0 r)\pi_n^{|m|}(\cos\theta) + B_n^m \xi_n(k_0 r)\tau_n^{|m|}(\cos\theta)\right] e^{(im\varphi)} \tag{25}$$

$$H_r^s = -\frac{H_0}{k_0^2 r^2}\sum_{n=1}^{\infty}\sum_{m=-n}^{+n} c_n^{pw} B_n^m n(n+1)\xi_n(k_0 r)P_n^{|m|}(\cos\theta)e^{(im\varphi)} \tag{26}$$

$$H_\theta^s = \frac{H_0}{k_0 r}\sum_{n=1}^{\infty}\sum_{m=-n}^{+n} c_n^{pw}\left[m A_n^m \xi_n(k_0 r)\pi_n^{|m|}(\cos\theta) - B_n^m \xi_n'(k_0 r)\tau_n^{|m|}(\cos\theta)\right] e^{(im\varphi)} \tag{27}$$

$$H_{\varphi}^{s}=\frac{iH_0}{k_0 r}\sum_{n=1}^{\infty}\sum_{m=-n}^{+n}c_n^{pw}\left[A_n^m\xi_n(k_0r)\tau_n^{|m|}(\cos\theta)-mB_n^m\xi_{nn}'(k_0r)\pi_n^{|m|}(\cos\theta)\right]e^{(im\varphi)} \tag{28}$$

Where $\xi_n(k_0r)$ is Ricatti-Hankel functions, and scattering coefficients A_n^m and B_n^m can be expressed by traditional Mie scattering coefficients a_n, b_n and BSCs $g_{n,TM}^m$, $g_{n,TE}^m$:

$$A_n^m=a_n g_{n,TM}^m,\qquad B_n^m=b_n g_{n,TE}^m \tag{29}$$

with

$$a_n=\frac{-m_1\psi_n'(x)\psi_n(y)+m_2\psi(x)\psi_n'(y)}{-m_1\xi_n^{(1)'}(x)\psi_n(y)+m_2\xi_n^{(1)}(x)\psi_n'(y)} \tag{30}$$

$$b_n=\frac{-m_2\psi_n'(x)\psi_n(y)+m_1\psi(x)\psi_n'(y)}{-m_2\xi_n^{(1)'}(x)\psi_n(y)+m_1\xi_n^{(1)}(x)\psi_n'(y)} \tag{31}$$

$$x=m_2k_0a,\qquad y=m_1k_0a \tag{32}$$

Substituting Eqs. (14) - (19) and (23) - (28) into Eqs. (11) - (12), and after some algebra, we can get the formula for RPFs which can be characterized by radiation pressure cross section (RPCS):

$$\mathbf{F}(\mathbf{r})=\frac{2_2I_0}{c}\left[\hat{\mathbf{e}}_{\mathbf{x}}C_{pr,x}(\mathbf{r})+\hat{\mathbf{e}}_{\mathbf{y}}C_{pr,y}(\mathbf{r})+\hat{\mathbf{e}}_{\mathbf{z}}C_{pr,z}(\mathbf{r})\right] \tag{33}$$

where RPCS $C_{pr,i}$ $(i=x,\ y,\ z)$ has a longitudinal cross section $C_{pr,z}$

$$\begin{aligned}C_{pr,z}=&\frac{\lambda^2}{\pi}\sum_{n=1}^{\infty}Re\left\{\frac{1}{n+1}(A_ng_{n,TM}^{0}g_{n+1,TM}^{0*}+B_ng_{n,TE}^{0}g_{n+1,TE}^{0*})+\sum_{m=1}^{n}\left[\frac{1}{(n+1)^2}\frac{(n+m+1)!}{(n-m)!}\right.\right.\\&\times(A_ng_{n,TM}^{m}g_{n+1,TM}^{m*}+A_ng_{n,TM}^{-m}g_{n+1,TM}^{-m*}+B_ng_{n,TE}^{m}g_{n+1,TE}^{m*}+B_ng_{n,TE}^{-m}g_{n+1,TE}^{-m*})\\&\left.\left.+m\frac{2n+1}{n^2(n+1)^2}\frac{(n+m)!}{(n-m)!}C_n(g_{n,TM}^{m}g_{n,TE}^{m*}-g_{n,TM}^{-m}g_{n,TE}^{-m*})\right]\right\}\end{aligned} \tag{34}$$

and two transverse cross section $C_{pr,x}$ and $C_{pr,y}$

$$C_{pr,x}=\mathrm{Re}(C)\qquad C_{pr,y}=\mathrm{Im}(C) \tag{35}$$

where

$$C = \frac{\lambda^2}{2\pi}\sum_{n=1}^{\infty}\left\{-\frac{(2n+2)!}{(n+1)^2}F_n^{n+1} + \sum_{m=1}^{n}\frac{(n+m)!}{(n-m)!}\frac{1}{(n+1)^2}\right.$$
$$\times\left[F_n^{m+1} - \frac{n+m+1}{n-m+1}F_n^m + \frac{2n+1}{n^2}(C_n g_{n,TM}^{m-1} g_{n,TE}^{m*}\right. \tag{36}$$
$$\left.\left. - C_n g_{n,TM}^{-m} g_{n+1,TE}^{-m+1*} + C_n^* g_{n,TE}^{m-1} g_{n,TM}^{m*} - C_n^* g_{n,TE}^{-m} g_{n,TM}^{-m+1*})\right]\right\}$$

with

$$F_n^m = A_n g_{n,TM}^{m-1} g_{n+1,TM}^{m*} + B_n g_{n,TE}^{m-1} g_{n+1,TE}^{m*} + A_n^{m*} g_{n+1,TM}^{-m} g_{n,TM}^{-m+1*} + B_n^{m*} g_{n+1,TE}^{-m} g_{n,TE}^{-m+1*} \tag{37}$$

$$A_n = a_n + a_{n+1}^* - 2a_n a_{n+1}^* \tag{38}$$

$$B_n = b_n + b_{n+1}^* - 2b_n b_{n+1}^* \tag{39}$$

$$C_n = -i(a_n + b_{n+1}^* - 2a_n b_{n+1}^*) \tag{40}$$

Note that substituting Eq. (13) into Eqs. (11) - (12) shows that the total RPF can be devided into thress parts:

$$<\mathbf{F}> = <\mathbf{F}_i> + <\mathbf{F}_{mix}> + <\mathbf{F}_s> \tag{41}$$

where $<F_i>$ depends only on the incident fields, $<F_s>$ is associated with the scattered fields, and $<F_{mix}>$ involves the interactions of the incident beam with the scattered field. After a great deal of algebra, we can get that $<F_i>=0$, which can be understood by the momentum conservation law for monochromatic fields in free space. The RPCS for $<F_{mix}>$ and $<F_s>$ can be directly given using Eqs. (34) - (40) by changing Eqs. (38) - (40) using

$$\begin{aligned} A_n &= a_n + a_{n+1}^* \\ B_n &= b_n + b_{n+1}^* \\ C_n &= -i(a_n + b_{n+1}^*) \end{aligned} \tag{42}$$

for $<F_{mix}>$, and

$$\begin{aligned} A_n &= -2a_n a_{n+1}^* \\ B_n &= -2b_n b_{n+1}^* \\ C_n &= 2i a_n b_{n+1}^* \end{aligned} \tag{43}$$

for $<F_s>$.

3.2.2. Beam shape coefficients for CVB

This section is devoted the derivation of BSCs for CVB using ILA. In the ILA, the beam shape coefficients $g_{n,TM}^{m}$ and $g_{n,TE}^{m}$ are obtained respectively from the radial component of electric and magnetic field E_r and H_r according to [34]

$$g_{n,TE}^{m} = \frac{Z_n^m}{2\pi H_{B0}} \int_0^{2\pi} \overline{H_r}(r,\theta,\phi) e^{-im\phi} d\phi \tag{44}$$

$$g_{n,TM}^{m} = \frac{Z_n^m}{2\pi E_{B0}} \int_0^{2\pi} \overline{E_r}(r,\theta,\phi) e^{-im\phi} d\phi \tag{45}$$

with

$$Z_n^m = \begin{cases} \dfrac{2n(n+1)i}{2n+1} & m = 0 \\ \left(\dfrac{-2i}{2n+1}\right)^{|m|-1} & m \neq 0 \end{cases} \tag{46}$$

$\overline{E_r}$ and $\overline{H_r}$ are respectively the localized fields of E_r and H_r, and they are obtained by changing kr to $(n+1/2)$ and θ to $\pi/2$ in their expression. For a radially polarized Bessel-Gaussian beam, the localized radial component of electric field are derived from Eq. (2):

$$\begin{aligned} \bar{E}_{rad}^{r} &= E_0\bar{\Omega}_0\left[-(\rho_n\cos\phi - \xi_0)\cos\phi + (\rho_n\sin\phi - \eta_0)\sin\phi\right] \\ &= E_0\bar{\Omega}_0\left[\rho_n - \rho_0\sin(\phi+\phi_0)\right] \end{aligned} \tag{47}$$

with

$$\bar{\Omega}_0 = -\sqrt{2}\exp\left[-(\rho_n^2 + \xi_0^2 + \eta_0^2)\right]\exp\left[2\rho_n(\xi_0\cos\phi + \eta_0\sin\phi)\right] \tag{48}$$

$$\rho_n = \frac{kr}{kw_0} = \frac{1}{kw_0}\left(n + \frac{1}{2}\right) \tag{49}$$

$$\xi_0 = \rho_0\sin\phi_0, \quad \eta_0 = \rho_0\cos\phi_0 \tag{50}$$

Substituting Eq. (47) into Eq. (45), and considering the formula of Bessel function

$$J_n(x)=\frac{1}{2\pi}\int_{-\pi}^{\pi}e^{i(x\sin\theta-n\theta)}d\theta=\frac{1}{2\pi}\int_{0}^{2\pi}e^{i(x\sin\theta-n\theta)}d\theta \tag{51}$$

we can obtain the final expression of BSCs

$$g_{n,TM}^{m,rad}=\frac{1}{2}Z_n^m\overline{\Omega}_n e^{im\phi_0}\left[2\rho_n J_m(-2i\rho_n\rho_0)+i\rho_0(J_{m-1}(-2i\rho_n\rho_0)-J_{m+1}(-2i\rho_n\rho_0))\right] \tag{52}$$

Here we only derive $g_{n,TM}^{m,rad}$, and $g_{n,TE}^{m,rad}$ can be derived in the similar way.

3.2.3. Debye series expansion

GLMT is a rigorous electromagnetic theory, and can exactly predict the RPF exerted on a sphere by focused beam. Whereas the solution is complicated combinations of Bessel functions, and the mathematical complexity obscures the physical interpretation of various features of RPF. The DSE, which is a rigorous electromagnetic theory, expresses the Mie scattering coefficients into a series of Fresnel coefficients and gives physical interpretation of different scattering processes. The DSE is an efficient technique to make explicit the physical interpretation of various features of RPF which are implicit in the GLMT. The DSE is firstly presented by Debye in 1908 for the interaction between electromagnetic waves and cylinders. Since then, the DSE for electromagnetic scattering by homogeneous, coated, multilayered spheres, multilayered cylinders at normal incidence, homogeneous, multilayered cylinder at oblique incidence, and spherical gratings are studied. In our previous work, DSE has been employed to the analysis of RPF exerted on a sphere induced by a Gaussian and Bessel beam.

As shown in Fig. 3, when an incoming spherical multipole wave, which is

$$\Psi=\xi_n^{(1)}(m_2kr)P_n^m(\cos\theta)\begin{Bmatrix}\cos m\phi\\ \sin m\phi\end{Bmatrix}, \tag{53}$$

encounters the interface of the sphere at *r*=*a*, portion of it will be transmitted into the sphere, and another portion will reflected back. The transmitted and reflected waves are respectively:

$$\Psi_1=T_n^{21}\xi_n^{(1)}(m_1kr)P_n^m(\cos\theta)\begin{Bmatrix}\cos m\phi\\ \sin m\phi\end{Bmatrix} \qquad r\le a \tag{54}$$

$$\Psi_2=\left[\xi_n^{(1)}(m_2kr)+R_n^{212}\xi_n^{(2)}(m_2kr)\right]P_n^m(\cos\theta)\begin{Bmatrix}\cos m\phi\\ \sin m\phi\end{Bmatrix} \qquad r\ge a \tag{55}$$

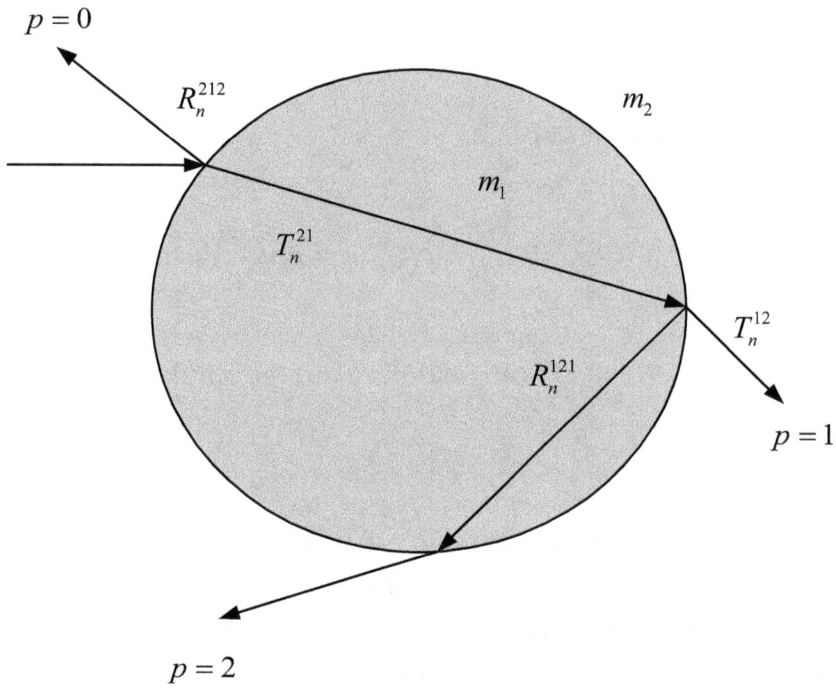

Figure 3. Debye model of light scattering by a sphere

Applying the boundary conditions, which reqires continuity of the tangential components of filds at the interface, to the incident, transmitted and reflected waves,we can obtain:

$$T_n^{21} = \frac{m_1}{m_2} \frac{2i}{D_n} \tag{56}$$

$$R_n^{212} = \frac{\alpha \xi_n^{(1)'}(\kappa_2)\xi_n^{(1)}(\kappa_1) - \beta \xi_n^{(1)}(\kappa_2)\xi_n^{(1)'}(\kappa_1)}{D_n} \tag{57}$$

with

$$D_n = -\alpha \xi_n^{(2)'}(\kappa_2)\xi_n^{(1)}(\kappa_1) + \beta \xi_n^{(2)}(\kappa_2)\xi_n^{(1)'}(\kappa_1) \tag{58}$$

$$\kappa_j = m_j ka \tag{59}$$

$$\alpha = \begin{cases} 1, & for\ TE \\ \dfrac{m_1}{m_2}, & for\ TM \end{cases}, \quad \beta = \begin{cases} \dfrac{m_1}{m_2}, & for\ TE \\ 1, & for\ TM \end{cases} \tag{60}$$

Similarly, the consideration of outgoing multipole waves can get

$$T_n^{12} = \frac{2i}{D_n} \tag{61}$$

$$R_n^{121} = \frac{\alpha \xi_n^{(2)'}(\kappa_2)\xi_n^{(2)}(\kappa_1) - \beta \xi_n^{(2)}(\kappa_2)\xi_n^{(2)'}(\kappa_1)}{D_n} \tag{62}$$

Substituting all Fresnel coefficients into

$$\left(1 - R_n^{121}\right)\left(1 - R_n^{212}\right) - T_n^{21} T_n^{12} \tag{63}$$

and after much algebra, we get

$$\left.\begin{matrix} a_n \\ b_n \end{matrix}\right\} = \frac{1}{2}\left[1 - R_n^{212} - \frac{T_n^{21} T_n^{12}}{1 - R_n^{121}}\right] = \frac{1}{2}\left[1 - R_n^{212} - \sum_{p=1}^{\infty} T_n^{21}\left(R_n^{121}\right)^{p-1} T_n^{12}\right] \tag{64}$$

where the prime indicates the derivative of the function with respect to its argument. $\xi_n^{(1)}(\cdot)$ and $\xi_n^{(2)}(\cdot)$ are respectively the spherical Ricatti-Hankel functions of first and second kinds. The definition of all Fresnel coefficients and Debye term p are given in Fig. 3. For convenience, we note p = - 1 and p = 0 respectively for the diffraction and direct reflection. In our previous work, we have theoretically and numerically proved that when p ranges from 1 to , the Eq. (64) is identical to the traditional Mie scattering coefficients. Here we provide the DSE for homogeneous spheres, and the DSE for multilayered spheres can be found in our previous work.

4. Numerical results and discussions

In this section, the GLMT and DSE will be employed to analyze the RPF exerted on a homogeneous sphere induced by a radially polarized vector Bessel-Gaussian beam. Xu et al. used GLMT to analyze the RPF exerted on a slightly volatile silocone oil of refractive index , which can be levitated in the air by a beam of wavelength . We first use GLMT to analyze the RPF exerted on such oil induced by vector Bessel-Gaussian beam, and DSE will be employed to the study of the contribution of various scattering process to RPF. In our calculation, the radius of the particle is .

We first explore the influence of beam center location on the RPF. In our calculation, we assume the beam center is located on the x axis so that . Fig. 4 gives the transverse RPCS ∞ versus m_1=1.5 for various beam-waist radius λ =0.5μm. Here we consider a=2.5μm and y_0=z_0=0, which are respectively larger, equal and smaller than the radius of the particle. We can find that the particle can not be trapped at the beam center $C_{pr,x}$ for all beams. This results from the fact all beams have null central intensity. It is worth pointing out that a stable trap corresponds to a particle position where the RPF is zero and its slope is positive. All curves have two equilibrium points, which are symmetric with respect to beam axis (x_0). This is decided by the intensity maxima of beams. So a vector Bessel-Gaussian beam can simultaneously trap more than one particles. We can also find that the interval of equilibrium points increases with the increasing of w_0. This can be easily explained from the fact that the interval of intensity peaks increases with the increasing of w_0=5μm, 2.5μm.

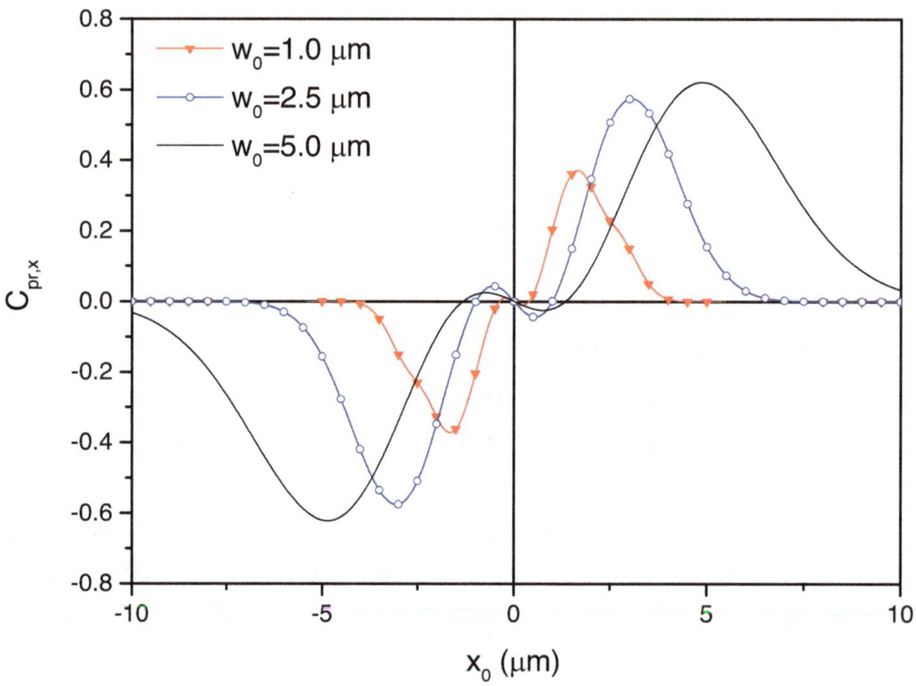

Figure 4. Transverse cross-section 1μm versus x_0=0 with parameter x_0=0. w_0, w_0, $C_{pr,x}$, x_0, w_0 and λ =0.5μm

To clarify the physical explanation of some features of RPCS, it is necessary for us to consider the contribution of each mode p to RPCS. The contribution of each p mode to RPCS can be computed separately by considering a single term in Eq. (64). Now we consider the contribution of a single p mode to transverse RPCS m_1=1.5. Here we set beam-waist radius m_2=1.

It is shown in Fig. 5 the transverse RPCS y_0=0 versus a=2.5μm with parameter p_{max}=∞. In the calculation, the beam-waist radius is assumed $C_{pr,x}$. Comparison of Fig. 5 with Fig. 4 shows that when w_0=5μm the results obtained by DSE are identical to those by GLMT. In fact, when

$C_{pr,x}$ is large enough, the difference between two theories should be negilible. For example, if x_0, the results of DSE is very close to GLMT results. Special attention should be paid to the case of p_{max}. Fig. 5 shows that when w_0=5μm, the agreement between the results of GLMT and DSE is already good. This concludes that main contribution of RPF comes from the scattering processes of diffraction ($p_{max} \to \infty$), specular reflection (p_{max}) and direct transmission (p_{max}=100).

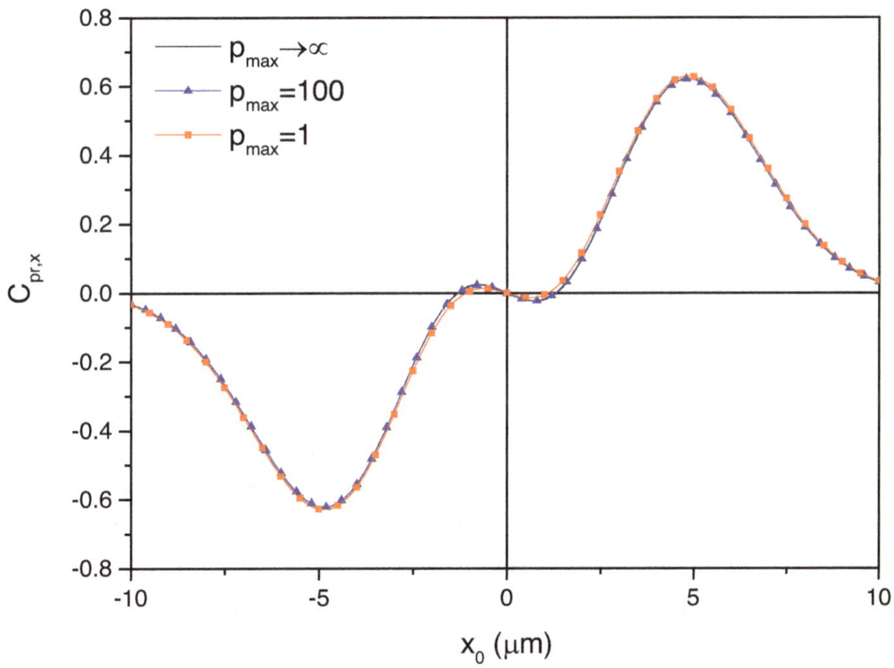

Figure 5. Transverse cross-section p_{max}=1 versus p_{max}=1 with parameter $p=-1$. p=0, p=1, $C_{pr,x}$, x_0, p_{max} and λ=0.5μm.

To clarify the physical explanation of some phenomena of RPF, it is necessary to consider the contribution of each mode p to RPCS, which can be computed separately by considering a single term in Eq. (64). Now we consider the contribution of a single p mode to transverse RPCS m_1=1.5. Fig.6 depicts the transverse RPCS m_2=1 versus y_0=0 for a=2.5μm and w_0=5μm. Generally, the RPF at $C_{pr,x}$ is zero for any mode p because of the symmetry. The magnitude of $C_{pr,x}$ for x_0 is much greater than that for p=1. This validates that the transverse RPCS p=2 is dominated by the contributions of direct transmission (x_0=0). Note that the curve for $C_{pr,x}$ has two equilibrium points at about p=1, while the curve for p=2 has only one points at $C_{pr,x}$. Near the beam axis (p=1), the curvesfor p=2 has positive slope, while that for x_0= ±1.3μm has negative one. To explain such phenomena, we must consider the integral effect of all intensity peaks.

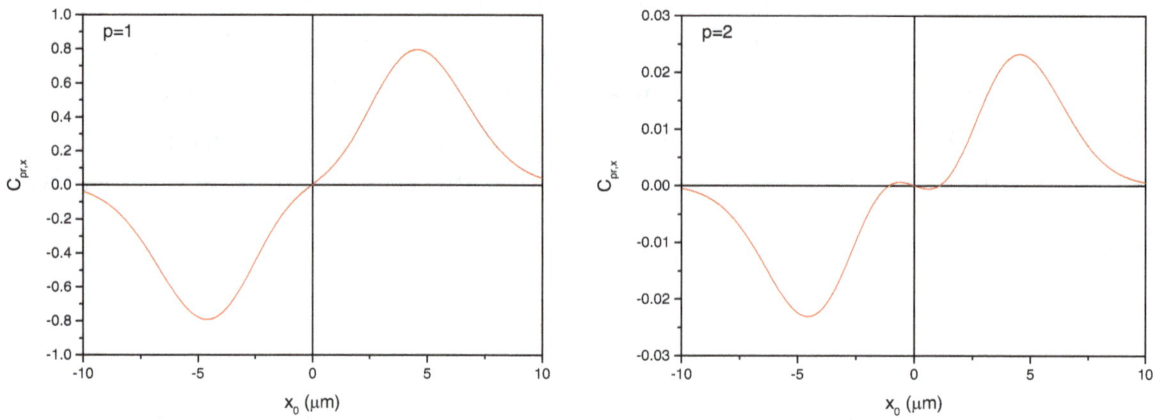

Figure 6. Transverse cross-section $p=1$ versus $x_0=0$ corresponding to single mode *p*. $x_0=0$, $p=1$, $p=2$, $C_{pr,x}$ x_0 and $\lambda=0.5\mu m$.

5. Conclusions

Rigorous theories including GLMT and DSE for RPF exerted on spheres induced by CVB is derived. The incident beam is described by a set of BSCs which is calculated by integral localized approximation, and the scattering coefficients are expanded using Debye series. For very small particles, namely Rayleigh particles, an approximation model is also given. Such thoery can be easily extended to the RPF exerted on multilayered sphere, and also to the RPF induced by other beams. Debye series is used to isolate the contribution of various scattering process to the RPF. The results are of special importance for the improvement of optical tweezers system.

Acknowledgements

The authors acknowledge support from the Natural Science Foundation of China (Grant No. 61101068), the National Science Foundation for Distinguished Young Scholars of China (Grant No.61225002), and the Fundamental Research Funds for the Central Universities.

Author details

Renxian Li*, Lixin Guo, Bing Wei, Chunying Ding and Zhensen Wu

*Address all correspondence to: rxli@mail.xidian.edu.cn

School of Physics and Optoelectronic Engineering, Xidian University, China

References

[1] Dholakia,, Dholakia, K.K., Reece, P.. Optical micromanipulation takes hold. Nano Today 2006;1(1):18–27.

[2] Neuman,, K.C.,, Block, S.. Optical trapping. Rev Sci Instrum 2004;75:2787.

[3] Enger, J., Goksor, M., Ramser, K., Hagberg, P., Hanstorp, D.. Optical tweezers applied to a microfluidic system. Lab Chip 2004;4(3):196–200.

[4] Domachuk, P., Cronin-Golomb, M., Eggleton, B., Mutzenich, S., Rosengarten, G., Mitchell, A.. Application of optical trapping to beam manipulation in optofluidics. Opt Express 2005;13(19):7265–7275.

[5] Ashkin, A.. Forces of a single-beam gradient laser trap on a dielectric sphere in the ray optics regime. Biophys J 1992;61(2):569–582.

[6] Roosen, G.. A theoretical and experimental study of the stable equilibrium positions of spheres levitated by two horizontal laser beams. Opt Commun 1977;21(1):189–194.

[7] Yao, X., Li, Z., Cheng, B., Zhang, D.. Analysis and calculation of the optical force on a double-layer dielectric sphere. Acta Optica Si 2000;10:13051310.

[8] Chaumet, P.C., Rahmani, A., Nieto-Vesperinas, M.. Optical trapping and manipulation of nano-objects with an apertureless probe. Phys Rev Lett 2002;88(12):123601.

[9] Barton, J. P., Alexander, D.R., Schaub, S.A.. Theoretical determination of net radiation force and torque for a spherical particle illuminated by a focused laser beam. J Appl Phys 1989;66(10):4594–4602.

[10] Ren, K.F., Gréhan, G., Gouesbet, G.. Prediction of reverse radiation pressure by generalized Lorenz-Mie theory. Appl Opt 1996;35(15):2702–2710.

[11] Ren, K., Gréhan, G., Gouesbet, G.. Radiation pressure forces exerted on a particle arbitrarily located in a Gaussian beam by using the generalized Lorenz-Mie theory, and associated resonance effects. Opt Commun 1994;108(4-6):343–354.

[12] Lock, J.A.. Calculation of the radiation trapping force for laser tweezers by use of generalized Lorenz-Mie theory. i. localized model description of an on-axis tightly focused laser beam with spherical aberration. Appl Opt 2004;43(12):2532–2544.

[13] Lock, J.A.. Calculation of the radiation trapping force for laser tweezers by use of generalized Lorenz-Mie theory. ii. on axis trapping force. Appl Opt 2004;43(12):2545–2554.

[14] Gouesbet, G., Maheu, B., Gréhan, G.. Light scattering from a sphere arbitrarily located in a Gaussian beam, using a bromwich formulation. J Opt Soc Am A 1988;5(9): 1427–1443.

[15] Polaert, H., Gréhan, G., Gouesbet, G.. Forces and torques exerted on a multilayered spherical particle by a focused Gaussian beam. Opt Commun 1998;155(1-3):169–179.

[16] Xu, F., Ren, K.F., Gouesbet, G., Cai, X., Gréhan, G.. Theoretical prediction of radiation pressure force exerted on a spheroid by an arbitrarily shaped beam. Phys Rev E 2007;75(2):026613.

[17] Li, R., Han, X., Shi, L., Ren, K., Jiang, H.. Debye series for Gaussian beam scattering by a multilayered sphere. Appl Opt 2007;46(21):4804–4812.

[18] Li, R., Han, X., Ren, K.F.. Debye series analysis of radiation pressure force exerted on a multilayered sphere. Appl Opt 2010;49(6):955–963.

[19] Bo Sun, Yohai Roichman, and David G. Grier, Theory of holographic optical trapping, Optics Express 2008, 16: 15765-15776

[20] Anna T. O'Neil, Miles J. Padgetta, Axial and lateral trapping efficiency of Laguerre–Gaussian modes in inverted optical tweezers, Optics Communications 2001, 193: 45–50

[21] Garces-Chavez, V., McGloin, D., Melville, H., Sibbett, W., Dholakia, K., et al. Simultaneous micromanipulation in multiple planes using a self-reconstructing light beam. Nature 2002;419(6903):145–147.

[22] Arlt, J., Garces-Chavez, V., Sibbett, W., Dholakia, K.. Optical micromanipulation using a Bessel light beam. Opt Commun 2001;197(4-6):239–245.

[23] Garc´es-Ch´avez, V., Roskey, D., Summers, M., Melville, H., McGloin, D., Wright, E., Dholakia, K.. Optical levitation in a Bessel light beam. Appl Phys Lett 2004;85:4001.

[24] Cizm´ar, T.. Optical traps generated by non-traditional beams. Ph. D. thesis, Masaryk University in Brno; 2006.

[25] Ambrosio, L., Hern´andez-Figueroa, H., et al. Gradient forces on double-negative particles in optical tweezers using Bessel beams in the ray optics regime. Opt Express 2010;18(23):24287– 24292.

[26] Ambrosio, L., Hern´andez-Figueroa, H., et al. Radiation pressure cross sections and optical forces over negative refractive index spherical particles by ordinary Bessel beams. Appl Opt 2011;50(22):4489–4498.

[27] Milne, G., Dholakia, K., McGloin, D., Volke-Sepulveda, K., Zem´anek, P.. Transverse particle dynamics in a Bessel beam. Opt Express 2007;15(21):13972–13987.

[28] Preston, T.C., Mason, B.J., Reid, J.P., Luckhaus, D., Signorell, R.. Size-dependent position of a single aerosol droplet in a Bessel beam trap. Journal of Optics 2014;16(2): 025702.

[29] Q. Zhan, Cylindrical vector beams: from mathematical concepts to applications, Advances in Optics and Photonics 2008, 1:1-57.

[30] Q. Zhan, Trapping metallic Rrayleigh particles with radial polarization, Opt. Express 2004, 12: 3377-3382.

[31] Y. Kozawa and S. Sato, Optical trapping of micrometer-sized dielectric particles by cylindrical vector beams, Opt. Express 2010, 18: 10828-10833.

[32] L. Huang, H. Guo, J. Li, L. Ling, B. Feng, and Z.-Y. Li, Optical trapping of gold nanoparticles by cylindrical vector beam, Opt. Lett. 2012, 37:1694-1696.

[33] G. Volpe, D. Petrov, Generation of cylindrical vector beams with few-mode fibers excited by Laguerre–Gaussian beams 2004, Opt. Commun. 237: 89–95.

[34] Ren, K.F., Gouesbet, G., Gr´ehan, G.. Integral localized approximation in generalized lorenz-mie theory. Appl Opt 1998;37(19):4218–4225.

2

Optimization of WDM-POF Network for In-Car Entertainment System

Mohammad Syuhaimi Ab-Rahman , Hadi Guna and Norhana Arsad

1. Introduction

Recent advances in In-Car Entertainment System will play an important role on automotive industry. It was assumed that in the year 2015, every new car, especially built in Europe will be equipped with Internet connection. As cars become connected to the Internet, the demand for Internet-based entertainment and applications and services increases [1].

Polymer optical fibers (POFs) are in a great demand for the data transmission and processing of optical communications compatible with the Internet, which is one of the fastest growing industries in automotive field. POFs become replacement for copper cable technology for future IVI system.

In this chapter reports experimental demonstration of a POFs based solutions in wavelength division multiplexing (WDM) network and some effects due to the placement of color filters as a demultiplexer for the In-Car Entertainment System. A *weakly fused* (WF) and a *highly fused* (HF) POF star coupler based on fusing and combining a group of POFs are designed and experimentally investigated. A low-cost demultiplexer is realized by using color filters. The specialized designed plastic-based interference films are used to filter out any other wavelength (color) that is not within the range. Variation of temperature were applied directly to the fused tapered region and an LED fiber source is launched into the fiber input, while the optical power deviation is measured at each output port. The effect of the return optical power and the coupling ratio to the temperature variation of 20 °C to 125 °C were investigated. The excess loss, EL_0 for the WF and the HF coupler at 20 °C is ~ 2 dB and ~ 8 dB, respectively.

Media Oriented Systems Transport (MOST) is one of the advance IVI system provider company which also utilized a POF-LED technology to transmit numerous signals represent a different data transmission via time division multiplexing (TDM) network (refer Fig. 1).

Figure 1. Recent technology on ring topology IVI system with single line of optical fiber cable facing main problem with no backup line when failure occurred.

In our research, we offered a wavelength division multiplexing (WDM) communication network over POF due to the rapid increase of traffic demands [2, 3]. WDM is the network that allows the transmission of multimedia data in IVI system over multiple wavelength (color) and thus greatly increases the POF's bandwidth. Beside, this network proposes a backup path in order to mitigate a serious breakdown in TDM-based network in IVI system.

Refer to Fig. 2, the proposed WDM-POF system, three unit of transmitters with different color of LED will carry single information simultaneously. For example in IVI network, red LED with 650nm wavelength modulated with video signal while blue ($\lambda 1$), green ($\lambda 2$), and yellow ($\lambda 3$) lights carry ethernet, audio and RF signal, respectively. The light has to be combined by the multiplexer (MUX) at the sending side. And to separate the wavelength channels at the receiver side, a wavelength demultiplexer (DEMUX) has to be used.

Demultiplexing, perform the reverse process with the same WDM techniques, in which the data stream with multiple wavelengths decomposed into multiple single wavelength data streams. POF coupler has similar function, operates to combine a number of optical data pulses

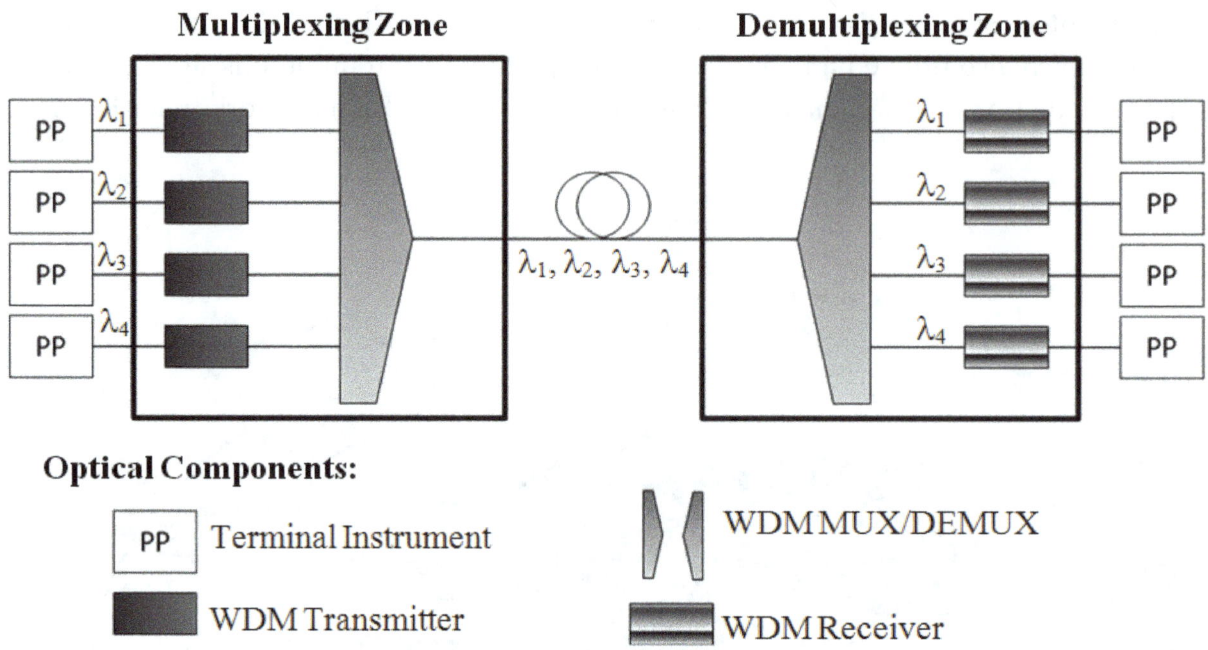

Figure 2. Simple schematic of a 4-channels WDM system.

as a single coupled signal. Hence, the development of MUX based on POF coupler is possible. A low-cost solution for POF-WDM system application will be presented.

A novel fused POF couplers has been fabricated by a fusion technique, as an effective transmission media to split and couple numerous different wavelengths which represents different signals. These novel coupler, however, suffer from several disadvantages. The high cost of the available couplers was raised as a challenge to the development of WDM systems in short-haul networks [5-8]. In addition, from the stand point of device design, the diameter of the fused tapered region, where stress is concentrated, is too small in conventional couplers. The structure causes a high incidence of fiber fracture, which results in poor reliability [9-12].

Thus a cost-effective 3×3 POF couplers based on a fused tapered structure to address these drawbacks of conventional couplers is demonstrated. The coupler is fabricated by a new and simple fabrication method, using a Bunsen burner and a metal tube. In this study, two types of fused couplers method focused of weakly fused (WF) and highly fused (HF). The WF coupler is, however, not considered to be a low-loss device, as the excess loss of the coupler was high, 12 to 22 dB.

The HF coupler is then developed to be the successor of the WF coupler. The excess loss of the HF coupler is very low, 0.3 to 5 dB. The device is developed as an optical switch which optical power can be switched completely from one fiber to another fiber at a temperature increase of T=55°C [13]. The switching characteristic can be achieved by varying the refractive index of the cladding at the coupling region of the coupler by temperature.

One of the aim of this chapter is to examine and optimize the feasibility of both methods of 3×3 POF couplers as thermal optical switches to be integrate in WDM-POF-based network for IVI system. The investigation is also to determine whether the thermal treatment is required to improve the quality and shift the device specification. Hence, a study of thermal effect on both polymer-based WF and HF couplers by varying the temperature of a hot plate from 20 °C to 125 °C is studied. The fused tapered fiber in the coupling region is exposed to the hot plate surface and optical power is launched into the input fiber of the coupler. In this temperature-dependence experiment, we investigate a relationship between temperature and several parameters such as coupling ratio, insertion loss and excess loss of the couplers.

In this chapter, red LED (650nm) has been utilized to transmit Ethernet data while green LED (520nm) can transmit a video image generated from CCTV network or DVD player, and blue LED with 470nm wavelength represents an audio transmission system for home networking. Refer to Fig. 3, special polymer color filters has been located between the coupler and receiver-end to ensure the entire WDM system can select a single signal as desired [4].

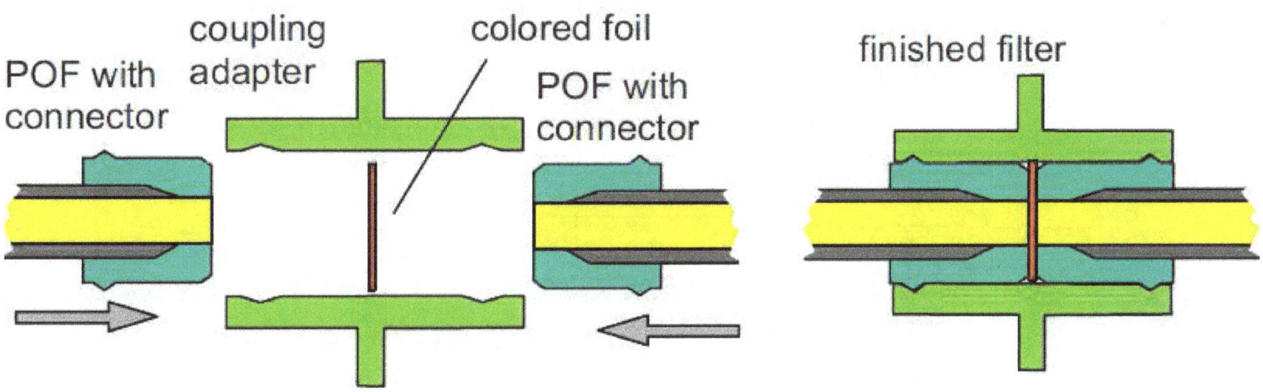

Figure 3. Each of those color filter lets pass exactly one wavelength and reflects all the others. By using several of those filters, the different wavelengths can be sorted out in a very low-cost way.

The performance of the novel coupler either with or without attachment of color filter can be evaluated in terms of insertion loss (IL). Some experiments in order the optimized the performance of WDM-POF based system for IVI system need to be conduct to minimized the value of the insertion loss (IL) in the network. The insertion loss (IL) is the amount of power loss that arises in the fiber optic line from input to the output of the fabricated coupler, expressed below,

$$IL_{port}(dB) = 10\log\frac{P_{o\,port}}{P_i}$$

Some wavelengths interfere with their reflected parts constructively, whereas others interfere destructively. Those wavelengths that interfere constructively can pass the filter, whereas the others get reflected. Besides the material parameters, the incident angle plays a major role, as each layer gets relatively thicker when tilting the filter. Color filters are manufactured for a long time and therefore high quality filters are readily available. They usually have sizes of several square millimeters. Color filters therefore are a valid choice to build DEMUX.

2. WDM-POF integrated network for in-car entertainment system

Adapting the fused tapering technique for conventional multimode fiber, we successfully established fabrication process for 1×3 POF twisted and fused couplers to be used as a MUX and DEMUX in IVI system. The 1×3 low cost coupler is an optical device, which ended by 3 number of POF output ports, while the other side ended by one POF port.

Similar to common coupler, it is also possible to work bidirectional, whereby it works from the 3 ports into 1 port (for coupling signal purpose), or vice versa (for splitting signals purpose). Optical 3×3 coupler has been symmetrically cut into two part to generate a pair of 1×3 couplers by the jointing of three polymethylmethacrylate (PMMA) POF [14]. Other specification for the design, the input POF is designed and fabricated to be twisted and fused shape as the fabrication process and 1×3 POF coupler is illustrated in Fig. 4.

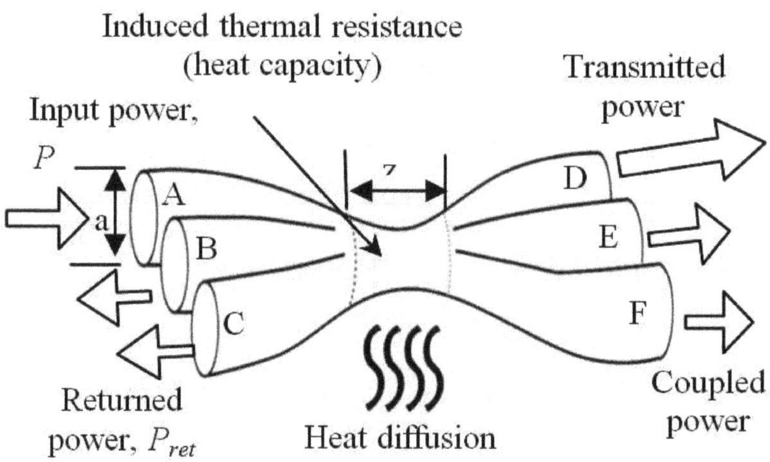

Figure 4. Thermal resistance at the coupling region of POF coupler in which all fibers are fused together.

Standard multimode SI-POF is used with its core diameter of 980 μm and cladding thickness of 10 μm. To obtain the results, DEMUX is realized using a special color filter attached using epoxy resin to the end of the connectors. The components are chosen because they are low cost and are easily found in the market.

Almost similar to POF material itself, the color filters are comprised of two types of plastic. More than 65% of the line is made from co-extruded polycarbonate plastic. The remainder of the line is deep dyed polyester [15, 16]. Filters create color by subtracting certain wavelengths of color. Thus, a red filter absorbs blue and green, allowing only the red wavelengths to pass. The process is subtractive not additive, so the light source must emit a full spectrum.

The swatch book provides detailed information on the spectral energy curve of each filter. The curve describes the wavelengths of color transmitted through each filter. For example, Supergel 342 transmits approximately 40% of the violet and blue energy of the spectrum and 75% of the orange and red energy. It absorbs all energy in the yellow and green range [15, 16].

After putting the resin onto the filter to be attached to the socket, the component then is hold together tightly for about two minutes to assure that no gap or air bubbles all over and also to assure the strong bond. This part has to be done gently since the epoxy resin has to be avoided covering the fiber's surface as much as possible so that any power losses can be minimized when the measurement is taken. However, since the edge of the socket is quite thin and sharp, the spread of the epoxy resin to the fiber surface cannot be 100% avoided.

After the fabrication is done, readings and measurements are taken for insertion loss for each of the fiber using a power meter. In this experiment, a lot of samples were fabricated to get the optimal results and to see which of the color filters that shows the most transmission and gives least losses. The length of the POFs is fixed at 3 meters long.

In this study, for fused plastic optical fiber, the optical loss is categorized as extrinsic loss due to the physical change of POF, LED projection to POF and the core-to-core connection and [17, 18]. It is obtained that the physical change of POF caused by fabrication process, where by diameter of POFs increasingly decrease to approach 1 mm and the POFs finally has twisted and fused shape. In characterization process, optical loss may present through the direct LED projection to POF surface. Besides, optical loss may also present through the connection between the fused tapered POF and POF cable [17].

3. Results and discussion

Comparison for the optical line either using the color filter or not, has been analyzed. The insertion loss of as much as 21 samples for the output terminal over POF has been utilized in 3-channels IVI system through red filter (internet data) have been visualized in Fig. 5. From the results gathered, it is seen that when all the components are set up and red LED (650nm) was injected, the insertion loss measured by the power meter shows small increase of losses when the film is attached to the socket.

This is also true when blue and green LEDs are injected. For the characterization of same film using different sources, we take a red film A (filter labeled *#4690*) as the primary filter and it is injected with all three LEDs, red, green and blue transmitters. As the results depict, the red filter injected with red transmitter shows a small increase of losses which averagely *IL*<~3 dB compared to the initial loss before the film is attached to the fiber. Same goes to the power output when the film is attached to the fiber.

The effect of plastic-based filter attachment at the receiver-end did not indicate a significant deviation in power efficiency since both PMMA and Co-extruded Polycarbonate for fiber core and plastic filter's material have the same refractive index which approaching 1.59 [16].

Study on the saturation level of each color filters – red, green and blue – has been carried out. As much as eleven best samples chosen with different level of color saturation labeled from 1 to 11 which sample number 1 indicate the darkest (more saturated) color filter while number 11 with the lightest color (less saturated). Each of different color filter has been injected by LED in a range of blue to red light (470~650nm) and the result can be obtained in Fig. 6.

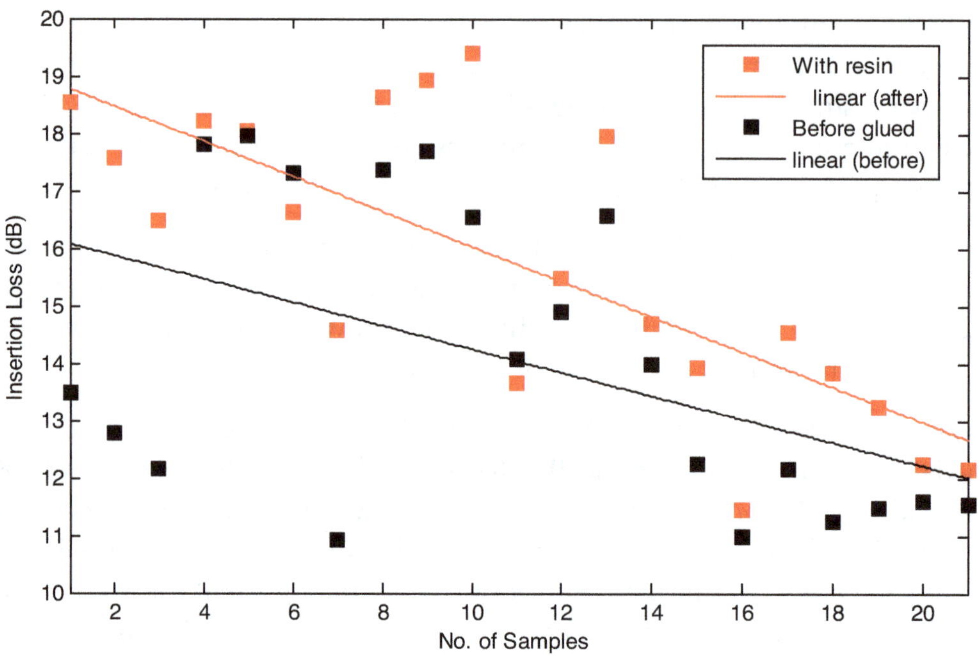

Figure 5. Effect on resin for demultiplexer filters approximately 1 dB insertion loss occurred on the measurement between before and after connector glued by resin.

From the result in Fig. 6(a), it is observed that sample 1 to 7 shows highest losses and decrease of efficiency. Since sample 1 to 7 being among the darkest film color meaning that only small or narrow transmission percentage of red LED or transmitter is allowed to get through. This phenomena also found in Fig 6(b), for the first six sample has quite high losses especially when it was injected by LED in blue to green range (470-580nm). However, red filter in Fig 6(c) did not indicate a significant deviation of loss by variying the saturation of the filter.

Compare with others, red filters block most of the green and blue LED transmission since it is clear that only red wavelength (λ=600~650nm) will be allowed to get through the film. The film filters out any other wavelength transmission that is not within the range. This concept is used as the primary idea for designing the demultiplexer.

It has been proven that, when blue filter was injected by all three optical source (blue, green and red color) all the data were more fluctuated right after the light hit the filter through the fiber. Same goes to green filter, all light sources will fluctuated the efficiency of the data transmission. This fluctuation generated as an effect from the SED percentage of each filters and also came from the intensity of each light source. The higher intensity of the light source transmitted the more oscillated graph plotted and the less significant of SED percentage deviation the less effect on fluctuation.

For the temperature-dependence experiment, two samples of 3×3 fused couplers with multimode SI PMMA POF are tested: WF and HF couplers. For the WF coupler, the fabrication method includes three processes: fiber bundle configuration, fabrication of spiral fiber and fiber tapering. Firstly, a fiber bundle consists of N core unjacketed polymer fibers (where N=3) is inserted into the metal tube and placed in symmetrical coordination. The fibers bundle is exposed to the heat of the flame repeatedly twisted while pulled continuously to produce a spiral fiber with a particular number of twists in the centre of the fiber bundle. The length of the fused region is limited to 8 cm. While the indirect heating process continues, fiber bundle is pulled and twisted. When the POFs reach the melting point, the fiber bundle is pulled to fuse the fibers, and the fused tapered fiber is formed. Each fiber ports (output and input) are insulated with PVC sleeve.

The HF coupler fabrication method includes four processes, as a new process was introduced to enhance the coupling characteristic of the fused tapered fiber by removing twisting effects on the fused fiber. A twisting effect implies that each fiber is not melted sufficiently to combine with each other. Firstly, a fiber bundle consisting of N core unjacketed polymer fibers (where N=3) are inserted into the metal tube and placed in symmetrical coordination. The fibers bundle are pulled and exposed to the heat of the flame repeatedly and continuously to produce a spiral fiber with a particular number of twists in the centre of fiber bundle. The length of the fused region is limited to 5 cm.

The fiber bundle is pulled and twisted from both sides repeatedly and continuously over a long fusion time, $t_f < 65$s, to reach a sufficiently high melting temperature to remove twisting effects. While indirect heating process continued, fibers bundle is pulled and twisted. When the POFs reach the sufficient melting point, the fiber bundle is pulled to fuse the fiber, and the fused tapered fiber is formed. Each fiber ports (output and input) are insulated with a PVC sleeve.

The experimental setup has been setup, which consisted of a digital hot plate, an AF-OM110A power meter and an LED fiber source with a 650 nm wavelength. Refer to Fig. 4, each fibers on the left side of the coupler are defined as Port A, B, or C, whereas on the other side, each fiber ports are termed as Port D, E, or F. In the experimental setup, for each coupler, the hot plate is exposed directly to the centre region (the fused and tapered fiber) and the temperature of the hot plate is varied from 30 °C to 125 °C to investigate the light propagation behavior and power loss for each fiber port. The LED source is launched into a single input fiber, while each output fibers was connected into an optical power meter using a suitable adapter to measure the output power. In the meantime, another optical power meter is also placed at the end of each input fibers to measure the returned power for both fiber ports. By using the optical power acquired from the measurement, several parameters were calculated for both couplers, such as the excess loss and the coupling ratio.

When the heating temperature is increased to $T > 85$°C, the behavior of power dissipation at the other input port differs from that in the first case described above. This phenomenon reveals that the induced thermal resistance in the centre of the coupler reaches a very high resistivity, where large heat capacity stored in the coupling region is sufficient to oppose optical power transferred between input and output fibers. In addition, there is a possibility that the optical power is switched incompletely from output to input fibers. While output power P_o in through-put and cross-coupled fibers remains fall to zero, the returned optical power P_{ret} in other input fiber increase drastically. In this case, the optical power tends to be reflected and coupled to other input fibers, instead of output ports. As the fused tapered fiber

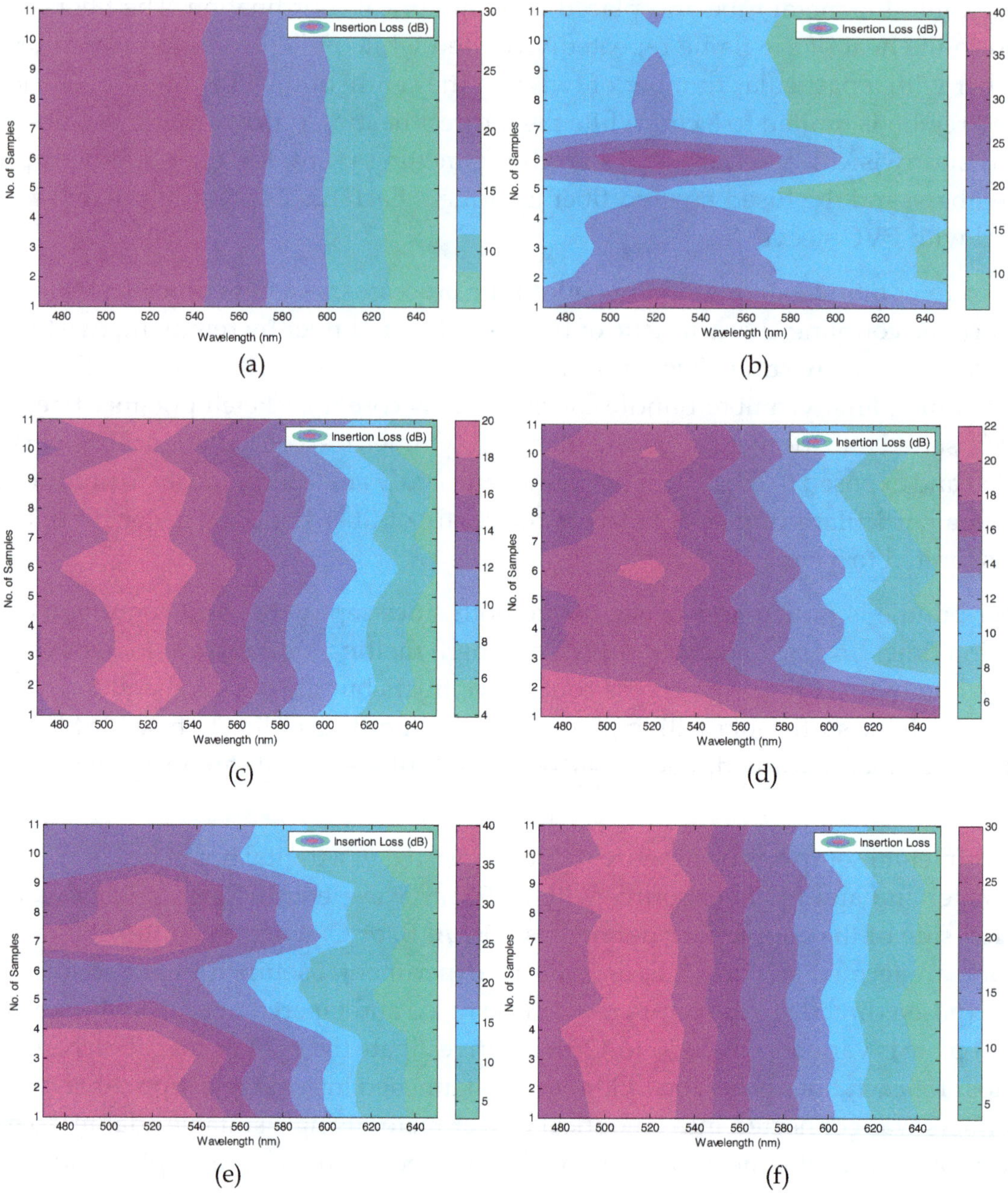

Figure 6. Comparison on before (left side) and after (right side) effect on injection of different saturation of color filter. Three basic colors have been investigated for (a) blue, (b) green and (c) red filter, before (left side) and after (right side) injected with LED in visible light range (450~650 nm).

is overheated ($T >> 85°C$), however, the returned optical power begins to break down as the fused fiber in the coupling region is approaching damage threshold at which the coupler lifetime begins to decrease. The same behavior is also exhibited by using the optical power at the output fibers.

The heating temperature of the hot plate T, is increased gradually from 30°C to 125°C to examine the thermal effect each port of the fibers. Previous experimental results show that the refractive index variation with temperature is analyzed for large-core PMMA POF [13]. The following function is used to describe the variation of the refractive index by performing a nonlinear least-square fit: $n(T)=n_0+aT+bT^2$. The symbol-α=dn/dT (°C^{-1}) denotes the thermal dispersion of the refractive index, where α=-5 × 10^{-4} for PMMA and α=-4 × 10^{-4} for fluorinated polymer [19, 20], which are the materials of the core and the cladding, respectively. At room temperature (T=20°C), the refractive index of the core is higher than that the cladding. For typical POF, it has been found that the index for the PMMA core and the fluorinated polymer cladding at 20°C are n_{co}=1.490 and n_{cl}=1.402, respectively [21]. The refractive indices of the cladding and the core at the fused region exhibit different temperature-dependence behavior. As the heating temperature is increased (T > 20°C), the difference in refractive index between the core and the cladding, $\Delta n=n_{co}$ - n_{cl} becomes smaller. In comparison, the percentage of the core index reduction is higher than in the cladding material, as the thermal dispersion dn/dT for the PMMA core is greater. As calculated, at temperature T=100°C, the refractive indices for both material would be equal ($n_{co}=n_{cl}$) and the difference of the two indices would be zero ($\Delta n=0$). At this point, the fiber stops confining light to the core, and the output light intensity drops to near zero. In practice, optical loss and/or power dissipation increase gradually; they do not cut off abruptly.

While heat energy is supplied from a hot plate to the coupling region at the fused coupler and an LED fiber source with wavelength of 650 nm is injected into an input fiber, thermal resistance would be induced at the centre of the fused coupler to oppose light propagation from a single input fiber into multiple output fibers. The symbols T_R and T_m denote room temperature and melting point temperature, respectively. It is known that T_R=20°C and T_m=85°C. In the experiment, there are two cases of power dissipation occurring within different behavior.

Firstly, when the temperature is increased to T > 20°C, a small heat capacity is stored in the fused and tapered fiber, while heat energy is distributed along its taper length z. In this case, the heat capacity accumulated in the coupling region represents thermal resistance. As the induced thermal resistance is not sufficient to block optical power transmission completely, optical power can pass through coupling region to output fibers with low power intensity. As the coupling region is heated directly, the change in the refractive indices of the core and the cladding result an optical power loss in each of the fiber output port.

The optical power was measured for one directions, the LED fiber source was injected into Port A, while the optical power meter was positioned at the ends of the output fibers (Ports D, E, and F). The room temperature T=20 was taken as a reference value before the experiment was started. In the experiment, the temperature of the hot plate was increased in steps of 5 °C, with a time delay of 60 – 70 s to reach a stable condition. Fig 7 shows the influence of varying temperature T from 30 °C to 125 °C on output power Po for both fused couplers (HF and WF) wherein the LED fiber was injected into the input fiber from the left side to the right side of the coupler.

As shown in Fig 7(a) and 7(b), in each fiber port, output power decreases as temperature rises. Both types of fused polymer couplers will be damaged when the heating temperature increased to T=125°C. In the case of the HF coupler, as shown in Fig 7(a), optical power decreased gradually when temperature varied from 30°C to 95°C. At 30°C, the differences between the transmitted power at Port D and the coupled power at other output fibers were significantly large. However when the temperature of the hot plate increased to *T*>> 95°C, however, the difference became relatively small (Δ ≈ 0) such that the optical power in each output fiber fell down drastically to zero. The temperature point at 95°C thus defined as the damage threshold, because the coupler lost temperature stability at this point. As the melting point of the polymer material at 85°C, it is believed that the polymer fiber suffered from excessive bond rupture at 95°C.

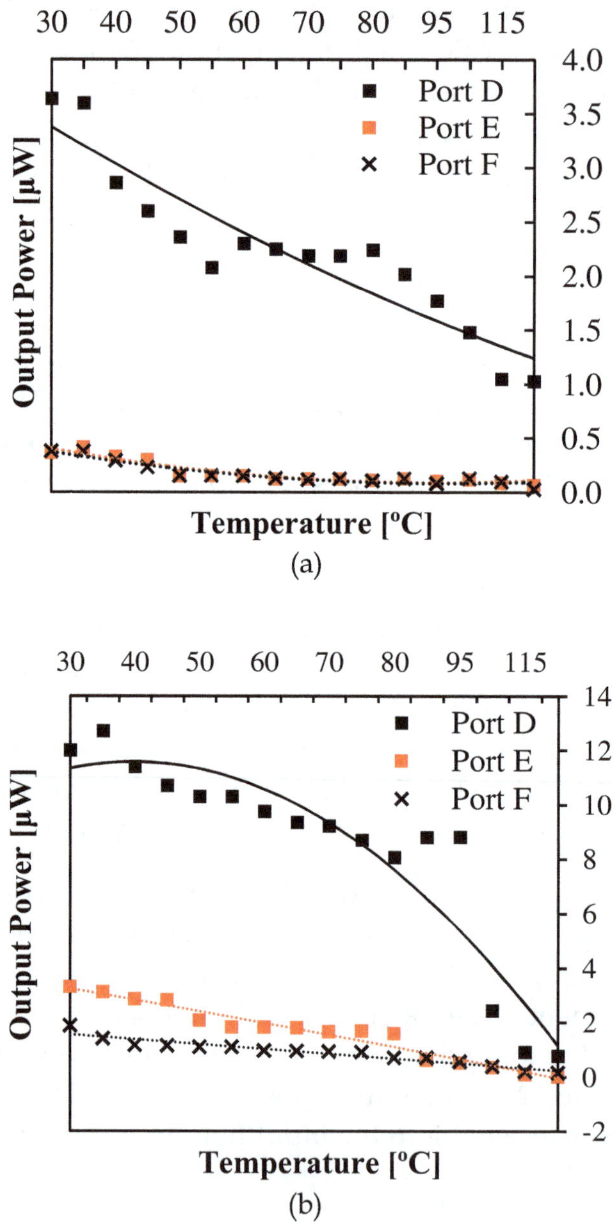

Figure 7. The relationship between temperature variation (30 °C to 125 °C) and optical power for (a) the HF coupler and (b) the WF coupler.

In the case of the WF coupler, as shown in Fig 7(b), the optical power reduction in the cross-coupled fibers (Ports E and F) was not significant, as the power intensities for both cross coupled ports were too small, less than 0.5 μW at 30°C. It was found that the throughput port D decreased with similar behavior as that seen in the WF coupler, with the power falling to zero suddenly when the temperature of the hot plate increased to *T*>> 95 °C.

However, for temperature variations from 30°C to 85°C, the downward slope for the WF coupler (−dP/dT) was greater than for the HF coupler. It is believed that the geometrical taper design (in the coupling region) influenced the −dP/dT slope. As the twisting effect was featured in the fused tapered fiber in the centre of the WF coupler, it is believed that the large fiber imperfection in the fused fiber region changes the total optical transmission characteristic of the polymer fiber. Therefore, the twisting effect is considered to be a minor factor in determining the power loss in the WF coupler.

Another effect to take into account is the heating time delay that occurred during power measurement for each fiber port. The delay caused the temperature of the polymer material to increase and thus resulted in optical loss. At *T*>> 95°C, the results show that the optical power P_0 in each output fiber (Ports D, E, and F) decreased drastically to zero with high −dP/dT slope. In this case, it was believed that the refractive indices of the core and the cladding were equal, $(n_{co}=n_{cl})$. As the PMMA core has a large dn/dT coefficient, the core index reduced faster than the cladding index, and as a result, the indices for the core and the cladding became equal at a certain point before the coupler damage occurred.

In addition to the measurement presented above, the returned power P_{ret} was measured at the other input fibers to investigate the thermal switching behavior the couplers. During the measurement, the optical power meter was placed at the ends of the two other input fibers.

Fig 8 shows the temperature dependence of the coupling ratio for both couplers in their throughput and cross-coupled fiber ports for both directions of lightguide propagation. The coupling ratio in the throughput port is defined as $CR_T\ (\%)=P_t\,/\,P_s$, whereas the cross-coupled port is $CR_c\ (\%)=P_c\,/\,P_s$. The symbols P_s and P_t denote the transmitted power at the throughput port and the coupled power at the throughput port, respectively.

As both the WF and the HF multimode PPMA POF fiber couplers are wavelength independent, their coupling ratios are not periodic functions. For an ideal wavelength-independent 3×3 coupler, it is assumed that the fused fiber in the coupling region has a strong coupling; the output power ratio in the throughput fiber and the cross-coupled fibers at room temperature, *T*=20 °C, is thus equal to 33 % and 66 %, respectively. For practical couplers, however, the couplers are afflicted with large error in the power ratio, as the coupling ratio at the throughput outlet is significantly higher than the power ratio at the cross-coupled port.

For the WF coupler, as shown in Fig 8, the ratio error ε at 20 °C is less than ± 50 %, whereas for the HF coupler, ε=± 36 %. It is believed that, for both couplers, the fused fibers in the coupling zone were not completely fused. In the other words, the fibers were not coupled to each other with a 100 % coupling ratio. The WF coupler has a lower coupling efficiency than the HF coupler, as the HF coupler's coupling ratio error is smaller. It is known that the HF coupler

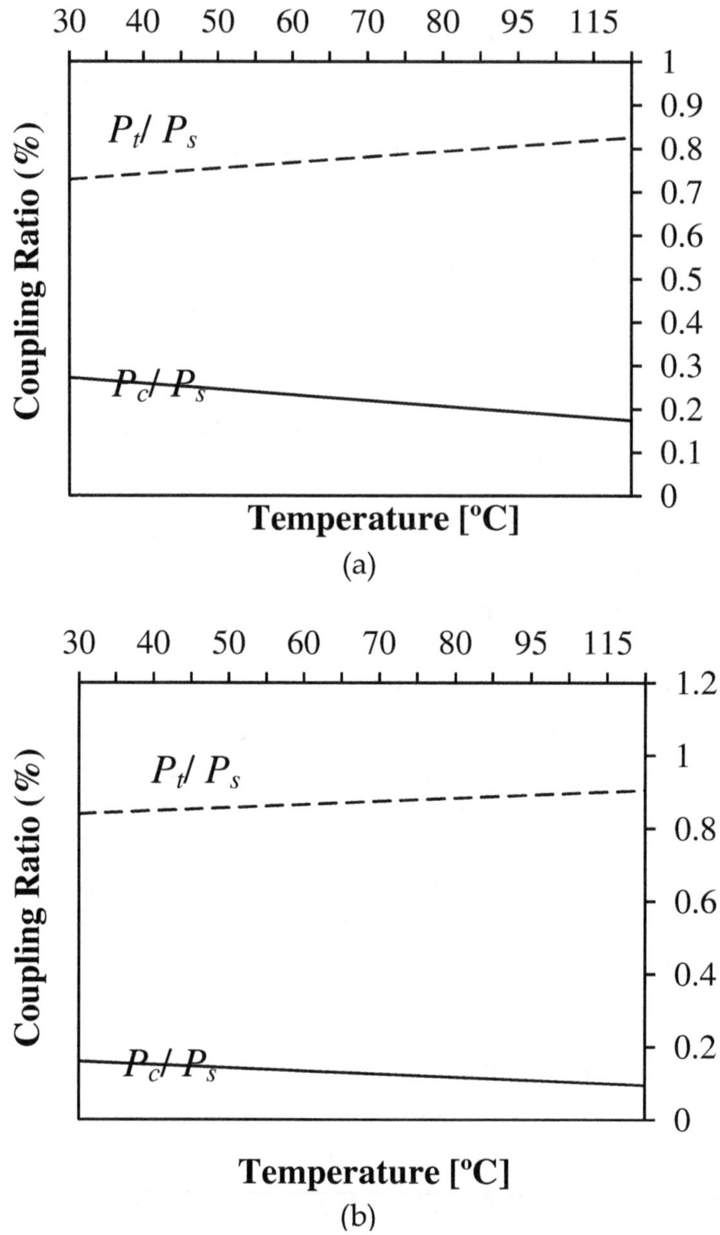

Figure 8. Coupling ratio variations with temperature increase for (a) HF and (b) WF coupler.

was fabricated to have a plane surface on its fused tapered fiber, with no twisting effect featuring in the structure. The twisting effect is an undesired design characteristic; as the fibers in the coupling region may not be highly fused and thus have low optical power coupled.

As mentioned before, the refractive index of the fused tapered fiber decreases with temperature. The thermal change of the refractive index in the core and the cladding in the coupling region will result in the variation of the relative phase velocity of the interaction light modes, and the coupling ratio of the fused coupler will be thus influenced by rising temperature [22].

As shown in Fig 8, the variation of the coupling ratio for both couplers is a linear function. The experimental result shows that, for both couplers, the ratio of the transmitted power increases with rising temperature from 30 °C to 125 °C, while the ratio of the coupled power decreases in the meantime.

Fig 8(a) indicates that, for the HF coupler, the average ratio of the coupled power reduces from 35 % to 23 %, whereas the ratio of the output power in the throughput fiber increases linearly from 68 % to 74 %. Fig 8(b) shows that, in contrast to the HF coupler, the WF coupler's power ratio in the cross-coupled fiber decreases linearly from 18% to 16%. Moreover, the ratio of the transmitted power increases from 82% to 86%. As the $\Delta CR/\Delta T$ slope is small, it can be seen that the influence of the temperature effect on the coupling ratio is minor, and the couplers are thermally stable with respect to their coupling ratios.

In this study, at room temperature (T=20 °C), extrinsic loss is considered as major contributing factor to optical loss for both the HF and WF couplers, as the fused fiber in the coupling region was degraded by structural fiber imperfections. These fiber imperfections include the change in fiber diameter from 2.8 mm to ~1 mm, twisting effects and polymer degradation via liquefaction.

In normal conditions without temperature influence, the power splitting performance of the HF coupler is more significant than that of the WF coupler, as the fused fibers in HF coupler suffer from relatively few imperfections. In spite of the different levels of power loss for the two couplers, the fiber imperfections that are characteristic of the fused fiber regions can be considered a design constant, as the excess losses of both couplers decrease with similar $\Delta EL/\Delta T$ slopes.

On the other hand, the efficiencies of both couplers lie at the same point when the temperature has been increased to T>95 °C, as both couplers are close to the ends of their lifetimes. Both couplers are destroyed when the temperature is increased to 125 °C. It is realised that the highest temperature to which polymer material may be exposed to the heat while still retaining its structural integrity is ~85 °C and that the glass transition temperature for POF is 90 °C. The experimental results show that each coupler lost its temperature stability when its fused tapered fiber is overheated to T >95 °C, bringing the device close to the end of its lifetime. It is known that supplying thermal energy of 290 kJ/mol – 375 kJ/mol causes the bond rupture of the polymer chains and thus changes the total performance characteristics of the fused coupler [23, 24].

In the experiment, 95 °C is thus defined as the PMMA POF damage threshold. In the case of polymer material, the dn/dT coefficient is negative because the material's thermal expansion is higher than the temperature coefficient of the electronic polarisability [25]. Hence, the thermal expansion coefficient is dominant. As the dn/dT coefficients for the PMMA core and the cladding are negative, the refractive indices of both materials decrease with increasing temperature. It is realised that the dn/dT coefficient for the PMMA core is higher than that of the cladding. As a result, the refractive index of the core decreases more rapidly, and the indices of both materials become nearly equal when the temperature increases to 95 °C. At T>>100 °C, the output power in each fiber port suddenly breaks down to zero, as no power reflection and/or transmission occurs through the medium of the fused fibers.

4. Conclusion

The combination of WDM with POF will broaden the horizon of low cost optical customer premises networks [26]. A technique has been used for fabricating the optical coupler based on POFs technology using multimode SI-POF type with 1 mm core size. Fabrication and characterization stages have been carried out to develop the coupler [14]. A technique also has been used to develop a demultiplexer for short-haul communication based on plastic optical fibers. This experiment shows the transmission of multiple signals with different wavelengths carried through one fiber. The concept of multiplexer and demultiplexer are the basic of this system. The system only utilizes three colors for the transmitters and also the filters for the demultiplexer which are blue, green and red (λ=430, 570 and 650nm). Light source from the red, green and blue transmitters are combined by using multiplexer. In order to separate the combined signals, special separators – called demultiplexers (DEMUX) – are utilized. These DEMUX are realized by employing the principle of the Color filters.

Filters play an important role in giving a higher insertion loss from the WDM-POF system, but the quality of a number of output port is not badly destructed due to the color band gap from the filter itself, speed rate of the Internet still stable and the resolution of the video image is quite good. Some parameters, such as optical output power and power losses on the devices were observed, and not to mention about the effect of filter placement and the efficiency of the handmade $1 \times N$ coupler itself.

Red LED with a 650 nm wavelength has been injected to different Color filters for the purpose of characterization test in order to analyze the level of power efficiency of the demultiplexer. Analysis shows that efficiency maintains for filter of the same wavelength as the transmitter while other range of wavelengths will mostly be filtered out or blocked. This main idea is fully utilized for the designing of demultiplexer for WDM-POF-based IVI-Systems applications. Final analysis shows that efficiency of the filter can reach up to 70%. Improvement of performance can be made through practice. Although the setup IVI system exhibits very high attenuation of the transmission, this concept of handmade optical coupler and demultiplexer has been tested for sending data for video, audio and Ethernet and the output shows successful performance.

In the temperature-dependence experiment, it was proven that thermal resistance exists in the fused tapered fiber at the centre of the coupler. The thermal resistance of the fused fiber is dependent on the heat capacity stored in the coupling region. As the heat capacity of the fused tapered fiber reaches its level of saturation, the internally induced thermal resistance is sufficient to block light propagation from the input fibers. Thus, some portion of the total input power is reflected along the opposite path to the two other input fibers. The resultant light guide propagation is called thermal switching.

Hence, the obtained result reveals that WDM-POF has great potential to be employed as economical wavelength divisions multiplexer because it is able to couple different wavelengths with main advantages that are low optical loss and low cost. An intensive study suggested in order improving the homogeneity of this prototype. In fact, fusion technique afflicted with some disadvantages has no consistency of producing coupler as it was almost not possible to fabricate POF coupler with good performance consistently. This WDM-POF technology can be improved gradually through experience and practice. This device is highly recommended for WDM-POF system as it is not as costly as other commercial POF coupler. Furthermore, the fabrication and installation process is simple, easy and suitable to be used for WDM-POF based IVI-system application.

Acknowledgements

This research has been conducted in Computer & Network Security Laboratory, Universiti Kebangsaan Malaysia (UKM). This project is supported by Ministry of Science, technology and Environment, Government of Malaysia, 01-01-02-SF0493 and Prototype Research Grant Scheme PRGS/1/11/TK/UKM/03/1. All of the handmade fabrication method of POF coupler, 1×N handmade™-POF coupler and also the low cost WDM-POF network solution were protected by patent numbered PI2010700001.

Author details

Mohammad Syuhaimi Ab-Rahman , Hadi Guna* and Norhana Arsad

*Address all correspondence to: hadi_guna87@yahoo.com

Universiti Kebangsaan Malaysia, Selangor Darul Ehsan, Malaysia

References

[1] Europe, E. Cars get internet connection, study says, 2008.

[2] Hashimoto, M., et al. Design and Control System over WWW for Regional CWDM Optical IP Networks with Reconfigurable Optical Add/Drop Multiplexers. International Journal of Innovative Computing, Information and Control, 4(5): 1299-1313, 2008.

[3] Zahilah, R., et al. Lightpath Route Management System for IP-Over-CWDM Networks With ROADMS, Based On A ROADM Graph, International Journal of Innovative Computing, Information and Control, 7(5(A)): 2485–2501, 2011.

[4] Ziemann, O., P. Zamzow, and W. Daum. POF Handbook: Optical Short Range Transmission Systems, Springer, 2008.

[5] Chandrappan, J., et al. Optical Coupling Methods for Cost-Effective Polymer Optical Fiber Communication, Components and Packaging Technologies, IEEE Transactions 32: 593-599, 2009.

[6] Haupt, M. and U.H.P Fischer. Design and development of a MUX/DEMUX element for WDM over POF, in International Students and Young Scientists Workshop in Photonics and Microsystems 2017: 27-31.

[7] Koonen, A.M.J., et al. Cost optimization of optical in-building networks, in 37th European Conference and Exhibition Optical Communication (ECOC) 2011: 1-3.

[8] Kagami, M. Visible Optical Fiber Communication, R&D Review of Toyota CDRL. Toyota Central R&D Labs, Japan, 2005.

[9] Imoto, K. et al. New biconically tapered fiber star coupler fabricated by indirect heating method. Journal of Lightwave Technology, 5(5): 694 – 699, 1987.

[10] David, S., F.M. Antonio, and J.A. Valenzuela. Fused fiber optics couplers, in The International Society for Optical Engineering Proceedings of SPIE p. 330-333 (2001).

[11] S. Ci-jun, D. Ji-an and Z. Jue. Development of a Novel Optical Fiber Coupler, in Sixth International Conference on Intelligent Systems Design and Applications 2006, pp: 183-186.

[12] Pal, B.P. Fabrication and Modeling of Fused Biconical Tapered Fiber Couplers. Fiber and integrated optics, 22(2): 97-117, 2003.

[13] Diemeer, M.B.J., W.J. DeVries and K.W. Benoist. Fused coupler switch using a thermo-optic cladding. Electronics Letters, 24(8): 457-458, 1988.

[14] Ab-Rahman, M.S., H. Guna and M.H. Harun. 1×N Self-Made Polymer Optical Fiber Based Splitter for POF-650 nm-LED based Application. in International Conference on Electrical Engineering and Informatics 2009, Selangor, Malaysia.

[15] Rosco, ROSCOLUX Color Filter, Rosco, Editor. Rosco: Harbor View Avenue, Stamford. pp. 1-2, 2003.

[16] Rosco, Roscolux, in 1991-2010. Rosco Laboratories, 2010.

[17] Appajaiah, A., V. Wachtendorf and W. Daum, Climatic exposure of polymer optical fibers: Thermooxidative stability characterization by chemiluminescence. Journal of Applied Polymer Science, 103(3): 1593-1601, 2007.

[18] Kuzyk, M. Polymer fiber optics: materials, physics, and applications, CRC/Taylor & Francis, 2007.

[19] Bosc, D. Thermo-optical coefficient determination of index liquids used for optimization of optical integrated components. Optics Communications, 194(4-6): 353-357, 2001.

[20] Kim, K.-T and K.-H. Park. Fiber-Optic Temperature Sensor Based on Single Mode Fused Fiber Coupler. J. Opt. Soc. Korea, 12: 152-156, 2008.

[21] Zhao, L.J., et al., Wide-Range Temperature Dependence of Brillouin Shift in Optical Fiber, in 1st IEEE International Conference-Nano/Micro Engineered and Molecular Systems. NEMS 2006. pp. 1327-1330.

[22] Yang, S.-W and H.-C. Chang. Numerical modeling of weakly fused fiber-optic polarization beam splitters. I. Accurate calculation of coupling coefficients and form birefringence. Journal of Lightwave Technology, 16(4): 685, 1998.

[23] Appajaiah, A. Climatic Stability of Polymer Optical Fibers. Potsdam University: Berlin. pp. 13-31, 2004.

[24] Hsieh, C.S., T.L. Wu and W.H. Cheng. Optimum approach for fabrication of low loss fused fiber couplers. Materials Chemistry and Physics, 69(1): 199-203, 2001.

[25] Zieamann, O., et al., POF Handbook: Optical Short Range Transmission Systems. 2nd ed. Berlin: Springer, 2008.

[26] Lutz, D., et al. Wavelength Division Multiplex Instructional Lab System with Polymeric Fibers for use in Higher Education. in Proceedings of the Symposium on Photonics Technologies for 7th Framework Program 2006, Wroclaw.

3

The Influence of Nonlinear Effects Upon Oscillation Regimes of Erbium-Doped Fiber Lasers

Y.O. Barmenkov and A.V. Kir'yanov

1. Introduction

Erbium-doped fiber lasers (EDFLs) are contemporary sources of coherent radiation, attractive for numerous applications requiring both continuous-wave (CW) and pulsed operations, among which telecommunications in a wide wavelength range covering C and L bands ought to be emphasized. Pulsed operation, presenting big interest for practice, is attained in EDFLs by means of standard active and passive Q-switching and mode-locking techniques, capable of enforcing a laser to generate short pulses with durations ranged from hundreds fs to hundreds ns [1]. In the meantime, transients to CW lasing and a laser's relaxation frequency are also of close attention, e.g. when targeting the sensor applications [2, 3]. The detailed knowledge of the processes involved in Erbium-doped fibers (EDFs) to be used, when pumped, as a laser or amplifying medium in each of the referred regimes cannot be overestimated. The present work is a review of some of the most featuring nonlinear-optical effects that affect the oscillation regimes of EDFLs.

In spite of certain advantages (availability at the market, low cost, and easiness of handling), EDFLs demonstrate considerably lower efficiency as compared to the lasers based on Ytterbium-doped fibers. The basic cause is the multi-level energy scheme of Er^{3+} ions, which makes unavoidable absorption of photons at both the pump and laser wavelengths by the ions being at upper Er^{3+} levels, i.e. the "excited-state absorption" (ESA) [4, 5], including the state where 1.5-μm laser transition stems from. In other words, ESA presents a kind of "up-conversion" (UC) loss inherent to EDF; see section 2.

Another, nonlinear in nature, source of UC loss in EDFs and EDFLs on their base relates to so-called "collective" (concentration-related) effects arising in Er^{3+} ion pairs and in more complicate clusters (percentage of which grows with Er^{3+} concentration) [6]; see section 3. Appearance of the latter phenomenon is also associated with the multi-level energy scheme of Er^{3+} ions.

Eventually, the cases of low-and heavily-doped EDFs are to be carefully segregated and properly addressed if one looks for optimization of an EDFL.

One more kind of optical nonlinearity that arises in actively Q-switched (AQS) EDFLs is stimulated Brillouin scattering (SBS). Depending of the EDF length and Q-cell's modulation frequency EDFLs may operate in one of the two QS regimes: either in the "conventional" QS (CQS) one in which QS pulses are composed of several sub-pulses separated by a photon's round-trip time (with negligible pulse jitter) [7, 8], or in the essentially stochastic SBS-induced QS (SBS-QS) one where pulse amplitude is bigger by an order of value as compared with CQS but where pulses suffer severe jitter [9]. The results of an experimental study of the basins the CQS and SBS-QS regimes belong to and the basic spectral features of these regimes are discussed in section 4.

In section 5, the review's conclusions are formulated.

2. ESA in EDF at the pump and laser wavelengths

In this section, we report a study aiming at determination of the ESA's spectral dependence in EDF, covering the most important for applications spectral range, 1.48...1.59 μm, and at ~978 nm, the wavelength that usually is used for pumping EDFs by laser diodes (LDs). In the experiments discussed hereafter a low-doped silica EDF (*Thorlabs M5-970-125*, ~300 ppm of Er^{3+} concentration, Al-Ge-silicate host composition, NA=0.24, cut-off wavelength – 0.94 μm) was chosen to avoid possible effect of UC (observed in heavily-doped EDFs; see section 3). [This fiber belongs to M-series (fabricated through the modified chemical vapor deposition (MCVD) process), to be under scope in section 3 where Er^{3+} concentration effects are treated; "M5" signifies that small-signal absorption peaked at 978 nm is 5 dB/m.] The measurements were completed by modeling, a necessary chain in interpretation of the experimental results. The developed modeling also provides an estimate for the so-called ESA parameter, obtained at both pump and laser (signal) wavelengths, and thereafter allows determination of the EDF's net-gain coefficient, "deliberated" from the ESA interfere.

Figure 1 shows the fife level Er^{3+} ion energy scheme upon excitation at the pump ($\lambda_p \approx 980$ nm) and signal ($\lambda_s \approx 1.5$ μm) wavelengths, useful for the discussion presented in this section.

The equations that describe functioning of the Er^{3+} system, in accord to the scheme shown in Figure 1, at excitation at sole or at both excitation wavelengths (λ_s and/or λ_p) in the steady-state are written as follows:

$$\frac{\sigma_{12} I_s}{h\nu_s} N_1 - \frac{\sigma_{21} I_s}{h\nu_s} N_2 - \frac{\sigma_{24} I_s}{h\nu_s} N_2 - \frac{N_2}{\tau_{21}} + \frac{N_3}{\tau_{32}} = 0 \tag{1}$$

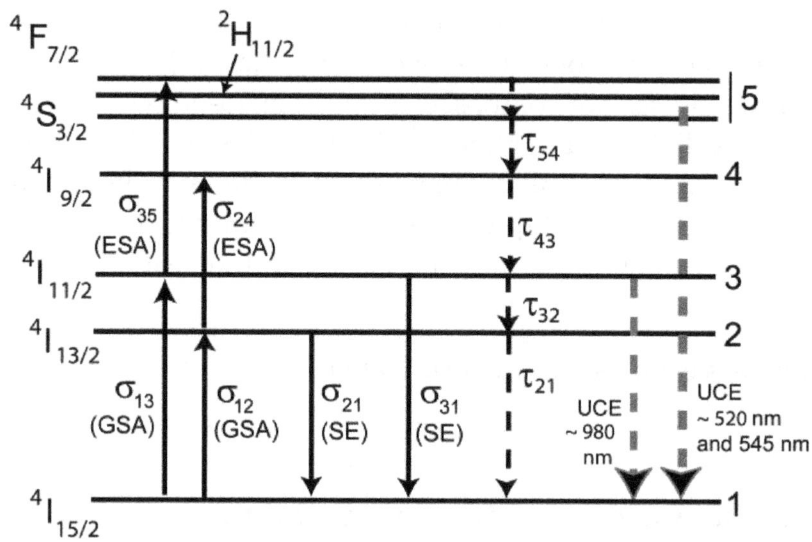

Figure 1. Fife-level scheme of Er^{3+} ion used in modeling. GSA and ESA indicate the ground-state and excited-state absorptions, UCE marks the up-conversion emission transitions (weak but detectable), σ_{ij} and τ_{ij} are, respectively, the cross-sections and decay times for the transitions between the levels *i* and *j*. Three closely-spaced energy levels $^4S_{3/2}$, $^2H_{11/2}$, and $^4F_{7/2}$ are regarded as an effective level "5".

$$\frac{\sigma_{13}I_p}{h\nu_p}N_1 - \frac{\sigma_{31}I_p}{h\nu_p}N_3 - \frac{N_3}{\tau_{32}} - \frac{\sigma_{35}I_p}{h\nu_p}N_3 + \frac{N_4}{\tau_{43}} = 0 \tag{2}$$

$$\frac{\sigma_{24}I_s}{h\nu_s}N_2 - \frac{N_4}{\tau_{43}} + \frac{N_5}{\tau_{54}} = 0 \tag{3}$$

$$\frac{\sigma_{35}I_p}{h\nu_p}N_3 - \frac{N_5}{\tau_{54}} = 0 \tag{4}$$

$$N_1 + N_2 + N_3 + N_4 + N_5 = N_0 \tag{5}$$

where N_0 is Er^{3+} concentration in the EDF core, N_i are the populations of the correspondent Er^{3+} levels (*i*=1 to 5), *h* is Plank constant, ν_p and ν_s are the frequencies of the pump and signal waves, and I_s and I_p are the pump and signal waves' intensities. The Er^{3+} parameters' values used in modeling are listed in Table 1.

Parameter	Value	Units
Low-signal absorption at 1531 nm (experimental data)	$\alpha_{s0} = 0.016$	cm^{-1}
Low-signal absorption at 977 nm (experimental data)	$\alpha_{p0} = 0.012$	cm^{-1}
Relaxation time for $^4I_{13/2} \rightarrow {}^4I_{15/2}$ transition [10]	$\tau_{21} = 10$	ms
Relaxation time for $^4I_{11/2} \rightarrow {}^4I_{13/2}$ transition[11]	$\tau_{32} = 5.2$	μs
Relaxation time for $^4I_{9/2} \rightarrow {}^4I_{11/2}$ transition [12]	$\tau_{43} = 5$	ns
Relaxation time for $(^4F_{7/2}/^2H_{11/2}/^4S_{3/2}) \rightarrow {}^4I_{9/2}$ transition [13]	$\tau_{54} = 1$	μs
Cross-section of $^4I_{15/2} \rightarrow {}^4I_{13/2}$ transition @ 1531 nm [14]	$\sigma_{12} = 5.1\times10^{-21}$	cm^2
Cross-section of $^4I_{15/2} \rightarrow {}^4I_{11/2}$ transition @ 977 nm [14]	$\sigma_{13} = 1.7\times10^{-21}$	cm^2

Table 1. EDF "M5" / Er^{3+} parameters used in modeling

2.1. Spectral features of ESA in low-doped EDF within the 1.48 to 1.59-μm range

The first experiment serves to reveal the existence of the ESA process in the EDF at pumping through $^4I_{15/2} \rightarrow {}^4I_{13/2}$ transition, refer to Figure 1. The UC emission (UCE) spectra were recorded in the wavelengths range near 980 nm ($^4I_{11/2} \rightarrow {}^4I_{15/2}$ transition). The pump source was a 12-mW narrow-line LD (*Anritsu Tunics Plus SC*), tunable through the spectral interval, λ_s=1.48...1.59 μm. Experimentally, light from the LD was launched into the EDF (length, 1 m) through a standard 980-nm / 1550-nm wavelength division multiplexer (WDM) while de-multiplexed backward emission from the EDF was registered by an optical spectrum analyzer (OSA, *Ando AQ6315A*). The recorded UCE spectra, at 10-mW in-fiber power and for four different excitation wavelengths λ_s, are shown in Figure 2(a). It is seen that UCE power depends on λ_s and that the maximal UCE signal spectrally matches the peak wavelength of the Er^{3+} ground-state absorption (GSA) contour. The appearance of the UCE spectra at λ_p ~980 nm is very similar to that of $^4I_{11/2} \rightarrow {}^4I_{15/2}$ emission, thus testifying for effective populating of $Er^{3+4}\,I_{11/2}$ state at pumping the EDF at λ_s=1.48...1.59 μm and so for the presence of the ESA process ($^4I_{13/2} \rightarrow {}^4I_{9/2}$), followed by fast non-radiative ($^4I_{9/2} \rightarrow {}^4I_{11/2}$) relaxation. Note that no UCE spectral components in the range below 960 nm were detected.

The second experiment, allowing us to reveal the spectral dependence of ESA, was arranged through measurement of the integral lateral emission power collected from the EDF's lateral surface, using a Si photo-detector directly placed above the fiber. Given Si is sensitive from the visible to near-IR (Si band-gap wavelength is ~1.1 μm), the Si photo-detector (PD) does register the UCE signal (centered at ~980 nm) and does not the spontaneous emission (SE) from level $^4I_{13/2}$ (~1.5 μm); a scattered pump-light component was found to be extremely weak. The results obtained at various pump powers are shown in Figure 2(b) (see curves 1–3). For comparison, we also plot in this figure (by curve 4) the integrated backward UCE power measured using OSA (from a 10-cm EDF sample). It is seen that the spectral dependence of the UCE signal is similar to the one of ESA power on the excitation wavelength. In fact, the shape of the presented dependencies is established by a convolution of the known GSA and yet unknown ESA spectra of Er^{3+} (notice that the latter depends on Er^{3+} ions population inversion).

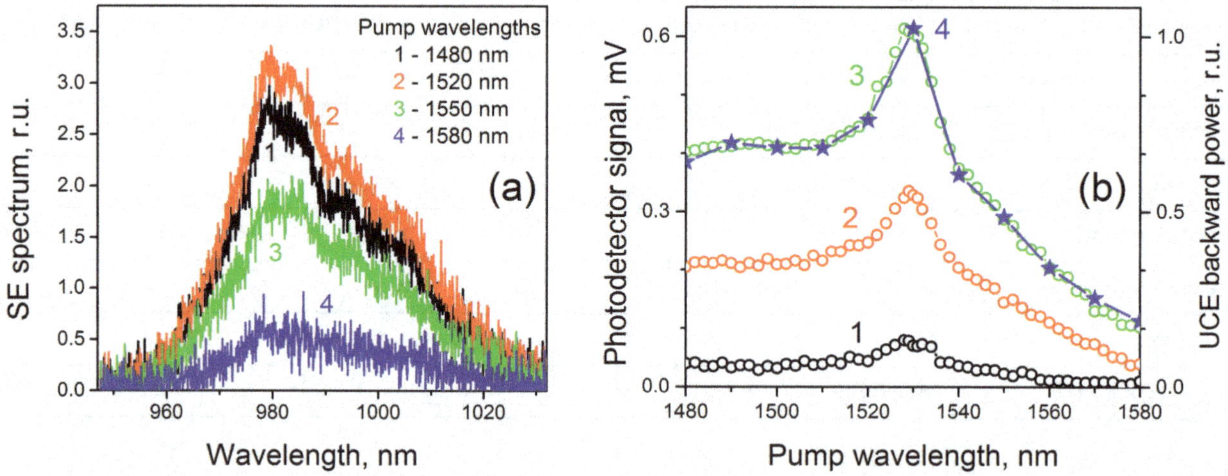

Figure 2. (a) UCE spectra at λ_s ~1.5-μm excitation. (b) Dependencies of PD signal (circles, left scale, curves 1 to 3) and normalized frontal UCE power (stars, right scale, curve 4) on excitation wavelength. Curves 1, 2, and 3 correspond to pump powers of 1, 5, and 10 mW; curve 4 is for 10-mW pump.

Considering the simplified Er^{3+} ion's model and the processes involved at excitation at λ_s ~1.5 μm (see Figure 1 and formulas (1-5)), we obtain simple formula for the normalized population of $^4I_{11/2}$ state, $n_3=N_3/N_0$ [11]:

$$n_3 = \frac{\varepsilon_s \gamma_s s_s^2}{1+\xi_s s_s + \varepsilon_s \gamma_s s_s^2}, \tag{6}$$

where $\varepsilon_s=\sigma_{24}/\sigma_{12}$ is the ESA parameter at signal wavelength, $\gamma_s=\tau_{42}/\tau_{21} \approx \tau_{32}/\tau_{21}$ ($\tau_{43} \approx 5$ ns is neglected in simulations given its smallness respectively the other decay times), $\xi_s=1+\sigma_{21}/\sigma_{12}$ is the spontaneous emission (SE) parameter at the signal wavelength, and $s_s=I_s/I_s^{sat}=P_s/P_s^{sat}$ is the saturation parameter for signal wave (with I_s being the signal wave intensity, $I_s^{sat}=h\nu_s/\sigma_{12}\tau_{21}$ the saturation intensity, $P_s=A_s I_s$, $P_s^{sat}=A_s I_s^{sat}$, P_s and P_s^{sat} the signal wave power and saturation power, respectively, and A_s the area of the Gaussian distribution of the signal wave in the EDF core). Notice that since $\varepsilon_s\gamma_s s_s \ll \xi_s$ ($\gamma_s \approx 7\times10^{-4}$ and $\varepsilon_s \le 0.6$, see below), the term with a second power on s in the denominator of (6) is omitted in further calculus. From formula (6), we find the normalized population n_3 averaged over the fiber core area if the pump wave intensity is taken to obey the Gaussian law, $s_s(r)=s_{s0}\exp[-2(r/w_s)^2]$:

$$\bar{n}_3 = \frac{1}{\pi a^2}\int_0^a n_3(r)\times 2\pi r dr = \frac{\varepsilon_s \gamma_s s_{s0}}{2\xi_s}\left(\frac{w_s}{a}\right)^2 \ln\left[\frac{1+\xi_s s_{s0}}{1+\xi_s s_{s0}(1-\Gamma_s)}\right], \tag{7}$$

where s_{s0} is the on-axis saturation parameter that depends on the pump power, P_s, as $P_s=I_s^{sat}$ $s_{s0}(\pi w_s^2/2)$, a is the EDF core radius, w_s is the modal radius of the Gaussian wave, and Γ_s is the mode's to fiber core's radii overlap factor at λ_s. The parameter ξ_s comprises the ESA parameter ε_s, the measured EDF low-signal absorption coefficient $\alpha_{s0}=\Gamma_s\sigma_{12}N_0$, and the full-saturated fiber gain $g_s=\Gamma_s\sigma_{21}N_0-\Gamma_s\sigma_{24}N_0=g_{s0}-\varepsilon_s\alpha_{s0}$ (g_{s0} is the EDF net gain at λ_s):

$$\xi_s = 1+\frac{\sigma_{21}}{\sigma_{12}} = 1+\frac{g_{s0}}{\alpha_{s0}} = 1+\frac{g_s}{\alpha_{s0}}+\varepsilon_s. \tag{8}$$

Apparently, UCE power depends on n_3 that in turns depends on ε_s. It is worth wising that the latter can be easily found from the UCE lateral signal, registered using PD (see Figure 2(b)). The results of calculation of the ESA parameters are plotted in Figure 3(a) by filled symbols (left scale), where the ε_s-spectrum is normalized on the GSA peak value (at λ_s=1.531 μm). It is seen that the ESA parameter weakly depends on λ_s within the spectral interval 1.48 μm to 1.55 μm whilst it sharply grows as λ_s approaches ~1.6 μm; this behavior reminiscences the trend reported in [15].

The method to find the absolute values of the ESA parameter ε_s is based on the measurements of the EDF's transmission coefficient $T(\lambda_s, P_s)$, i.e. in function of excitation power, and subsequent comparison of the experimental dependencies with the simulated ones, obtained after integrating, through the fiber length, of the saturation parameter $s_{s0}(z)$, with the latter being obtained from the equation for the excitation wave propagating in the fiber (z is the direction of light propagation) (see [11]):

$$\frac{ds_{s0}}{dz} = -\alpha_{s0}\frac{\varepsilon_s}{\xi_s}\Gamma_s s_{s0} - \alpha_{s0}\frac{\xi_s-\varepsilon_s}{\xi_s^2}\ln\frac{1+\xi_s s_{s0}}{1+\xi_s s_{s0}(1-\Gamma_s)}. \tag{9}$$

Note that the contribution of amplified SE (ASE) to the output spectra was as low as 0.1–1.9% respectively to the overall transmitted power, provided a short (50 cm) EDF piece has been used.

The results of fitting of the experimental EDF transmission curves within the interval 1.48...1.59 μm are shown in Figure 3(a) by asterisks (see right scale). It is evident that these results, as the ones regarding the UCE experiments, almost coincide. The solid curve shown in this figure is the best fit of the experimental data by the polynomial regression.

The knowledge of the ε_s-spectrum for the 1.48...1.59 μm band allows one to find the spectral dependence of the net-gain coefficient g_{s0} in the EDF, which is the real gain in the fiber, in contrast to the gain coefficient g_s, the quantity commonly but uncritically dealt with in experiments and simulations with EDF and EDFL[13]: In fact g_{s0} is diminished by the ESA parameter's value. Figure 3(b) demonstrates the spectral dependences of the three coefficients (g_{s0}, g_s, and α_{s0}), where the spectra for g_s and α_{s0} were measured using standard methods (see

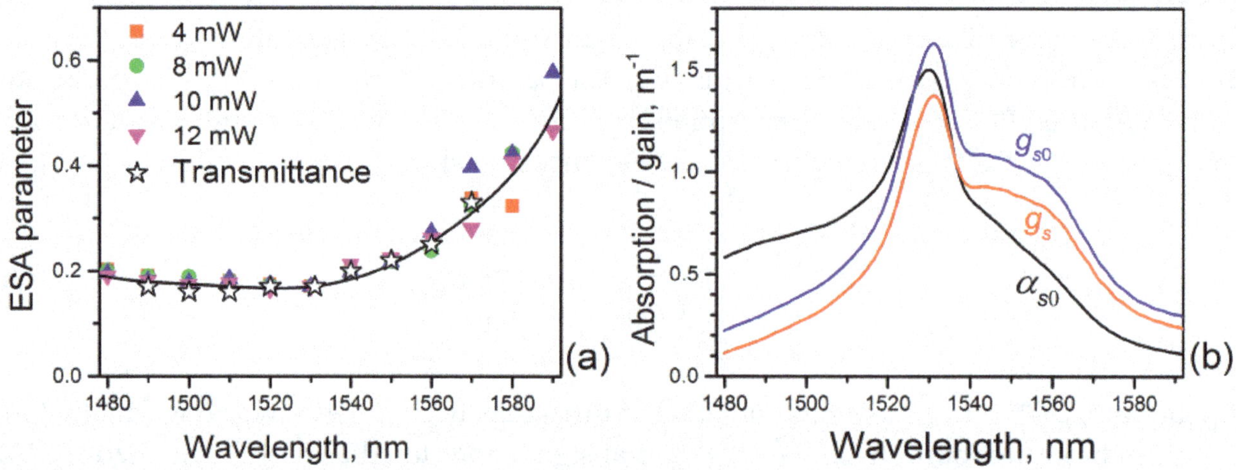

Figure 3. (a) ESA parameter ε_s calculated from the measurements of PD signal (filled symbols) and the ones of the EDF transmission coefficient (stars); solid line is the best polynomial fit. (b) Spectral dependencies of the EDF's small-signal absorption coefficient α_{s0}, full-saturated gain coefficient g_s, and net-gain coefficient g_{s0}.

e.g. Ref. 7) and the spectrum for g_{s0} was calculated using the g_s-and α_{s0}-spectra and the fitted ε_s-spectrum, taken from Figure 3(b). Worth noticing, the data obtained above are universal for any silica-based EDF weakly doped with Er^{3+}, where the "concentration" effects (to be discussed in section 3) are negligible.

2.2. ESA in low-doped EDF at 977 nm

We report here the results of a study of the other ESA process, $^4I_{11/2}\rightarrow{}^4F_{7/2}$ (see Figure 1), which takes the place when EDF is excited simultaneously at the two GSA wavelengths: λ_p ~977 nm (through $^4I_{15/2}\rightarrow{}^4I_{11/2}$ transition) and λ_s ~1531 nm (through $^4I_{15/2}\rightarrow{}^4I_{13/2}$ transition), a situation normally encountered in a diode-pumped (at λ_p~980 nm) EDFL or EDF-based amplifier (EDFA). Upon simultaneous excitation of EDF at these two wavelengths, population of $^4I_{11/2}$ state increases, resulting in significant growth of the pump-induced ESA loss (when $^4I_{11/2}\rightarrow{}^4F_{7/2}$ transition gets "switched on", see Figure 1). On the contrary, without the presence of a signal wave in the fiber, almost all ions will be at the upper laser level $^4I_{13/2}$, which leads to a lower absorption at λ_p.

The first experiment was focused on the demonstration of the presence of "pump-ESA", when EDF is pumped simultaneously at λ_p=977 nm and at λ_s=1531 nm. It was aimed to reveal of whether UCE is observed (through transitions $^2H_{11/2}\rightarrow{}^4I_{15/2}$ and $^4S_{3/2}\rightarrow{}^4I_{15/2}$ "5"→"3", see Figure 1), as following the ESA process at λ_p. Experimentally, an EDF piece was pumped, using 980 nm / 1550 nm WDM supporting up to 1 W of optical power, by the pump and signal beams from the same EDF's side. These beams were delivered from a standard fibered LD (λ_p=977 nm) and from another semiconductor laser with wavelength λ_s=1531 nm, followed by an EDFA. Figure 4 presents two photographs of lateral emission from the EDF sample when the fiber was pumped, respectively, (*i*) solely at wavelength λ_p (Figure 4(a)) and (*ii*) simultaneously at wavelengths λ_p and λ_s (Figure 4(b)); the incidence pump and signal powers were P_p=P_s=260 mW. One can readily see bright UCE (in the green spectral range, ~520–560 nm) when the EDF is pumped at both the wavelengths, and very weak UCE when it is pumped at λ_p only.

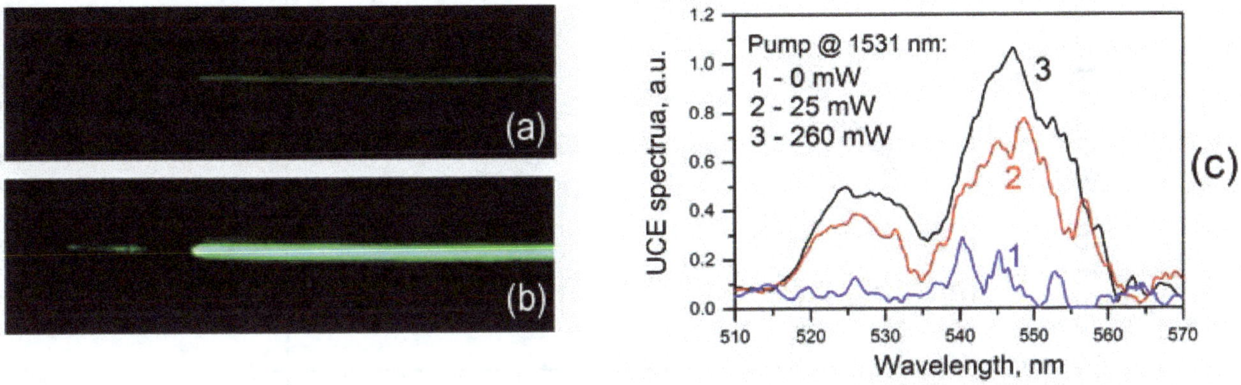

Figure 4. (a) and (b) Photographs of ~520/~545 nm UCE obtained from the EDF's surface at in-core excitation (a) only at λ_p=977 nm and (b) simultaneously at λ_p=977 nm and λ_s=1531 nm. (c) Frontal UCE spectra recorded from a 5-cm EDF sample at in-core excitation at pump wavelength (P_p=260 mW) and variable signal powers (P_s=0,25, and 260 mW).

The UCE spectra, obtained using a 5-cm EDF sample, are shown in Figure 4(c). The spectra were recorded for three signal powers (0, 25, and 260 mW) while fixed pump power (260 mW). One can see from the figure that even low-power (25 mW) signal radiation tremendously enhances UCE (compare curves 1 and 2) and that at further increasing signal power, up to P_s=260 mW, growth of the UCE power becomes slower, demonstrating a saturating behavior (compare curves 2 and 3). Thus, the signal at 1531 nm "ignites" the ESA process at λ_p and thereby increases the number of Er^{3+} ions in the ground state, resulting in an increase of the pump-light absorption. Consequently, there will be a non-negligible population of Er^{3+} ions on $^4I_{11/2}$ level and, as the fact of matter, a more effective ESA process at the pump wavelength, seen as growth of pump-induced loss.

Figure 5 shows the dependence of "green" UCE signal (transition "5"→"3", see Figure 1), detected from the EDF, on powers of pump and signal radiations launched into the fiber (in this case, again, the pump power was kept fixed, P_p=260 mW, and the signal power was varied, P_s=0...290 mW). Experimentally, lateral UCE power $P_{UCE}(P_s)$ was measured using a photo-multiplier with a cesium cathode from a short section (~1 mm) of the EDF near its splice with WDM. It is seen that the UCE's behavior demonstrates the already noticed aspects: At very low signal power UCE is extremely weak, whilst upon its increase UCE first strongly enhances and then gets saturated.

Since UCE power is proportional to population of the 5th level of Er^{3+} ions, the experimental data can be fitted well by a simulated curve of normalized population inversion n_5=N_5/N_0 (N_5 is the population of the 5th Er^{3+} ion's level) averaged over the fiber core area, which confirms the theory we built. Considering that both the excitation waves (at 977 nm and at 1531 nm) have Gaussian spatial distributions, one can obtain from the steady-state rate equations for the Er^{3+} ion (formulas 1-5) the following expression for the averaged population $\bar{n}_5$:

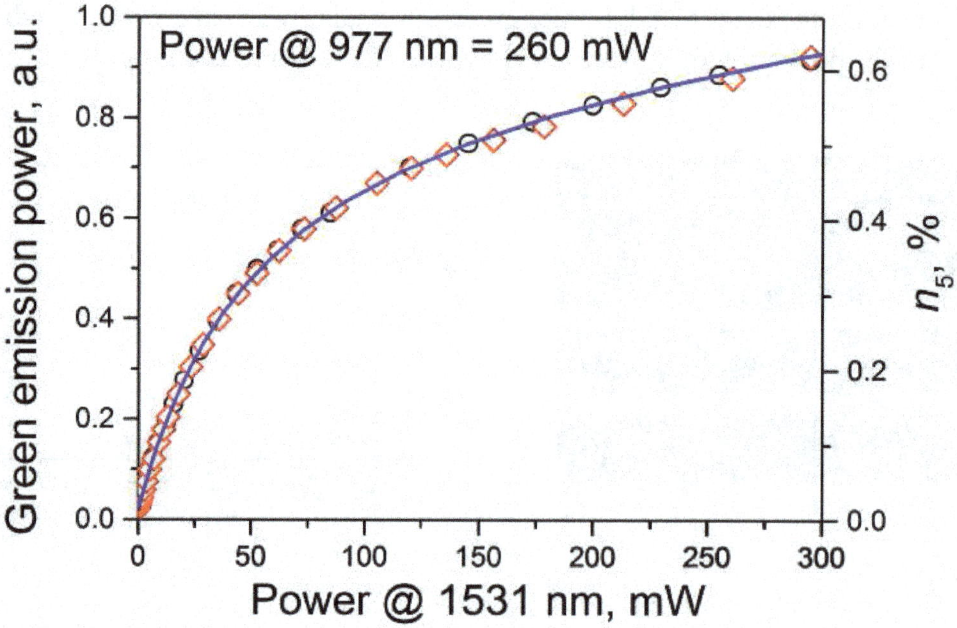

Figure 5. Dependence of UCE power on signal power (left scale): Circles and rhombs correspond to two different experimental series; plain curve is a theoretical fit of normalized inversion of the 5th Er^{3+} level (right scale).

$$\bar{n}_5 = \frac{\varepsilon_p \gamma_2}{a^2} \int_0^a \frac{1 - n_1(r) - n_2(r)}{1 + \varepsilon_p \gamma_2 s_p(r)} s_p(r) 2r dr \tag{10}$$

where the normalized populations $n_1(r)$ and $n_2(r)$ are taken from [16] (found from the same set of rate equations for the Er^{3+} ion); $\varepsilon_p = \sigma_{35}/\sigma_{13}$ is the ESA parameter at λ_p, $\gamma_p = \tau_{53}/\tau_{31}$ (with $\tau_{53} \approx \tau_{54}$ ≈ 1 μs and $\tau_{31} \approx \tau_{21} \approx 10$ ms, see Table 1), and $s_p = I_p / I_p^{sat}$ is the saturation parameter at the pump wavelength (I_p is the pump intensity, $I_p^{sat} = h\nu_p/\sigma_{13}\tau_{21}$, and $h\nu_p$ is the quanta energy at λ_p). As seen from Figure 5, the simulated curve for population of the 5th level ($\bar{n}_5$) fits well the experimental data, thus confirming correctness of the theory.

We found that the best way to find the ESA parameter at the pump wavelength, λ_p=977 nm, is to measure the EDF's nonlinear transmission coefficient at the pump wavelength in function of signal power, $T_p(P_s)$, and then compare this data with the simulated ones, obtained from the steady-state rate equations (formulas 1-5) for the fife level Er^{3+} energy diagram (see Figure 1). Considering the Gaussian radial intensity distributions for the pump and signal waves, the equations describing $T_p(P_s)$ at fixed pump power (P_p=260 mW in our case) take the form [16]:

$$\frac{dP_p(z)}{dz} = -\frac{4\alpha_{p0}}{\Gamma_p w_p^2}\left[\int_0^a [n_1(r, z) - (\xi_p - \varepsilon_p) n_3(r, z)] exp\left[-2\left(\frac{r}{w_p}\right)^2\right] r dr + \alpha_{BG}\right] P_p(z) \tag{11}$$

$$\frac{dP_s(z)}{dz} = \frac{4\alpha_{s0}}{\Gamma_s w_s^2}\left[\int_0^a [(\xi_s - \varepsilon_s) n_2(r, z) - n_1(r, z)] exp\left[-2\left(\frac{r}{w_s}\right)^2\right] r dr - \alpha_{BG}\right] P_s(z) \tag{12}$$

where α_{p0} is the small-signal EDF absorption and Γ_p is the pump wave to fiber core overlap factor, both at the pump wavelength, α_{BG} is the small background EDF loss (~3 dB/km for the EDF used), w_p is the radius of the Gaussian distribution of pump wave, and ξ_p=1+σ_{31}/σ_{13} is the SE parameter at the pump wavelength. The normalized populations of Er^{3+} levels n_1, n_2, and n_3 are found from the rate equations for the Er^{3+} ion (the routine is not discussed here because of its completeness; refer for details to [16]).

Actually, the coefficient adjacent to P_p from the left side (Equation (11)) is the EDF absorption coefficient at the pump wavelength, whereas the one adjacent to P_s from the left side (Equation (12)) is the EDF gain at the signal wavelength; both the coefficients depend on the pump and the signal powers. It is worth of mentioning that the set of equations (11) and (12), added by the corresponding boundary conditions, can be also used for modeling a CW EDFL.

Figure 6 shows the EDF transmissions measured for three different EDF lengths along with the best fits obtained using equations (7) at varying ε_p and ξ_p (and as the result their most relevant values found to be: ε_s=0.17 and ξ_s=1.08, both at 1531 nm). As seen from the figure, the best fitting of the experimental EDF transmission at the pump wavelength is provided at ε_p=0.95 and ξ_p=1.08. Thus, the ESA cross-section at the pump wavelength is strong enough (being less than the GSA one only slightly), which ought to affect (decrease) an EDFL's efficiency via the pump-induced loss distributed along the active fiber.

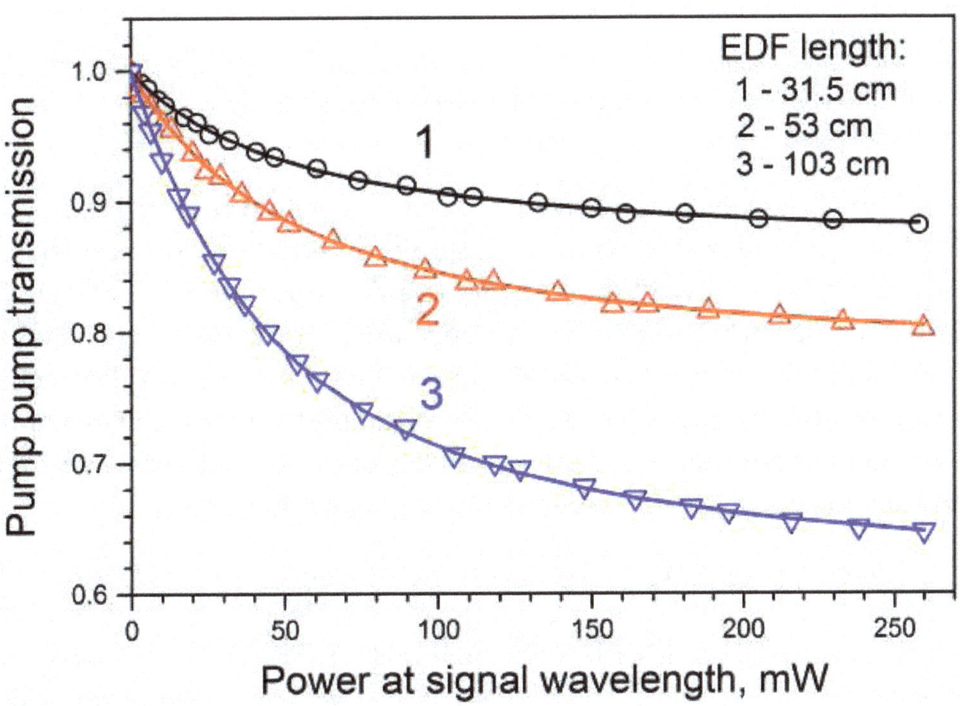

Figure 6. Experimental (symbols) and theoretical (curves) dependencies of pump (λ_p=977 nm) wave's transmission coefficient on signal (λ_s=1531 nm) wave's power; the data are obtained for three different EDF lengths (indicated in the inset). Theoretical curves are the best fits to the experimental data obtained at ε_p=0.95 and ξ_p=1.08.

2.3. Effect of ESA in EDF upon efficiency of CW EDFL.

We discuss hereafter the results of a theoretical study of influence of ESA inherent in Er^{3+} ions upon efficiency of an EDFL assembled in the linear (Fabry-Perot) configuration. The laser setup we shall deal with at modeling is shown in Figure 7. The low-doped EDF discussed above was considered to be an active medium and two fiber Bragg Gratings (FBGs) – to form selective couplers of the cavity, both centered at λ_s=1550 nm, the laser (signal) wavelength. The EDF length was chosen to be 4 m whereas the EDFL cavity length L_c, including the EDF piece and FBGs' tails, to be 6 m. Pump power on the EDF input was fixed in modeling at 200 mW.

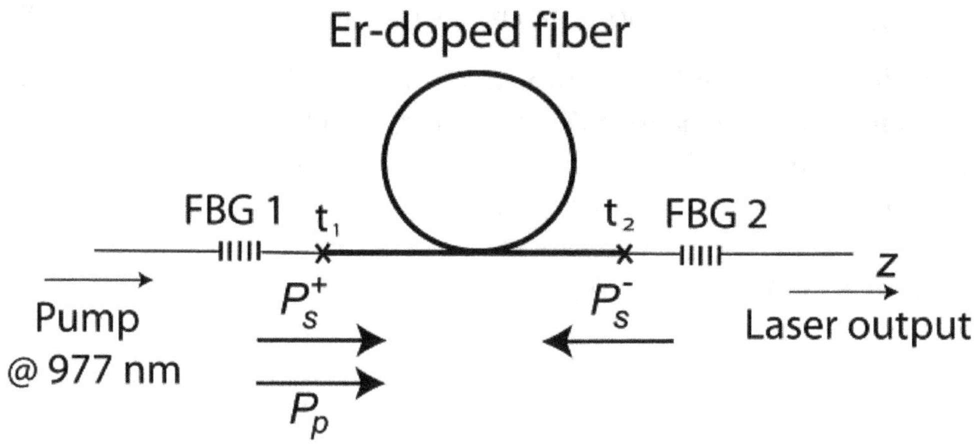

Figure 7. Sketch of the EDFL's geometry: $P_s^+$and P_s^- are the laser (signal) waves' powers propagating in the positive (+) and negative (-) z direction, respectively; P_p is the pump power propagating in the positive z direction; t_1 and t_2 are the transmissions on the fibers' splices.

We imply (see Figure 7) that the pump wave at 977 nm (absorption peak of $^4I_{15/2} \rightarrow {}^4I_{11/2}$ transition, see Figure 1) propagates rightward while the laser waves – rightward (marked by "+" superscript) and leftward (marked by "–" superscript), respectively. In the scheme, FBG1 plays the role of a rear 100% reflector (reflection R_1=1), transparent for the pump wave, whereas FBG2 forms an output coupler with reflection R_2 varied for optimizing the laser efficiency. To decrease loss on the fiber splices, both FBGs were considered to be written in a photosensitive fiber with waveguide parameters similar to the EDF's ones. At the laser wavelength the low-signal EDF absorption α_{s0} was taken to be 0.0069 cm^{-1}, according to [17].

The laser is simulated by applying the two contra-propagating laser waves' model discussed in details in [17] with taking into account Gaussian distributions of the laser and pump waves. In this model, the pump wave is described by equation (11) and the signal waves – by equation (12) with a small modification being that powers of the two contra-propagating signal waves are assumed to be governed by the equation:

$$\frac{d P_s^{\pm}(z)}{dz} = \pm \frac{4\alpha_{s0}}{\Gamma_s w_s^2}\left[\int_0^a [(\xi_s - \varepsilon_s)n_2(r,\ z) - n_1(r,\ z)]exp\left[-2\left(\frac{r}{w_s}\right)^2\right]rdr \mp \alpha_{BG}\right]P_s^{\pm}(z) \tag{13}$$

In equation (13) the superscripts "s" stand for the laser wavelength (1550 nm). The model includes also the two contra-propagating waves of SE (not shown in Figure 7) spectrally centered at λ_{se}=1531 nm (this wavelength corresponds to the GSA peak of $^4I_{15/2}\rightarrow{}^4I_{13/2}$ transition), powers of which obey the equation:

$$\frac{dP_{se}^{\pm}(z)}{dz} = \pm\left[g_{se}(z)P_{se}^{\pm}(z) \mp \frac{\Omega}{4\pi}\frac{\alpha_{se0}P_{se}^{sat}}{\Gamma_{se}}\int_0^a n_2(r, \ z)2rdr\right] \tag{14}$$

where the second term on the right side is the SE power generated by a short fiber section dz, α_{se0}=0.016 cm^{-1} is the low-signal absorption at λ_{se}, P_{se}^{sat} is the saturation power at 1531 nm, Γ_{se} is the overlap factor for the SE waves, $\Omega=\pi NA/n^2$ is the fraction of SE photons guided by the EDF core in each direction, and n is the modal refractive index at λ_{se}. The EDF gain at λ_{se} is written as

$$g_{se}(z) = \frac{4\alpha_{se0}}{\Gamma_{se}w_{se}^2}\left[\int_0^a\left[(\xi_{se} - \varepsilon_{se})n_2(r, \ z) - n_1(r, \ z)\right]exp\left[-2\left(\frac{r}{w_{se}}\right)^2\right]rdr - \alpha_{BG}\right] \tag{15}$$

The boundary conditions are written as:

$$P_p(z=0, \ t)=P_{p0} \tag{16}$$

$$P_s^+(z=0, \ t)=P_s^-(z=0, \ t)R_1t_1^2 \tag{17}$$

$$P_s^-(z=L_c, \ t)=P_s^+(z=L_c, \ t)R_2t_2^2 \tag{18}$$

$$P_{se}^+(z=0, \ t)=P_{se}^-(z=L_c, \ t)=0 \tag{19}$$

$$P_s^{out}(t)=P_s^+(z=L_c, \ t)(1 - R_2)t_2=0 \tag{20}$$

where P_{p0} is the pump power at the EDF input and P_s^{out} is the EDFL output power. To simplify calculations, we considered that t_1=t_2=0 (i.e. no loss on the fiber splices).

The EDFL's efficiency as a function of FBG2's reflectivity R_2, simulated using the laser model described above, is depicted in Figure 8(a). The EDFL was modeled for four different "versions" of the energy level system: Without considering all the ESA transitions (curve 1); with considering the pump ESA only (curve 2) and the signal ESA only (curve 3); with considering all the ESA transitions (curve 4).

The first important observation is that the optimal reflection of output FBG2, at which EDFL demonstrates the maximal efficiency, is drastically decreased when the ESA transitions are accounted for. For instance, the optimal reflection of FBG2 is ≈66% when only the EDF background loss (3.1 dB/km, see Table 1) is present (curve 1), whereas it is ≈11% when all kinds

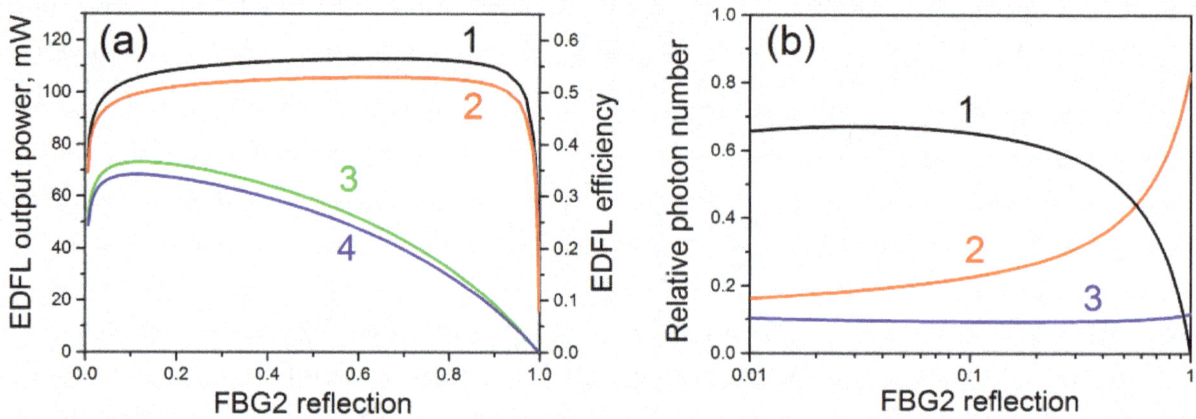

Figure 8. (a) EDFL output power and (b) fractions of the absorbed pump photons spent on the laser output (quantum efficiency) as functions of output (FBG2) reflectivity, R_2. The designations of curves 1 to 4 are given in the text. The small fraction of pump photons spent at the background loss and the ASE contribution are not shown.

of the ESA loss are accounted for (curve 4). The other important fact is that the range of FBG2's reflections, in which the output laser power varies within 10% (with respect to its maximum value if all the ESA transitions are considered), is relatively broad: ~2.5% to ~34%. At the optimal FBG2's reflectivity (R_2=11%) the laser efficiency reaches ~34% (implying all fiber splices are made lossless).

The fractions of pump and signal photons spent on the ESA transitions with respect to the absorbed by EDF pump photons are shown in Figure 8(b). It is seen that the contribution of the signal ESA loss is bigger when FBG2's reflection is bigger (curve 2). Furthermore, if reflection of the output coupler approaches 100%, the absorbed pump power is spent entirely on the ESA transitions (see curves 2 and 3) and no photons at the laser output are produced (see curve 1). At optimal FBG2's reflectivity (≈11%), about 23% of absorbed pump photons are spent on ESA at the laser wavelength and about 9.5% on ESA at the pump wavelength. Note that the sum of the relative photon numbers, as can be revealed from curves 1, 2 and 3, is approximately equal to 1 in the whole range of FBG2's reflections.

The reader is advised here to refer to [18] for comparison of the developed theory with some of the experimental data on laser efficiency of EDFLs based on the EDF of M-type with relatively high Er^{3+} concentrations.

Finally, we conclude that the ESA processes at the laser and pump wavelengths strongly affect an EDFL's efficiency and output coupler's optimal reflectivity, at which the laser output power is maximal.

3. Er^{3+} concentration effects in EDF

In this part, we shall discuss the Er^{3+} concentration effects in EDFs resulting in a reduced efficiency of EDF based lasers and amplifiers, which is associated with the phenomenon of

Er^{3+} ions' clustering that leads, in turn, to non-saturable absorption (NSA) through inhomogeneous up-conversion (IUC).

For our experiments we selected the most representative commercial EDFs fabricated through the MCVD and direct nanoparticle deposition (DND) processes; all the fibers under scope in this section are similar in the sense of Er^{3+} doped core's chemical composition being the most common alumino-silicate glass (in the case of MCVD-EDFs with addition of germanium). The first series of the EDFs (MCVD-based, "M"-series) includes two fibers: M5-125-980 and M12-125-980 (*Fibercore*), hereafter M5 and M12, and the second series (DND-based, "L"-series) – three fibers: L20-4/125, L40-4/125, and L110-4/125 (*Leikki / nLight*), hereafter L20, L40 and L110. These fibers have very similar waveguide parameters and differ mainly in Er^{3+} doping level. [Notice that the EDFs employed in the whole of above experiments, see Section 2, belong to M-series.]

3.1. Absorption and fluorescence spectra

The EDFs' absorption spectra are shown in Figure 9(a) where Er^{3+} transitions $^4I_{15/2} \rightarrow {}^4I_{11/2}$ (within a 940...1020 nm range with a peak at 978 nm) and $^4I_{15/2} \rightarrow {}^4I_{13/2}$ (within a 1400...1600 nm range with a peak at 1.53 μm) are featured. The spectra were obtained using a white light source with fiber output and OSA with 1-nm resolution. It is seen that the absorption spectra of the EDFs of both series have a very similar shape (given by similarity of core glass chemical compositions), differing only in intensity. The ratio of the peaks' magnitudes at 1.53 μm and at 978 nm was measured to be equal to ~1.6, for the M and L EDFs.

Figure 9(b) demonstrates the normalized fluorescence spectra for L fibers (to simplify the picture the spectra for M fibers are not shown), measured at the maximal pump power at 978 nm, P_p ~400 mW, within the 450...1650 spectral range (the area nearby the pump wavelength is cut out). In this experiment the lateral geometry, when fluorescence is captured by a multimode fiber patch cord from the lateral surface of the short fiber samples, was arranged. In spite of Er^{3+} concentration increases in the row of fibers L20 → L40 → L110, the Er^{3+} fluorescence band, centered at ~1.53 μm, is indistinguishable in shape. Although the 1.53-μm band dominates in the EDFs' fluorescence spectra, there also exist the spectral lines at its anti-Stokes side (~450...1100 nm), which evidences the presence of UC. Note that, in contrast to the ~1.53-μm band's stability against Er^{3+} concentration, the higher Er^{3+} content, the more intense the UCE (compare curves 1 – 3 in Figure 9(b)).

To understand the origin of UCE and the dependence of UC intensity upon Er^{3+} concentration in the EDFs, let's refer to Figure 10 in which the scheme of Er^{3+} energy levels and a sketch of the processes involved at the excitation at λ_p=978 nm are presented. UCE (shown in the figure by grey arrows) seems to be mostly associated to Er^{3+} ion clusters being in states $^4I_{11/2}$ and $^4I_{13/2}$, because the ESA process, equally acting for single and clustered Er^{3+} ions, is ineffective at 978-nm excitation.

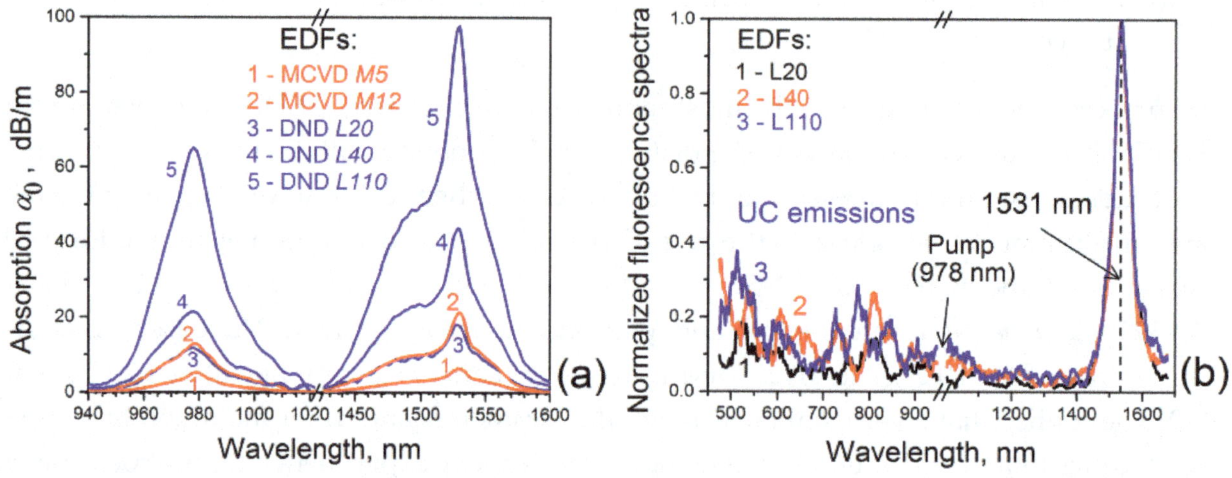

Figure 9. (a) Absorption spectra of the EDFs of L-(blue curves) and M-(red curves) series in the near-IR. (b) Fluorescence spectra of the EDFs of L-series in the VIS...near-IR spectral range at 978-nm pumping.

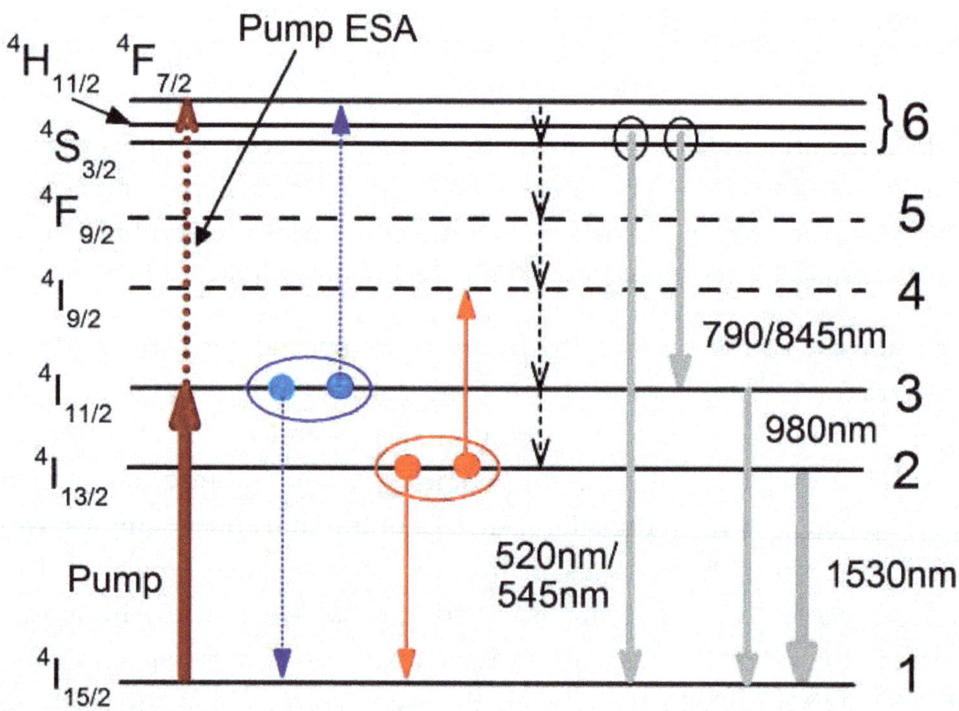

Figure 10. Scheme of Er^{3+} energy levels, applicable for the EDFs with high Erbium content. Functioning of Er^{3+} clusters (shown for simplicity as ion pairs) is sketched by the blue and red arrows for long-living manifolds $^4I_{11/2}$ and $^4I_{13/2}$; the black dotted arrows depict non-radiative relaxations; the grey arrows show the UC and "fundamental" (1.53-μm) emissions; the short-living levels are shown by dashed lines.

3.2. Fluorescence decay kinetics

Like at the fluorescence spectra' measurements discussed above, the kinetics of near-IR fluorescence at ~1.53 μm was detected using the lateral geometry. However, the pump light at 978 nm was in this case switched on / off by applying a rectangular modulation of LD current at Hz-repetition rate. The launched into the EDF samples pump power was varied between zero and ~400 mW. The fluorescence signal was detected either using an InGaAs PD with a Si filter placed between the fiber and a multimode patch cord delivering fluorescence to PD (being so the measurements above ~1.1 μm where the use of Si filter allows cutting off the pump light's spectral component), or using a fast Si-PD with no spectral filtering applied (being in fact the measurements below ~1.1 μm), placed directly above a slit segregating a portion of fluorescence from the EDF's surface. To diminish ASE and re-absorption on the results, we used short (~0.5 cm) EDF pieces.

Typical kinetics of the fluorescence signal, recorded after switching pump light at 978 nm off, are presented in Figure 11 for the heavier doped EDFs M12 (a) and L110 (b); the data were acquired using InGaAs-PD with Si filtering (transmission band above 1.1 μm). We don't present here the results for other, lower doped, EDFs as these showed similar but less featured trends in the decay kinetics.

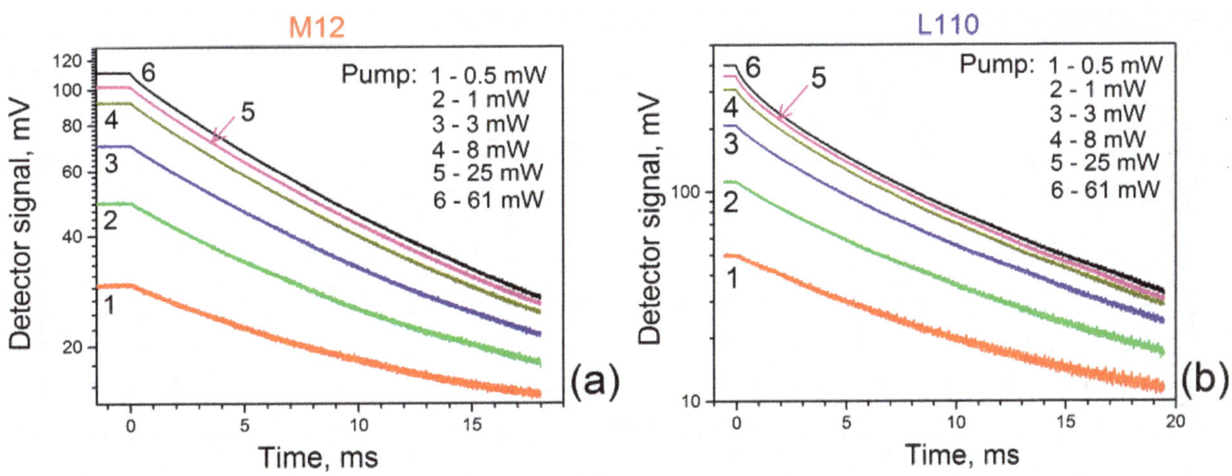

Figure 11. Fluorescence decay kinetics obtained for the EDFs M12 (a) and L110 (b). Curves 1 to 6 are captured for different pump levels (see the insets in the right upper corners). Zero-time corresponds to the moment when the pump light is switched off.

It is seen from Figure 11 that for these two EDFs fluorescence power, corresponding to 1.53-μm spectral band, is saturated (as is saturated GSA of Er^{3+} ions) yet at a few mW of pump power. However the key feature is deviation from the exponential law in the fluorescence kinetics in EDF L110 (see Figure 11(b)). A similar trend occurs but is less expressed in the rest of L and M fibers with lower Er^{3+} concentration; see e.g. Figure 11(a). Another fact deserving attention is the presence of a sharp drop in the fluorescence signal in fiber L110 at high pump powers, which happens after switching pump off (refer to curves 4–6 in Figure 11(b)). Such a feature is present but in a smaller degree also in fibers L40, L20, and M12 (having substantially lower Er^{3+} contents) and almost vanishes in fiber M5 (having the lowest Er^{3+} content). Note that similar fluorescence kinetics were observed in some of the earlier reports, see e.g. [4, 10].

Overview of the fluorescence decays for all the EDFs under scope is provided in Figure 12 (points). These data were obtained at maximal pump power, P_p at 978 nm (400 mW; the high pump power was found to be the right choice for minimizing spatial diffusion of excitation; see e.g. [10]). The dependences in the figure demonstrate the fluorescence decay "tails" obtained after cutting off the short initial segments just after switching pump off (~30 μs), which permits elimination of the influence of non-instant LD power decay (~8 μs).

As seen from Figure 12, 1.53-μm fluorescence decays get more and more deviated from the single exponential law when Er^{3+} concentration increases: The fibers with smaller contents of Er^{3+} ions (M5, M12, and L20) demonstrate decays nearly a single-exponent law whereas fibers L40 and L110 – the decays, apparently different from this law. These features, associated to the Er^{3+} concentration effect, can be addressed in terms of the IUC process – see Section 3.3 where the results of modeling of Er^{3+} fluorescence kinetics are presented. The modeling of the fluorescence kinetics allowed us to get, for each EDF, lifetime τ_0 and constant C_{HUP*} (characterizing the homogeneous UC process, HUC; see modeling below) and thereafter to build their dependences upon small-signal absorption value α_0 at 1.53 μm (and hereby upon Er^{3+} ions concentration, proportional to α_0).

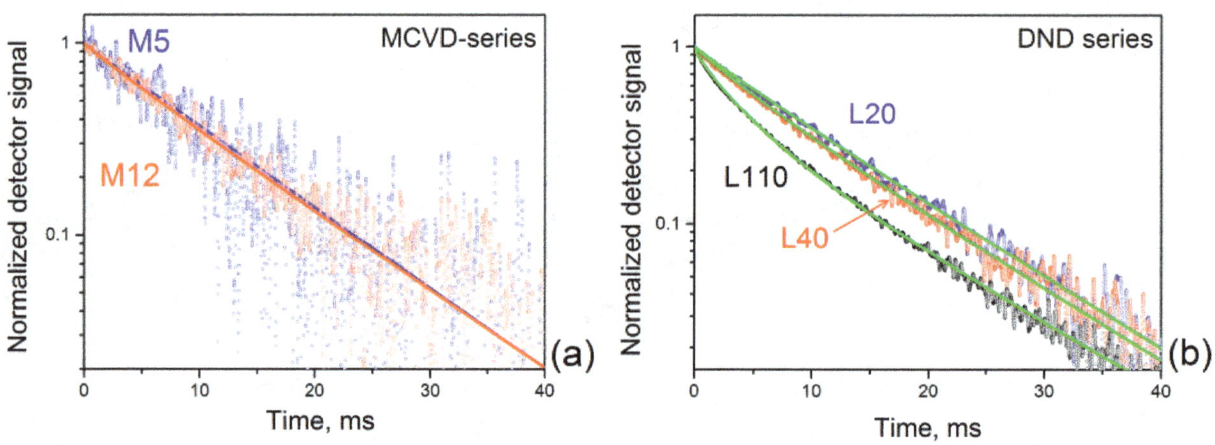

Figure 12. Normalized fluorescence decay kinetics obtained for the EDFs of L-(a) and M-(b) series; points – experimental data (using InGaAs-PD with Si filtering); plain curves are the theoretical fits made using formula (22).

Furthermore, Figure 13 demonstrates the results of the fluorescence kinetics' measurements in the optical band below 1.1 μm within a short (tens of μs) interval just after switching pump light off, for EDFs M12 (a), L40 (b), and L110 (c). The measurements were fulfilled using Si-PD without optical filtering at P_p=400 mW. As is seen from the figure, there are two "fast" components in the PD signal's decay: The shortest (≈8 μs) one, being in fact the setup's technical resolution and originating from the scattered pump light, and the longer one, measured by 21 μs to 26 μs for fibers M12, L40, and L110 (for fibers M5 and L20 this component was not resolved). A similar component was also detected in the 1.53-μm fluorescence kinetics; see Figure 11, which evidences its non-radiative nature. Note that there are known the processes in Er^{3+} -doped materials attributed by similar times [19, 20].

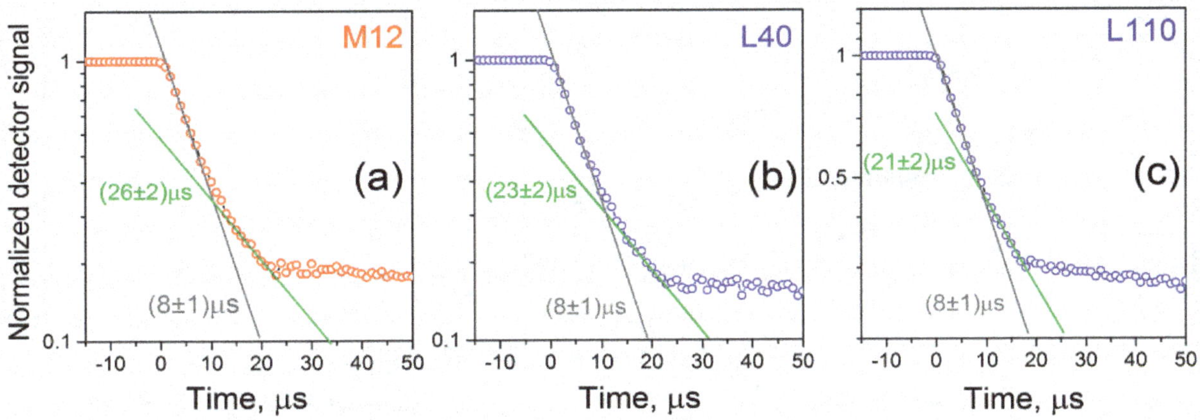

Figure 13. Fluorescence decay kinetics in the EDFs measured using Si-PD: M12 (a), L40 (b), and L110 (c); the short-time components of the fluorescence signals are specified in each plot.

We suggest that the found feature stems from a partial excitation relaxation in Er^{3+} clusters since it is present in the heavier doped EDFs but almost vanishes in the lower doped ones. The magnitude of the short-living component is a function of Er^{3+} concentration (and therefore of α_0), as seen when comparing the plots (a), (b), and (c) in Figure13: The higher Er^{3+} concentration the larger is relative (to the technical, i.e. originated from pump-light scattering) magnitude of this component.

3.3. Nonlinear absorption coefficient

The nonlinear absorption coefficient of a rare-earth doped fiber (e.g. EDF) as a function of pump power $\alpha(P_p)$ contains the useful information about GSA saturation and, consequently, about the fiber's potential as a laser medium. On the other hand, such effects deteriorating laser 'quality' of the fiber as ESA and concentration-related HUC / IUC (lifetime quenching and non-saturable absorption) ought to affect the behavior of $\alpha(P_p)$, too [18].

In the study to be reported hereafter, pump light was delivered to an EDF sample from the same LD operated at 978 nm (used in the measurements of fluorescence spectra and lifetimes); pump power was varied from ≈0.5 to 400 mW. We measured first the nonlinear transmission coefficient of the EDF sample with length L_0, which is defined as $T_{978}=P_P^{out} / P_P^{in}$ where P_P^{in} and P_P^{out} are the pump powers at the EDF's input and output. Then we made a formal re-calculation of the experimental transmission coefficient $T_{978}(P_P^{in})$ in the absorption coefficient, applying formula: $\alpha(P_P^{in})=-ln(T_{978}) / L_0$. The EDFs' lengths were chosen such that overall trends in the dependences $\alpha(P_p^{in})$ can be viewed within the whole range of pump powers. The ratio of the EDFs' lengths was such that the optical density (the product $\alpha_0 L_0$) is almost the same for all the samples, which is worth for estimation of the nonlinearity (saturation) of absorption at increasing Er^{3+} concentration in the fibers. Using the OSA, we checked the ratio of pump to ASE powers at the EDFs' outputs; it was found that the ASE contribution is negligible in all samples at P_p^{in}> 0.5 mW.

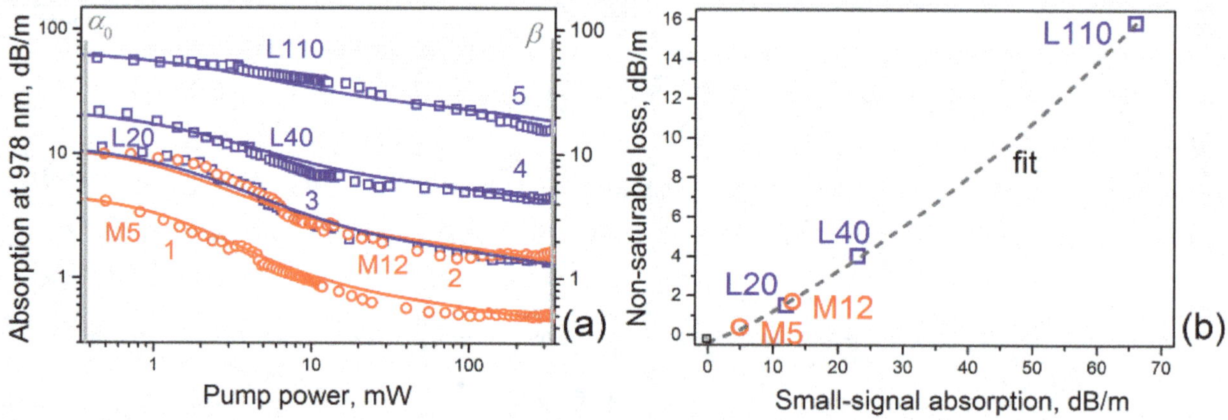

Figure 14. (a) Nonlinear absorption coefficients of the EDFs of L-and M-types vs. pump power at 978 nm. Symbols: experimental points; plain curves: theoretical fits obtained using Eqs. (25-27). Fiber lengths used in experiments and at modeling were, correspondingly: 188.6 (M5, curve 1), 59.4 (M12, curve 2), 43.5 (L20, curve 3), 22.4 (L40, curve 4), and 9.5 (L110, curve 5) cm. (b) Non-saturable absorption β vs. small signal absorption α_0, measured for the entire EDFs' set; the fitting curve is for guiding the eye.

The results obtained by applying the drawn procedure are shown in Figure 14(a) by symbols. Coefficients α_0 and β (saturated absorption at pump wavelength) marked in the upper left and right corners of the figure correspond to the limits of small-signal and saturated pump absorptions. First notice that absorption is "bleached" (in other words, transmission is "saturated") by a more or less similar manner for either fiber. However, as is also seen from the figure, the "residual" absorption (β) rises drastically with increasing Er^{3+} concentration in the fibers' sequences M5→M12 and L20→L40→L110. This trend points out that the ratio between the residual (β) and small-signal (α_0) absorptions is much bigger for the heavier doped fibers (M12, L40 and L110). In fact, pump-induced (looking as residual) absorption β is the measure of nonlinear absorption loss in EDF, as it stems from the modeling results (see Section 3.4 below). Furthermore, the dependence $\beta(\alpha_0)$ plotted in Figure 14(b) allows one to reveal that this pump-induced excessive loss in the EDFs appear as one of the most important Er^{3+} concentration effects.

3.4. Modeling

Firstly, the kinetics of near-IR (~1.53 μm) fluorescence decays obtained for the entire set of the EDF samples were modeled, which allows us to find Er^{3+} fluorescence lifetimes τ_0 and HUC coefficient C_{HUP}. The model implicitly implies, in accord to the definition of the HUC process, the only interactions between rather distant single Er^{3+} ions, not forming "chemical", tightly coupled, clusters, which are in turn supposed to interact via the IUC mechanism, discussed further. At this step of modeling we disregard the short-time features in the fluorescence decays, which are assumed to originate from Er^{3+} ions gathered into quenched (weakly fluorescing) clusters. That fact that in the experiments on fluorescence kinetics we used comparatively high pump powers (400 mW) allows us to neglect the intensity dependent HUC contribution and excitation migration effects.

For the normalized population density n_2^s of single (index "s") Er^{3+} ions being in the first excited (laser) state $^4I_{13/2}$, the following rate equation holds [21]:

$$\frac{dn_2^s}{dt} = -\frac{n_2^s}{\tau_0} - C_{HUP}\left(n_2^s\right)^2 \tag{21}$$

where $n_2^s = N_2^s / N_0^s$; N_2^s is the population density of single Er^{3+} ions in the excited state 2 ($^4I_{13/2}$), N_0^s is their concentration, and C_{HUP} [s^{-1}] is the UC parameter, being a product of the "volumetric" HUC constant C_{HUP}^* [$s^{-1}cm^3$] and concentration N_0^s: $C_{HUP} = N_0^s C_{HUP}^*$.

Assuming that pump power at 978 nm is high enough to achieve maximal populating of the excited state $^4I_{13/2}$, i.e. at "infinite" pump power, the part of Er^{3+} ions being in the excited state is $k=\sigma_{12}/(\sigma_{12}+\sigma_{21})$. Furthermore, since in our experimental circumstances (where near-IR fluorescence is detected at ~1.53 μm while excitation is at λ_p=978 nm) the SE process can be disregarded by means of formal zeroing cross-section σ_{21} in the dominator of this ratio (k=1). Implying that $n_2^s(t=0)=1$ and that pump is switched off at t=0, equation (21) is solved analytically, giving:

$$n_2^s(t) = \frac{e^{-\frac{t}{\tau_0}}}{1 + \tau_0 C_{HUP}\left(1 - e^{-\frac{t}{\tau_0}}\right)} \tag{22}$$

Formula (21) is a worthy approximation for fitting of the whole of experimental near-IR fluorescence decay kinetics, reported above for P_p ~400 mW (providing maximal population of manifold $^4I_{13/2}$). The modeling results obtained by using formula (22) are shown by plain curves in Figure 12 (the fitting procedure has been fulfilled until the residual sum R^2 exceeded 0.99) and are seen to be in good agreement with the experimental decay kinetics (points in the figure). The values of constants τ_0 (lifetime of single Er^{3+} ions, found to be ~10.8 ms for all the fibers under study), and C_{HUP} (an attribute of the HUC process, determined as the result of fitting) are plotted in Figure 15 in function of the small signal absorption α_0 at 978 nm.

As seen from Figure 15, the parameter C_{HUP} is proportional to Er^{3+} concentration (GSA α_0 at 978 nm). From this figure we found the value of HUC constant: C_{HUP}^*=2.7×10^{-18} $s^{-1}cm^3$. This value agrees well with the published data for EDFs of similar types; see e.g. Refs. [22-24]. Note that the quantity attributing the HUC phenomenon (C_{HUP}^* constant) should be proportional to the ESA cross-section (see e.g. [25]). As the latter does not depend on Er^{3+} concentration, C_{HUP}^* should be concentration-independent. Indeed, the dependence C_{HUP} vs. α_0 is seen from Figure 15 to be almost linear.

The next step in modeling is simulation of Er^{3+} clusters' contribution on the base of the experimental dependences of nonlinear absorption vs. pump power (see Figure 14(a)). A method to model nonlinear absorption of an EDF is based on the idea that ensemble of Er^{3+} ions in a fiber consists of two independent subsystems, assumed to be single ("s") and clustered

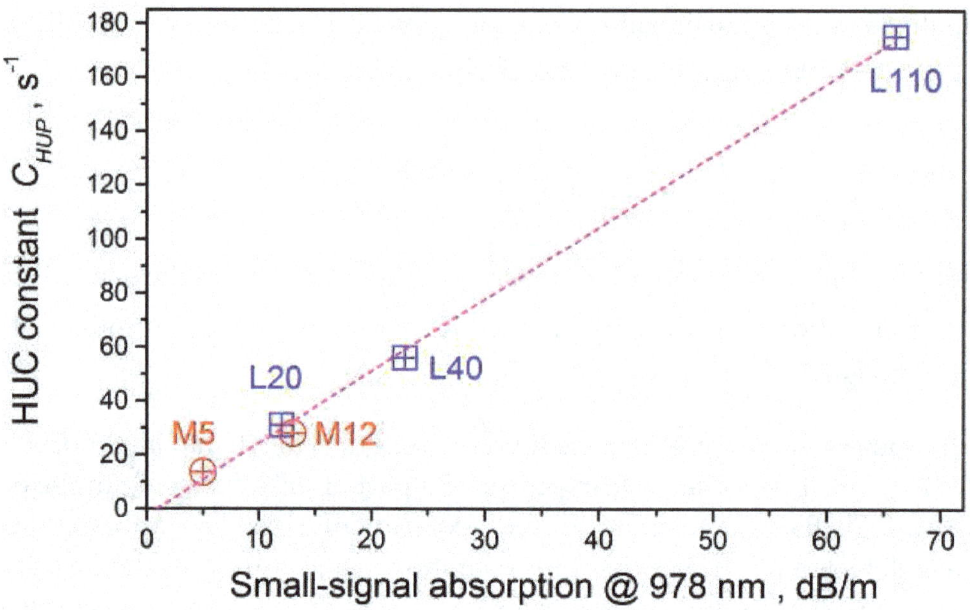

Figure 15. The values of HUC constant C_{HUP} vs. Er^{3+} concentration (in terms of α_0 at 978 nm), obtained for the entire set of EDFs as the result of modeling of the experimental fluorescence kinetic using formula (22); the fitting line in for guiding the eye.

("c") ions. Considering this hypothesis, we generalize the model developed in [21] for propagation of a pump wave through the system of single and paired resonantly absorbing and fluorescing centers (pairs are the simplest case of clusters). The model's generalization signifies here that the clusters' subsystem is meant to comprise an arbitrary number of centers (Er^{3+} ions in our case) whereas the other subsystem – to consist of single species [6, 26].

We assume that a cluster of Er^{3+} ions can occupy only one of the two permitted states – the state <11>, where all ions forming the cluster are in the ground state, and the state <12>, where one excepted ion (an acceptor of energy transferred from the adjacent donor ions within the cluster) is in the excited state. The latter holds because if other cluster's constituents absorb pump photons and move to the excited state, all them except one leave state 2 (down to state 1) immediately, whereas only this excepted one can stay in state 2.

We also take into account the presence of the short time τ_1 (together with much longer τ_0) for Er^{3+} clusters as an essential element of the model, coming from the experimental data.

We consider that the populations of single ($N_{1,2}^{S}$) and clustered ($N_{1,2}^{c}$) Er^{3+} ions, satisfy the following relations:

$$N_1^s + N_2^s = N_0^s = (1 - \chi\kappa)N_0 \tag{23}$$

$$N_1^c + N_2^c = N_0^c = \chi\kappa N_0 \tag{24}$$

where κ is the partial weight of clustered ions in ensemble, χ is the effective (averaged) number of Er^{3+} ions in a cluster, and N_0 is the total Er^{3+} ions concentration. The correspondent normalized population densities are defined as $n_{1,2}^{s,c} = N_{1,2}^{s,c}/N_0$, where the lower indices assign, correspondingly, the ground (1 or <11>) and the excited (2 or <12>) states, and $N_0^{s,c}$ are the concentrations of single ions and clusters, respectively.

The balance equations for pump power (P_p) and normalized dimensionless population densities of single and clustered ions being in metastable states 2 and <12>, $n_2^{s,c}$ ($0 \le n_2^{s,c} \le 1$), are as follows [27]:

$$\frac{dP_p(z)}{dz} = -\alpha_{p0}\left[1 - (1+\xi_s-\varepsilon_s)\left(n_2^s(z)+n_2^c(z)\right)\right]P_p(z) - \alpha_{BG}P_p(z) \tag{25}$$

$$\frac{\alpha_0}{h\nu_p N_0 \Gamma_p A_c}\left[\chi\kappa - (1+\xi_s)n_2^c(z)\right]P_p(z) - \left(\frac{1}{\tau_0}+\frac{1}{\tau_1}\right)n_2^c(z) = 0 \tag{26}$$

$$\frac{\alpha_{p0}}{h\nu_p N_0 \Gamma_p A_c}\left[1 - \chi\kappa - (1+\xi_s)n_2^s(z)\right]P_p(z) - \frac{n_2^s}{\tau_0} - C_{HUP}\left(n_2^s\right)^2 = 0 \tag{27}$$

where majority of the quantities have been designated above, parameter $A_c = \pi a^2$ is the EDF core area (a=1.5 μm is the core radius), and α_{BG}=0.03...0.1 dB/m, depending weakly upon the EDF type). In formulas (25-27) we omit the ASE contribution as negligible in our experiments at pump powers exceeding 0.5 mW (see above). Note that these formulas are written in a general form, applicable not only to the Er^{3+} ion but also to any other resonantly absorbing center, having a three equivalent level system and subjected to the aforementioned concentration effects.

The modeling results are plotted by plain curves 1 to 5 in Figure 16(a). It is seen that they fit well the whole of the experimental data for the EDFs of both (M and L) types. Thus, the IUC process, treated by us as mostly non-radiative relaxation within Er^{3+} ion clusters, is justified as the key mechanism responsible for nonlinear absorption (attributed by coefficient β). The dependence of β upon Er^{3+} concentration in terms of small signal absorption α_{p0} is shown in Figure 16(b) (see curve 1) as the result of modeling; the errors' bars in the curve show uncertainties of fitting the experimental data by the theory.

When making the numerical calculations, we found that, once searching for the best fit of the experiment by the theory, any χ-value (χ=2, 3, and so on) can be used, with κ-value being varied accordingly. Thus, we have concluded that the product $\chi\kappa$ (a relative number of clustered Er^{3+} ions in the system) serves an adjusting parameter at fitting rather than quantities χ and κ separately. Given by the modeling results, useful insight can be made to interrelation between the relative number of clustered Er^{3+} ions $\chi\kappa$ (modeling: Figure 16(a)) and the measured nonlinear (saturated) absorption β (experiment: Figure 14(a)). Placing on the same plot (at double logarithmic scaling) the dependences of these two quantities vs. small-signal absorption α_{p0}, we found that they have the slopes related as (~1.44/~0.63) ≈ 2.3; see Figure 16(b). The found slope's value signifies the average number of Er^{3+} ions in clusters, whose

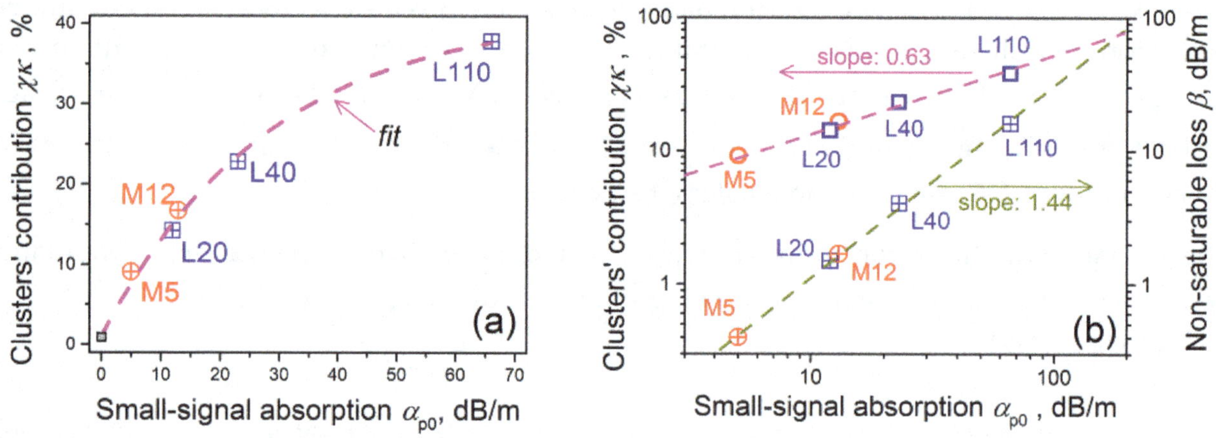

Figure 16. (a) Er^{3+} clusters' contribution $\chi\kappa$ vs. Er^{3+} concentration (in terms of α_{p0}) obtained for the entire set of EDFs of L-and M-types as the result of modeling the experimental dependences of nonlinear absorption coefficient using formulas (25-27); the fitting curve is for guiding the eye. (b) Er^{3+} clusters' contribution $\chi\kappa$ (left scale) and non-saturable absorption loss β (right scale) vs. Er^{3+} concentration (in terms of α_{p0}); the ratio of the slopes attributing the dependences reveals an effective number of Er^{3+} ions in clusters.

presence, according to the model, is responsible for NSA. This result deserves attention since it shows that NSA in the EDFs of both types mostly originates from paired Er^{3+} ions rather than from more complicate aggregates. Thus, in contrast to [28] where the role of "heavier" clusters in the IUC phenomenon is discussed, our results evidence for negligible contribution of Er^{3+} clusters, "heavier" than simple ion pairs.

Notice that the presence in EDFs of NSA at increasing Er^{3+} concentration affects net gain in heavily-doped Er^{3+} fibers, which becomes more and more limited (saturated) as it stems from the presence of single Er^{3+} ions being in the excited state, whereas the clustered ions, in their big part, i.e. ($\chi\kappa$–1) ions in each cluster, are always in the ground state. As a consequence, efficiency of an EDFL or EDFA is expected to drop down with increasing concentrations of Er^{3+} ions in the active fiber. Some recent studies confirm a severe character of the problem [29-30]. It seems that another problem could be encountered at the use of heavily-doped EDFs for pulsed operation where extensive heating via excitation relaxation within Er^{3+} clusters would affect the dispersive properties of the fibers and deteriorate the regime.

4. Effect of SBS upon operation of actively Q-switched EDFL

Actively Q-switched (AQS) EDFLs based on acousto-optics modulator (AOM), implemented in the Fabry-Perot geometry, usually produce Q-switch (QS) pulses with duration from a few to hundreds ns [1]. The QS pulses are normally composed of a few sub-pulses separated by round-trip time of photon inside the cavity [31, 32]; this AQS regime will be called further "conventional" (CQS). In the meantime, it is known that in certain conditions FLs demonstrate stochastic QS pulsing, which stems, as it will be clearly demonstrated below, from intra-cavity stimulated Brillouin scattering (SBS) [33]. Such pulses, referred further to as SBS-QS ones, are characterized by dramatic increasing of power as compared to CQS pulses but, at the same time, by perceptible jitters [34]. In this section, we show that in certain circumstances SBS-QS pulsing is inherent in AOM-based AQS EDFLs. We also demonstrate that the areas (basins) where CQS and SBS-QS regimes exist are defined by definite values of EDF length and AOM's repetition rate and that the most important condition for turning of the laser to one or another pulsing regime is absence or presence of spurious narrow-line continuous wave (CW) lasing in the intervals where the laser cavity is blocked (AOM is switched OFF).

4.1. Experimental setup

An experimental setup of the QS-EDFL is sketched in Figure 17. The laser cavity consists of a piece of a standard low-doped "M" EDF (*Thorlabs*, M5-980-125), two FBGs (1 and 2) centered at ~1549.4 nm (laser wavelength), which form Fabry-Perot cavity, and a standard down-frequency shifting AOM with fiber outputs (*Gooch & Housego*, operation frequency – 111 MHz), placed nearby FBG2. The full AOM's rise time was measured to be 50 ns, AOM gate was fixed at 2 μs in experiments. FBGs' reflection coefficients were ~ 30% (FBG1) and ~ 100% (FBG2). A long period grating (LPG) tuned to ~1533 nm was used as in-line stop-band filter for neutralizing fiber gain at Er^{3+} SE peak and thus avoiding spurious CW lasing at this wavelength, which might otherwise discharge EDF and thereby reduce QS pulse energy and, via interfering with targeted pulsed lasing at the wavelength selected by FBGs, produce instability of pulsing. The EDF was pumped by a fiber-coupled 976-nm LD through a 980/1550-nm WDM. To decrease the cavity loss, FBG1 and LPG were written in the EDF core after preliminary hydrogenation. The laser signal was registered by OSA with a 50-pm resolution or by 1.2-GHz PD used in-line with a 2.5-GHz oscilloscope. In experiments, pump power was fixed at 500 mW and AOM's repetition rate (f_{AOM}) was varied within a 0...30-kHz range.

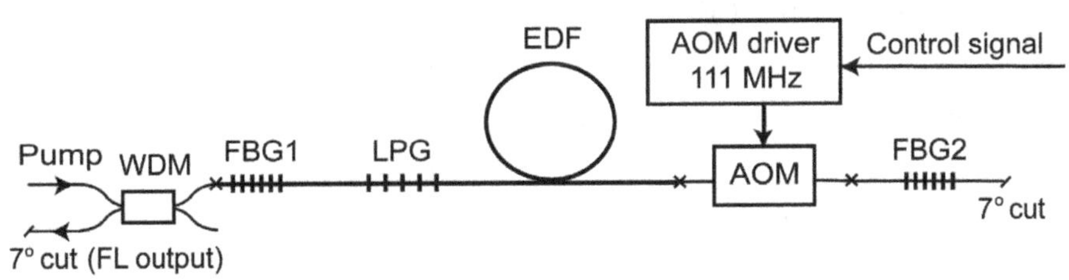

Figure 17. Experimental setup of the QS-EDFL (crosses indicate fiber splices).

4.2. Properties of CQS and SBS-QS regimes

As well-known, AQS FLs operated in CQS regime usually generate pulses consisting of train of sub-pulses (ripples), separated by a time interval equal to a photon's round-trip in the cavity. Such kind operation is fully described by the model of two contra-propagated waves in Fabry-Perot cavity, once considering the laser as a multi-pass amplifier of SE reflected several times by selective mirrors (FBGs) [32]. CQS is observed at any f_{AOM} when EDF length (L_{EDF}) is shorter than some specific value and at larger EDF when f_{AOM} is high. The common features of CQS pulsing observed experimentally and also modeled are as follows: Delay of a QS pulse with respect to the moment of AOM opening increases while its energy and power decrease with increasing AOM's repetition rate. Usually, the first detectable sub-pulse arises in a few round-trips of ASE after AOM got opened. For example, when L_{EDF}=8.8 m and f_{AOM}=8 kHz, the first visible sub-pulse appears at ~250 ns (~2.5 photon round trips; see Figure 18(a)). The RF (FFT) spectrum of pulse train measured at f_{AOM}=16 kHz (see Figure 18(b)) has three peaks centered at 0 MHz, ~10 MHz and ~20 MHz. Width of peak 0 relates to QS pulses width, whereas peaks 1 and 2 correspond to the first and the second harmonics of the round-trip frequency (an inverted interval between sub-peaks, or round-trip time), respectively.

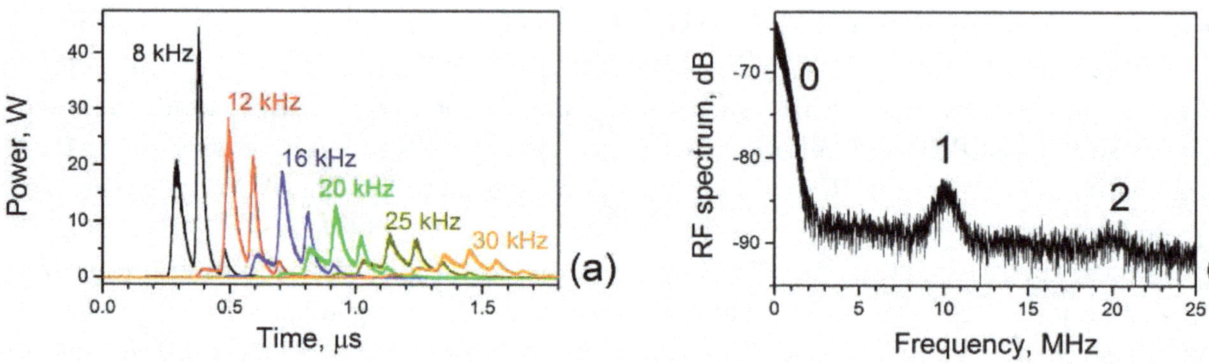

Figure 18. (a) CQS pulses registered on the EDFL output at L_{EDF}=8.8 m and at various f_{AOM}-values. Zero-time in both snapshots corresponds to the moment when AOM gets opened. (b) Averaged RF spectrum of CQS pulsing at f_{AOM}=16 kHz.

If the active fiber is long enough and AOM's repetition rate is not too high the QS EDFL turns into the regime of SBS-induced pulsing. This kind of pulsing is quite different as compared with CQS. Typical SBS-QS pulses are shown in Figure 19(a) for L_{EDF}=8.8 m and f_{AOM}=1 kHz. These pulses, as compared with CQS ones, arise earlier, approximately in ~180...280 ns after the moment of AOM's switching on; they are much narrower (~2.5...10 ns at 3-dB level); the pulses amplitude is more than by 10 dB higher as compared with the one at CQS while their envelop is apparently irregular. Emphasize that no SBS-QS pulses arisen within the intervals between the adjacent AOM's windows, in contrast to SBS-QS pulsing in an ytterbium-doped FL [34].

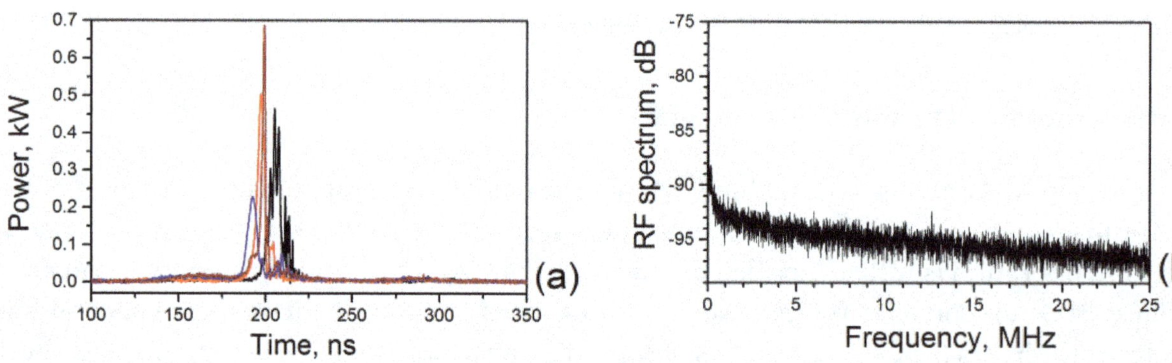

Figure 19. (a) CQS pulses registered on the EDFL output at L_{EDF}=8.8 m and f_{AOM}=1 kHz. (b) Averaged RF spectrum of CQS pulsing at f_{AOM}=1 kHz.

One more detail of SBS-QS is that pulses released in this regime suffer strong amplitude and timing jitters. Apparently, the presence of jittering is an indication of the stochastic nature of the SBS process involved. Furthermore, since the SBS-QS pulses are not composed of sub-pulses spaced by photon round-trip time, their RF (FFT) spectrum does not have peaks at the round-trip frequency (~10 MHz) and its harmonics (see Figure 19(b)).

4.3. Basins of CQS and SBS-QS regimes

To find basins of CQS and SBS-QS regimes existence, we measured the value of f_{AOM} at which the laser transits from one to another QS regime, for different L_{EDF}. The experimental points in the space (f_{AOM}, L_{EDF}) segregating CQS and SBS-QS operations were easily fixed since QS pulses captured at the laser output in these two regimes differ drastically in pulse amplitude, duration and shape (see snapshots in Figures 18(a) and 19(a)). We found that if L_{EDF} is less than or equal to a certain value (5.4 m in our arrangement) the laser operates in CQS regime, at any f_{AOM}. But if EDF is longer than 5.4 m, an operation regime depends on AOM's repetition rate: At low f_{AOM} the laser generates SBS-QS pulses while at high f_{AOM} it turns to CQS operation. The basins of CQS and SBS-QS regimes are illustrated in Figure 20. As seen from the figure, the laser operates in CQS and SBS-QS above and below the border line, respectively (this line schematically marks a transition between the regimes).

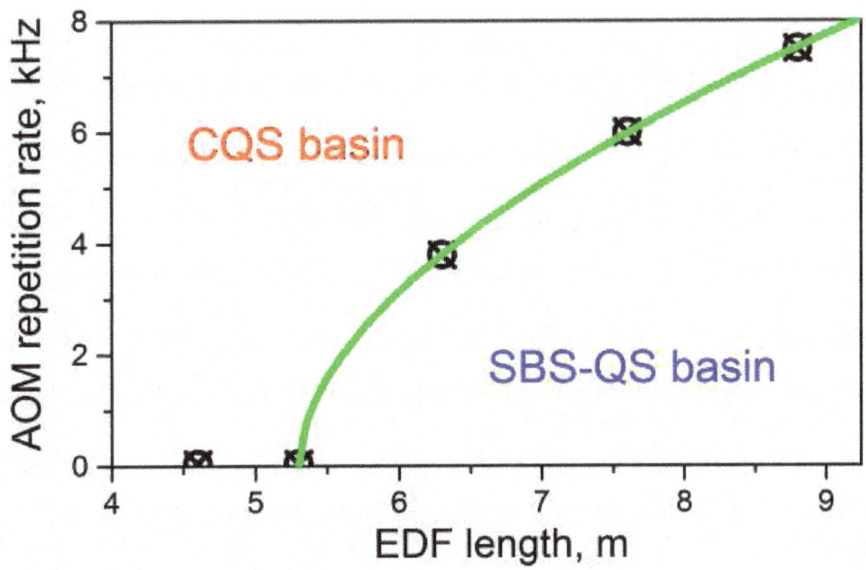

Figure 20. Basins of CQS and SBS-QS regimes; symbols label the experimental points corresponding to the border (solid line) between the basins.

The reason of the AQS-EDFL's switching to CQS or SBS-QS regime is the existence or the absence of spurious narrow-line CW lasing when AOM is blocked. In the last case the cavity is formed by the output reflector FBG1 and by a small reflection from closed AOM (~ –40 dB) ("bad" cavity). The overall loss of this cavity is estimated to be ~45-47 dB, revealing its very low Q-factor. At long EDF and low AOM repetition rates (the area below the border line in Figure 20) the EDF charge is sufficient to provide fiber gain capable of overcoming the cavity loss, which results in arising CW lasing. After switching AOM on, the CW wave starts to propagate along the main laser cavity (formed by FBG1 and FBG2) with simultaneous amplification of its power by the EDF until the latter reaches SBS-threshold and produces a "giant" SBS-QS pulse. Thus, CW spurious lasing arising in the "bad" cavity is a startup mechanism for SBS-QS pulsing. In the area above the border line in Figure 20 the spurious CW lasing is not established since the EDF cannot accumulate gain sufficient to overcome the "bad" cavity's loss.

To confirm the hypothesis that the mechanism "igniting" SBS-QS pulsing relates to arising of the narrow-line CW lasing in "bad" cavity (when AOM is closed), we fulfilled the experiments on measuring the laser's optical spectra.

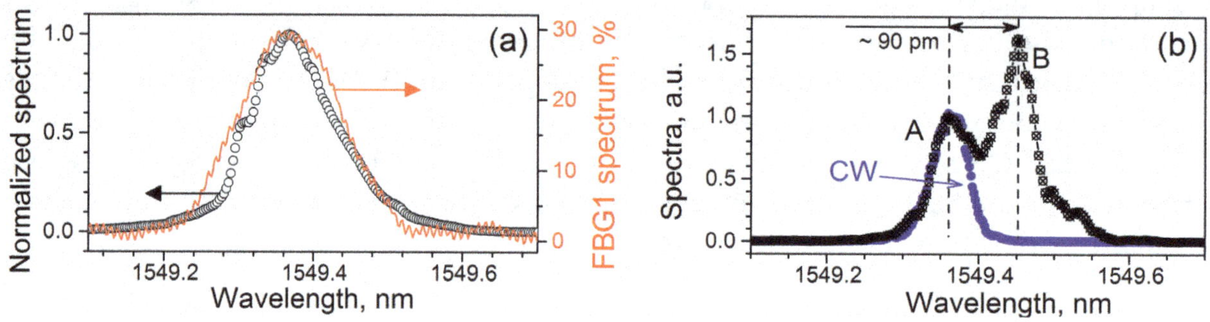

Figure 21. (a) Normalized AQS EDFL spectrum measured at f_{AOM}=8 kHz (open circles, left scale) and FBG1 spectrum (solid line, right scale). (b) Normalized AQS EDFL spectra measured at f_{AOM}=4 kHz (black crossed circles) and at AOM being always blocked (CW lasing, blue circles). In both cases L_{EDF}=7.6 m. OSA resolution is 50 pm.

Firstly, we compared the optical spectra of the laser operated in CQS and SBS-QS regimes; see Figure 21. Comparing the lasing spectra at f_{AOM}=8 kHz (CQS) and f_{AOM}=4 kHz (SBS-QS), one sees that in the former case (Figure 21(a)) the spectrum virtually repeats the reflection spectrum of FBG1 (the one of FBG2 is broader), whereas in the latter case (Figure 21(b)) it consists of two spectral lines A and B, spaced by ~90 pm (~11 GHz, a Brillouin shift at 1550 nm). Both laser lines A and B are narrower (~60 pm) than the lasing (in fact ASE) spectrum at CQS, ~160 pm. Supposedly, line A, centered at the FBG1's peak, corresponds to CW lasing arisen when AOM is in OFF state (between the adjacent AOM's gates) whereas line B – to SBS-QS pulsing. To provide more arguments in favor of this hypothesis, we plot in the same figure (Figure 21(b)) the spectrum of CW lasing, when the main cavity is always blocked (i.e. AOM is continuously in OFF state). It is seen that line A vastly reproduces the CW lasing spectrum. This allows us to reveal that the narrow line A is the signature of CW lasing and that it "ignites" the SBS-process; accordingly, the narrow line B, shifted by ~90 pm to the Stokes side, is the signature of SBS-lasing.

To shade more light on the scenario drawn above, we also measured the spectral width of spurious CW lasing arising when AOM is closed by employing another technique that utilizes a modified delayed self-heterodyne interferometer (DSHI, see Figure 22), described in details in [35-37]. To carry out this, the laser output signal was split into two beams, one of which being passed through an optical frequency shifter (AOM) and then through a long recirculating fiber delay line made using SMF-28 fiber (L_d=20 km in our case). As the result, optical frequency was shifted by f_{AOM}=111 MHz at each pass along the delay line. After that, the signals (un-delayed and delayed) were combined and registered by a fast PD, connected to a RF spectrum analyzer (RFSA). To increase the method's sensitivity, an EDFA was included into the DSHI scheme at the fiber delay line's exit. We should note that the multi-pass self-heterodyne scheme used for estimation the EDFL's line width was chosen because, for correct measurements, the path difference should be much higher that the light source's coherence length (~ 20 km, see below).

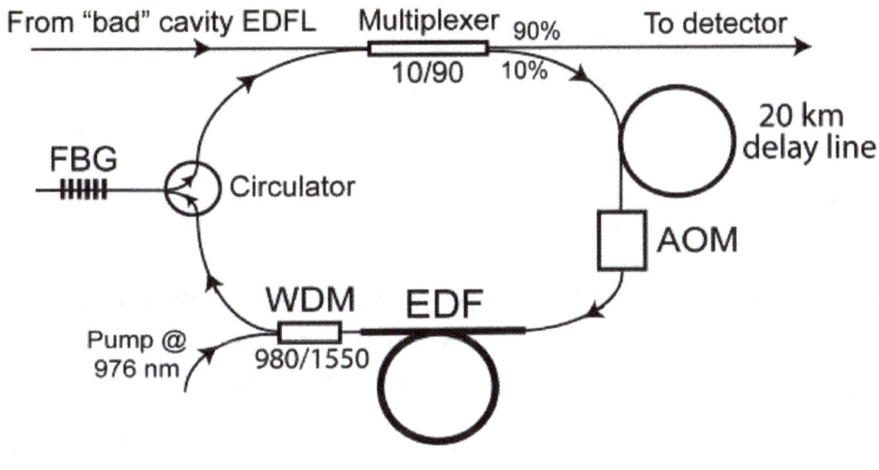

Figure 22. DSHI setup. AOM serves as a 111 MHz frequency shifter (always is open); 100%-FBG filters ASE produced by EDFA (reflects light only at the wavelength of the EDFL's "bad" cavity).

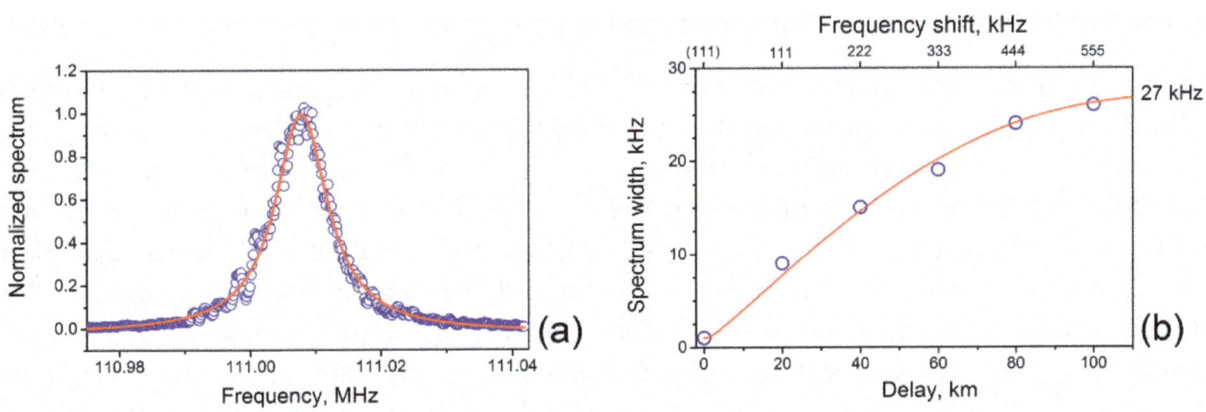

Figure 23. (a) An example of the CW EDFL RF-spectrum obtained from DSHI at 111 MHz; experimental data are shown by circles, solid line is a Lorentzian fit that gives 9.8-kHz width of the spectrum. (b) Spectrum width of CW EDFL measured using the DSHI technique at RF frequencies multiplied by AOM's frequency shift (111 MHz) when the cavity is blocked (circles). Solid line is a fit obtained using the theory presented in [37] (equation 16). The point at zero-delay was obtained with a 111-MHz frequency shift in the absence of delay line, which gives the RFSA's resolution (1 kHz).

Figure 23 shows the spectral width of the signals at the frequencies multiplied by the AOM's frequency shift, measured after fitting the DSHI signal by the Lorentzian law, in function of the delay-line length $L_d \times N$ (N is the number of passes through the line). As shown in [35-37], at high N a signal's width approaches the real value of an EDFL's optical width; in our case, this value is ~30 kHz, which corresponds to coherence length of ~20 km. The earlier reports [38, 39] also revealed a very narrow EDFL's line.

Consequently, a narrow-line CW laser wave developing in "bad" cavity and being a prerequisite of SBS-QS pulsing is highly coherent. This explains why SBS is unavoidably boosted by spurious CW lasing in the "blocked" cavity after the moment of AOM's opening, when the EDF is strongly inverted and thus strongly amplifying. Indeed, "unlimited" in length and therefore "uniform" Brillouin dynamic grating [40], induced in intra-cavity EDF (the fiber is always shorter than the estimated coherence length, ~20 km), is the main cause of SBS-QS pulsing.

5. Conclusions

In this Chapter, we reported some of the important nonlinear-optic features of EDFs, which, on one hand, impact efficiency of CW EDFLs on their base and, on the other hand, underlie the operation regimes established in EDFLs Q-switched using AOMs.

In particular, we showed that strong ESA transitions inherent in the Er^{3+} system at both the pump (~978 nm) and the laser (~1550 nm) wavelengths and ions' clustering inherent in the EDFs heavily doped with Er^{3+} cause unavoidable nonlinear losses that, in turn, strongly reduce efficiency of EDFLs, as compared to efficiency of Ytterbium-doped fiber lasers. We demonstrated as well that for making a correct numerical modeling of an EDFL one needs to consider all kinds of the nonlinear losses intrinsic in EDFs.

We also discussed in details the peculiarities of EDFLs operated in actively Q-switched regime using AOM. We demonstrated that the operation regimes of these lasers strongly depend on EDF length and AOM's repetition frequency. Specifically, at short EDF length or at high AOM's repetition frequency the laser operates in "conventional" Q-switching regime (being in fact multi-pass amplification of Er^{3+} SE) where pulses with relatively moderate power and relatively long in duration are composed of several, stable in time, sub-pulses, separated by a photon's round-trip time in the cavity. Furthermore, if EDF length is long enough and AOM's repetition frequency is not too high, the laser turns to the completely different pulsing regime, characterized by much shorter and much powerful pulses; however, pulses of this type are subjected to noticeable timing and amplitude jitters, originated from the stochastic in nature SBS process, ignited by spurious narrow-line CW lasing in "bad" (at closed AOM) cavity.

Acknowledgements

Authors acknowledge support from the CONACyT, Mexico (Project No. 167945).

Author details

Y.O. Barmenkov* and A.V. Kir'yanov

*Address all correspondence to: yuri@cio.mx

Centro de Investigaciones en Optica, Leon, Guanajuato, Mexico

References

[1] Richardson D. J., Nilsson J., Clarkson W. A. High power fiber lasers: Current status and future perspectives [Invited]. Journal of the Optical Society of America B 2010;27(11): B63–92.

[2] Barmenkov Yu. O., Ortigosa-Blanch A., Diez A., Cruz J. L., Andres M. V. Time-domain fiber laser hydrogen sensor. Optics Letters 2004;29(21): 2461-3.

[3] Arellano-Sotelo H., Barmenkov Yu. O., Kir'yanov A. V. The use of erbium fiber laser relaxation frequency for sensing refractive index and solute concentration of aqueous solutions. Laser Physics Letters 2008;5(11): 825-9.

[4] Laming R. I., Poole S. B., Tarbox E. J. Pump excited-state absorption in erbium-doped fibers. Optics Letters 1988;13(12): 1084-6.

[5] Miniscalco W. J. Erbium-Doped Glasses for Fiber Amplifiers at 1500 nm. Journal of Lightwave Technology 1991;9(2): 234-50.

[6] Blixt P., Nilsson J., Carlinas T., Jaskorzynska B. Concentration-dependent upconversion in Er^{3+} -doped fiber amplifiers: Experiments and modeling. IEEE Transactions Photonics Technology Letters 1991;3(11): 996-8.

[7] Myslinski P., Chrostowski J., Koningstein J. A. K., Simpson J. R. Self-mode locking in a Q-switched erbium-doped fiber laser. Applied Optics 1993;32(3): 286-90.

[8] Escalante-Zarate L., Barmenkov Y. O., Cruz J. L., Andres M. V. Q-switch modulator as a pulse shaper in Q-switched fiber lasers. IEEE Photonics Technology Letters 2012;24(4): 312-4.

[9] Barmenkov Y. O., Kir'yanov A. V., Cruz J. L., Andres M. V. Dual-kind Q-switching of erbium fiber laser. Applied Physics Letters 2014;104(9): 091124.

[10] Nykolak G., Becker P. C., Shmulovich J., Wong Y. H., DiGiovanni D. J., Bruce A. J. Concentration-dependent $^4I_{13/2}$ lifetimes in Er^{3+} -doped fibers and Er^{3+} -doped planar waveguides. IEEE Photonics Technology Letters 1993;5(9): 1014-6.

[11] Guzman-Chavez A. D., Barmenkov Y. O., Kir'yanov A. V. Spectral dependence of the excited-state absorption of erbium in silica fiber within the 1.48–1.59 μm range, Applied Physics Letters 2008;92(19): 191111.

[12] Bellemare A. Continuous-wave silica-based erbium-doped fibre lasers. Progress in Quantum Electronics 2003;27(4): 211–66.

[13] Krug P. A., Sceats M. G., Atkins G. R., Guy S. C., Poole S. B. Intermediate excited-state absorption in erbium-doped fiber strongly pumped at 980 nm. Optics Letters 1991;16(24): 1976-8.

[14] Barnes, W. L. Laming, R. I. Tarbox, E. J.; Morkel, P. R. Absorption and emission cross-section of Er^{3+} doped silica fibers. IEEE Journal of Quantum Electronics 1991;27(4): 1004-10.

[15] Bolshtyansky M., Mandelbaum I., Pan F. Signal Excited-State Absorption in the L-Band EDFA: Simulation and Measurements. Journal of Lightwave Technology 2005;23(9): 2796-9.

[16] Barmenkov Y. O., Kir'yanov A. V., Guzman-Chavez A. D., Cruz J. L., Andres M. V. Excited-state absorption in erbium-doped silica fiber with simultaneous excitation at 977 and 1531 nm. Journal of Applied Physics 2009;106(8): 083108.

[17] del Valle-Hernandez J., Barmenkov Y. O., Kolpakov S. A., Cruz J. L., Andres M. V. A distributed model for continuous-wave erbium-doped fiber laser. Optics Communications 2011;284(22): 5342–7.

[18] Kir'yanov A. V., Barmenkov Y. O., Il'ichev N. N. Excited-state absorption and ion pairs as sources of nonlinear losses in heavily doped Erbium silica fiber and Erbium fiber laser. Optics Express 2005;13(21): 8498-507.

[19] Kik P. G., Polman A., Exciton–Erbium interactions in Si nanocrystal-doped SiO_2. Journal of Applied Physics 2000;88(4): 1992–8.

[20] Kik P. G., Brongersma M. L., Polman A. Strong exciton-Erbium coupling in Si nanocrystal-doped SiO_2, Applied Physics Letters 2000;76(17): 2325–7.

[21] Kir'yanov A. V., Dvoyrin V. V., Mashinsky V. M., Barmenkov Y. O., Dianov E. M. Nonsaturable absorption in alumino-silicate bismuth-doped fibers. Journal of Applied Physics 2011;109(2): 023113.

[22] Berkdemir C., Ozsoy S. On the temperature-dependent gain and noise figure of C-band high-concentration EDFAs with the effect of cooperative upconversion. Journal of Lightwave Technology 2009;27(9): 1122–7.

[23] Jung M., Chang Y. M., Jhon Y. M., Lee J. H., Combined effect of pump excited state absorption and pair-induced quenching on the gain and noise figure in bismuth oxide-based Er^{3+} -doped fiber amplifiers, Journal of the Optical Society of America B 2011;28(11): 2667–73.

[24] Valles J. A., Bordejo V., Rebolledo M. A., Diez A., Sanchez-Marin J. A., Andres M. V. Dynamic characterization of upconversion in highly Er-doped silica photonic crystal fibers. IEEE Journal of Quantum Electronics 2012;48(8), 1015–21.

[25] Hehlen M. P., Cockroft N. J., Gosnell T. R., Bruce A. L., Nikolak G., Shmulovich J. Uniform upconversion in high concentration Er^{3+} -doped soda lime silicate and aluminosilicate glasses. Optics Letters 1997;22(11): 772–4.

[26] Myslinski P., Nguyen D., Chrostowski L., Effects of concentration on the performance of Erbium-doped fiber amplifiers. Journal of Lightwave Technology 1997;15(1): 112–9.

[27] Kir'yanov A. V., Barmenkov Y. O., Sandoval-Romero G. E., Escalante-Zarate L. Er^{3+} Concentration effects in commercial erbium-doped silica fibers fabricated through the MCVD and DND technologies. IEEE Journal of Quantum Electronics 2013;49(6): 511-21.

[28] An H. L., Pun E. Y. B., Liu H. D., Lin X. Z. Effects of ion clusters on the performance of heavily doped Erbium-doped fiber laser. Optics Letters 1998;23(15): 1197–9.

[29] Lim E.-L., Alam S. U., Richardson D. J., High energy in-band pumped Erbium doped fibre amplifiers. Optics Express 2012;20(17): 18803–18.

[30] Lim E.-L., Alam S. U., Richardson D. J. Optimizing the pumping configuration for the power scaling of in-band pumped Erbium doped fiber amplifiers. Optics Express 2012;20(13), 13886–95.

[31] Lim E. L., Alam S. U., Richardson D. J. The multipeak phenomena and nonlinear effects in Q-switched fiber laser, " IEEE Photonics Technology Letters 2011;23(23): 1763–5.

[32] Kolpakov S. A., Barmenkov Y. O., Guzman-Chavez A. D., A. V. Kir'yanov, Cruz J. L., A. Diez, Andres M. V. Distributed model for actively Q-switched erbium-doped fiber lasers. IEEE Journal of Quantum Electronics 2011;47(7): 928–34.

[33] Fotiadi A. A., Megret P., Self-Q-switched Er-Brillouin fiber source with extra-cavity generation of a Raman supercontinuum in a dispersion shifted fiber. Optics Letters 2006;31(11), 1621–3.

[34] Barmenkov Y. O., Kir'yanov A. V., Andres M. V. Experimental study of the nonlinear dynamics of an actively Q-switched ytterbium doped fiber laser. IEEE Journal of Quantum Electronics 2012;48(11): 1484–93.

[35] Tsuchida H. Simple technique for improving the resolution of the delayed self-heterodyne method. Optics Letters 1990;15(11): 640–2.

[36] Dawson J. W., Park N., Vahala K. J., An improved delayed self-heterodyne interferometer for linewidth measurements. IEEE Photonics Technology Letters 1992;4(9): 1063–5.

[37] Horak P., Loh W. H. On the delayed self-heterodyne interferometric technique for determining the linewidth of fiber lasers. Optics Express 2006;14(9): 3923–8.

[38] Chernikov S. V., Taylor J. R., Kashyap R., Coupled-cavity erbium doped lasers incorporating fiber grating reflectors. Optics Letters 1993;18(23): 2023–5.

[39] Choi K. N., Taylor H. F. Spectrally stable Er-fiber laser for applications in phase-sensitive optical time-domain reflectometry. IEEE Photonics Technology Letters 2003;15(3): 386–8.

[40] Song K. Y. Operation of Brillouin dynamic grating in single-mode optical fibers. Optics Letters 2011;36(23): 4686–8.

4

Self-Coherent Reflective Passive Optical Networks

S. Straullu, S. Abrate and R. Gaudino

1. Introduction

Fiber optic access networks are the necessary infrastructural approach for a real broadband delivery, allowing the fiber to arrive closer to the final customer, eventually up to the premises equipment; such infrastructures, depending on the depth of reach of the fiber, are usually referred to as FTTX (Fiber To The X), where X stands for H (Home), B (Building), C (Curb) or Cab (Cabinet). They are the basis for broadband access networks, enabling high-speed Internet access, digital TV broadcast (IPTV), video on demand (VOD) and other services. Comparing with copper technologies, like for example DSL (digital subscriber line), higher bandwidths (up to several 10 Gbit/s) and higher distances (up to several tens of km) are possible. FTTX infrastructures can address both residential and business access, and support mobile back-hauling.

Access networks can be divided in two main categories: Active Optical Networks (AONs) and Passive Optical Networks (PONs):

- AONs are a point-to-point (P2P) network structure, where all users have their own fiber optic line that is terminated on an optical access node, which can be designed differently depending on specifications;
- PONs feature a point-to-multi-point (P2MP) tree-like architecture (as shown in Fig. 1 below) to provide broadband access, based on a network composed by an Optical Line Terminal (OLT) at the service provider's Central Office (CO) and several Optical Network Units (ONUs) at the users' premises. PONs allow only passive elements along the Optical Distribution Network (ODN) that connects the OLT to the ONUs. The main advantages are that PONs do not require electrical power in the outside plant to power the distribution elements, thereby lowering operational costs and complexity and, moreover, they allow to highly reduce the number of required optical ports in the CO compared to P2P solutions.

The OLT is the interface between backbone and access networks, and it is also responsible for the enforcement of any media access control (MAC) protocol for upstream bandwidth arbitration. The ONU cooperates with the OLT in order to control and monitor all PON transmission and to enforce the MAC protocol for upstream bandwidth arbitration; it also acts as the residential gateway, coupling the ODN with the in-home network.

The ODN consists of the distribution fibers and all the passive optical distribution elements, mainly optical splitters and/or wavelength-selective elements (WDM filters), that are located in sockets or cabinets. The splitting ratio in most cases is between 1:8 and 1:128, and can be performed in lumped or cascaded elements. The ODN power budget is usually taken as a reference for PON reach calculations, and is computed as the difference between the transceiver back-to-back power budget (i.e. OLT transmitter directly coupled into ONU receiver or vice versa) and the passive optical equipment necessary inside the OLT and the ONU to perform all multiplexing of PON signals into a single fiber. Hence, the ODN power budget considers the remaining power budget that can be spent for the distribution fibers and the distribution elements in the remote nodes.

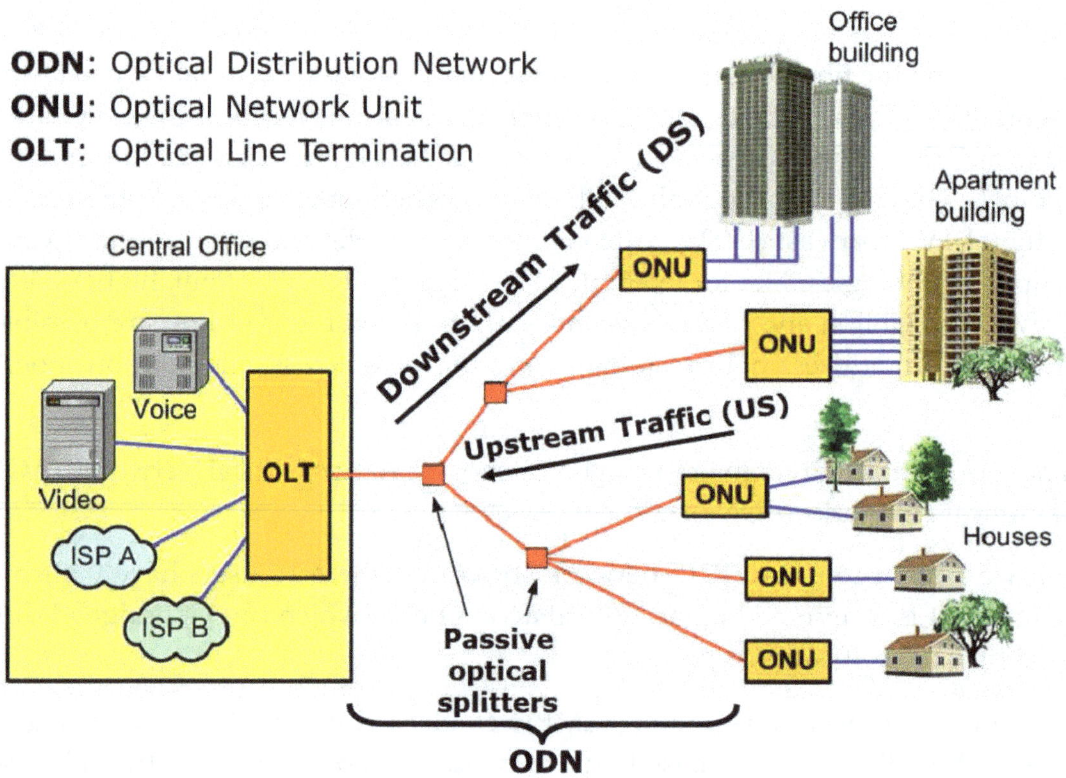

Figure 1. Passive Optical Network architecture

During the last 20 years, the Full Service Access Network (FSAN) and the Ethernet in the First Mile alliance (EFM) working groups, in cooperation with the International Telecommunications Union (ITU) and the Institute of Electrical and Electronics Engineers (IEEE) standardization bodies, defined several PON standards, as summarized in Figure 2.

For the purposes of this Chapter, we will not go through all these standards, since none of these has been approached with self-coherent reflective architectures that will be presented in this Chapter, but we will mention only the most recent one, called NG-PON2 (Next Generation PON 2), in progress of standardization at the time of writing and with still open technological issues. In particular, we will address only physical layer aspects. The definition of such standard started in 2010 by FSAN.

Since about the 70% of the total investments in deploying PONs is due to the optical distribution networks, it is crucial for the NG-PON2 evolution to be compatible with the deployed networks and to reuse the outside plant. Moreover, NG-PON2 technology must outperform existing PON technologies in terms of ODN compatibility, bandwidth, capacity and cost-efficiency. For this reason, the initial "wish list" for NG-PON2 was originally very demanding, requiring not only an increase of bit rates with respect to previous generations, but also in terms of total number of users per PON tree and reach.

The major original general requirements of NG-PON2 can be summarized as follows:

- increase the aggregate rate to 40 Gbit/s in downstream or upstream;
- increase the reach to 40-60 km (or more);
- increase the splitting ratio up to 256 users (or more);
- compatibility with the attenuation classes defined in XG-PON, as reported in Table 1;
- increase the effective bit rate per user (in both directions) to a value closer to 1 Gbit/s per ONU.

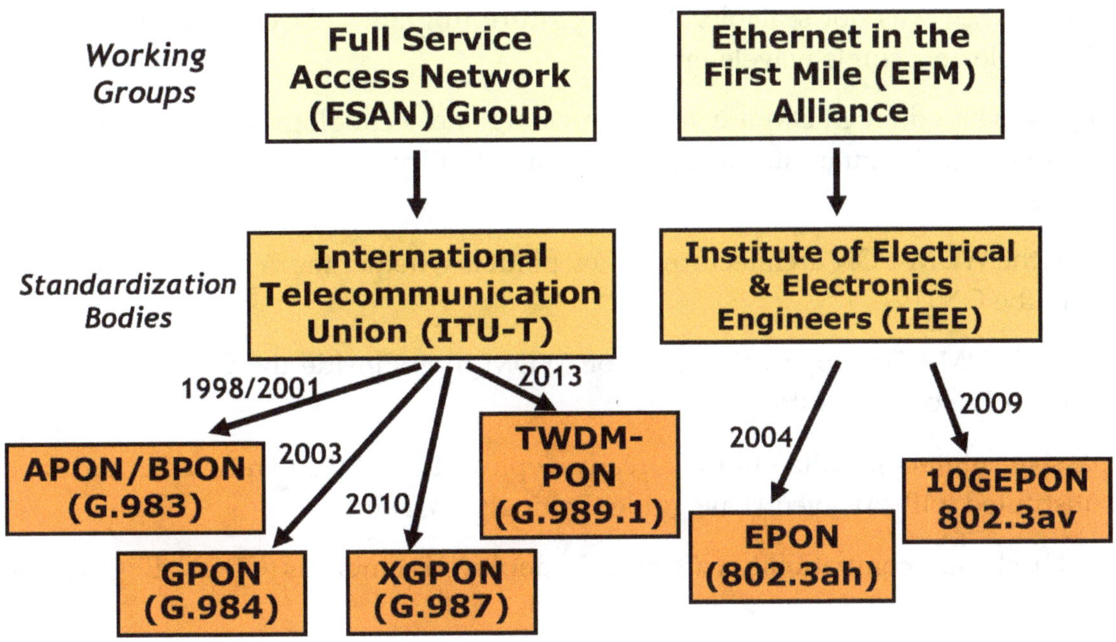

Figure 2. PON standards

	'Nominal1' class (N1 class)	'Nominal2' class (N2 class)	'Extended1' class (E1 class)	'Extended2' class (E2 class)
Minimum loss	14 dB	16 dB	18 dB	20 dB
Maximum loss	29 dB	31 dB	33 dB	35 dB

Table taken from G.987.2

Table 1. Classes for optical XG-PON path loss

Many PON technologies have been proposed to provide broadband optical access beyond 10 Gbit/s, like for example:

- the 40 Gigabit time-division multiplexed PON (XLG-PON) proposal [1] which increases the single carrier serial downstream bit rate of a 10 Gigabit PON (XG-PON) to 40 Gbit/s;
- a set of orthogonal frequency-division multiplexed (OFDM)-based PON proposals which employ quadrature amplitude modulation and fast Fourier transform to generate digital OFDM signals for transmission [2, 3];
- a group of WDM-PON proposals which provide a dedicated wavelength channel at the rate of 1 Gbit/s to each ONU with different WDM transmit or receive technologies [4, 5];
- the time-and wavelength-division multiplexed PON (TWDM-PON) proposal which stacks multiple XG-PONs using WDM [6].

The first proposal, based on a single wavelength (per direction) at very high bit rate, was finally discarded, since it was perceived as too expensive. Indeed, it had to include a duo-binary modulation and a downstream transmission around 1300 nm to avoid dispersion compensation. Moreover, it was not scalable, thus no future upgrades above 40 Gbit/s downstream seemed feasible on a single wavelength.

The proposals based on OFDM are very interesting and technologically advanced solutions, but they do not easily guarantee a complete compatibility with existing deployed networks and devices.

Regarding the WDM-PON architectures, a key point is whether the WDM filters are inside the ODN or in the ONU.

The "pure" WDM-PON approach, based on WDM filters inside the ODN, generates some constraints to Telecom operators:

- all passive splitters installed in the already deployed PON (brown field scenario) need to be substituted with Arrayed Waveguide Grating (AWG) filters;
- limited backward compatibility with already deployed standards (GPON, EPON, XG-PON, etc.);
- a dedicated wavelength per user is a "circuit" per individual user, thus no sharing or statistical multiplexing is possible.

For these reasons, this solution seems to be out of question for NG-PON2 standard. Among all of the aforementioned proposals, TWDM-PON has attracted the majority support from global vendors and was selected by the FSAN community in the April 2012 meeting as a primary solution to NG-PON2. TWDM-PON increases the aggregate PON rate by stacking XG-PONs via multiple pairs of wavelengths; since an XG-PON system offers the access rates of 10 Gbit/s in downstream and 2.5 Gbit/s in upstream, a TWDM-PON system with four pairs of wavelengths will then be characterized by:

- 4 wavelengths (upgradable to 8) at 100 GHz WDM spacing:
- 1595-1600 nm for downstream (up to 1605 nm with 8 wavelengths);
- 1535-1540 nm for upstream (up to 1545 nm with 8 wavelengths);
- 40 Gbit/s in downstream (10 Gbit/s per wavelength) and 10 Gbit/s in upstream (2.5 Gbit/s per wavelength);
- up to 64 ONUs (16 per wavelength);
- up to 40 km;
- up to 35 dB of power budget;
- full backward compatibility.

The basic TWDM-PON architecture is shown in Figure 3. Four XG-PONs are stacked by using four pairs of wavelengths (e.g., wavelength pairs of $\{\lambda 1, \lambda 5\}$, $\{\lambda 2, \lambda 6\}$, $\{\lambda 3, \lambda 7\}$ and $\{\lambda 4, \lambda 8\}$, in Figure 3). ONUs are equipped with tunable transmitters and receivers. The tunable transmitter is tunable to any of the four upstream wavelengths; the receiver is tunable to any of the four downstream wavelengths. In order to achieve power budget higher than that of XG-PON, optical amplifiers (OAs) are used at the OLT side to boost the downstream signals as well as to pre-amplify the upstream signals. The ODN remains passive since OAs are placed at the OLT side, together with WDM Mux/DeMux.

Therefore, the TDWM-PON key features are:

- colorless ONUs;
- splitter-based ODN.

Most of the TWDM-PON components are commercially available in access networks today. Comparing to previous generations of PONs (e.g., GPON, XG-PON), the only, but technically very challenging, new components in TWDM-PON are the tunable receivers and tunable transmitters at the ONU.

For what concerns the tunable receiver, it should tune its wavelength to any of the TWDM-PON downstream wavelengths by following the OLT commands. This function can be implemented by using candidate technologies such as:

- thermally tuned Fabry–Perot (FP) filter [7];
- angle-tuned FP filter, injection-tuned silicon ring resonator [8];

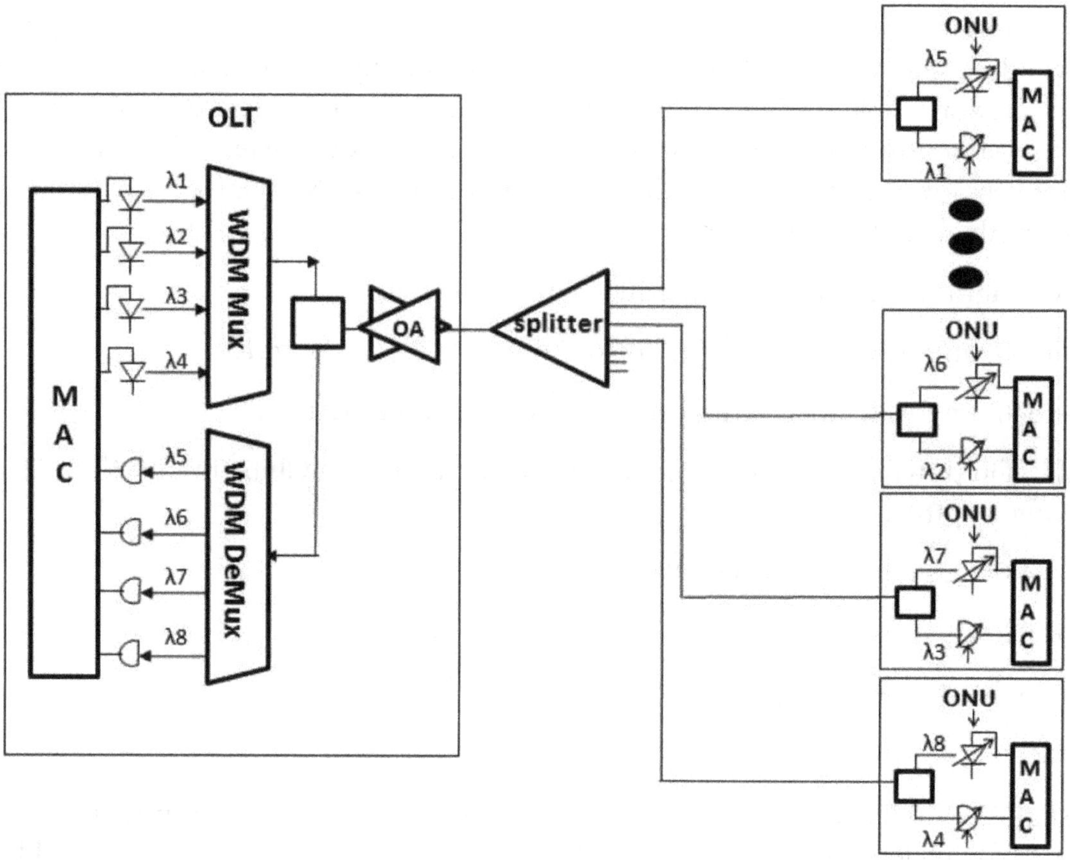

Picture taken from [6]

Figure 3. TWDM-PON system diagram

- liquid crystal tunable filter [9];
- thermally tunable FP detector [10].

The ONU tunable transmitter can tune its wavelength to any of the upstream wavelengths. The related implementation technologies are:

- distributed feedback (DFB) laser with temperature control (TC) [11];
- DFB laser with partial TC [12];
- multi-section distributed Bragg reflector laser (electrical control) without cooling [13];
- external cavity laser (ECL) with mechanical control without cooling [14];
- ECL with thermo/electro/piezo/magneto-optic control without cooling [15, 16].

All these solutions are today under consideration for NG-PON2, but none has yet completely demonstrated to be achievable at the (very low) target prices for ONU. Purpose of this Chapter is to describe a solution for the upstream transmission that avoids the need for a tunable laser at the ONU side: it is based on self-coherent reflective PON architectures as a possible technological approach to the NG-PON2 requirements. In the following we will then first show the concept of the reflective PON, describing the key components it needs, the problems it solves and the limitations it encounters; then, we will propose the self-coherent detection enhancement and will give an overview of the latest research results available in literature.

2. The reflective PON approach

A very basic schematic view of a WDM-PON architecture is depicted in Figure 4. It makes use of integrated multichannel transmit (TX) and receive (RX) arrays in the OLT, and tunable laser (T-TX) and filters (TOF) in the ONUs. From the OLT, all downstream wavelengths are broadcast via the ODN; therefore, a tunable filter is necessary at each ONU in order to select the correct wavelength. Similarly, each ONU has to be provided with a tunable laser for the correct upstream wavelength selection.

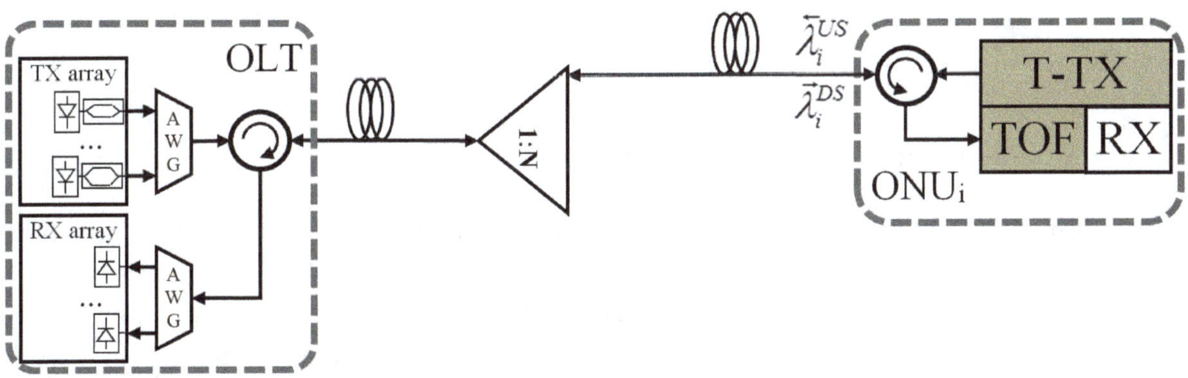

Figure 4. WDM-PON architecture

One of the few physical layer problems over which FSAN is still working today before the final NG-PON2 Recommendation release, is related to these two devices at the ONU. Indeed, they should both have a precision compatible with a 100 GHz wavelength grid, and be able to tune on (at least) four wavelengths, adopting some protection systems during their switch-on time, in order to avoid interferences with other channels (due to the uncontrolled wavelength transmission at the laser power on); moreover, they should operate on a very wide temperature range and they should have a target price compatible with equipment to be installed at the customer premises. Moreover, in the longer term, if more than four wavelengths will be used (for instance for WDM overly), this issue will be particularly critical.

An alternative solution to overcome this problem is represented by reflective PON architectures, whose key idea is to generate the unmodulated upstream wavelengths at the OLT side, and modulate them in reflection at the ONU side, as schematically depicted in Figure 5. With respect to the setup depicted in Figure 4, the OLT is equipped by an additional multichannel transmit array, for the unmodulated seed signals generation, and the tunable laser at the ONU side is replaced by a reflective transmitter (R-TX). This way, the costs and the technical issues (the wavelength control in particular) related to the tunable lasers are moved from the ONU to the OLT, where they can be more easily managed.

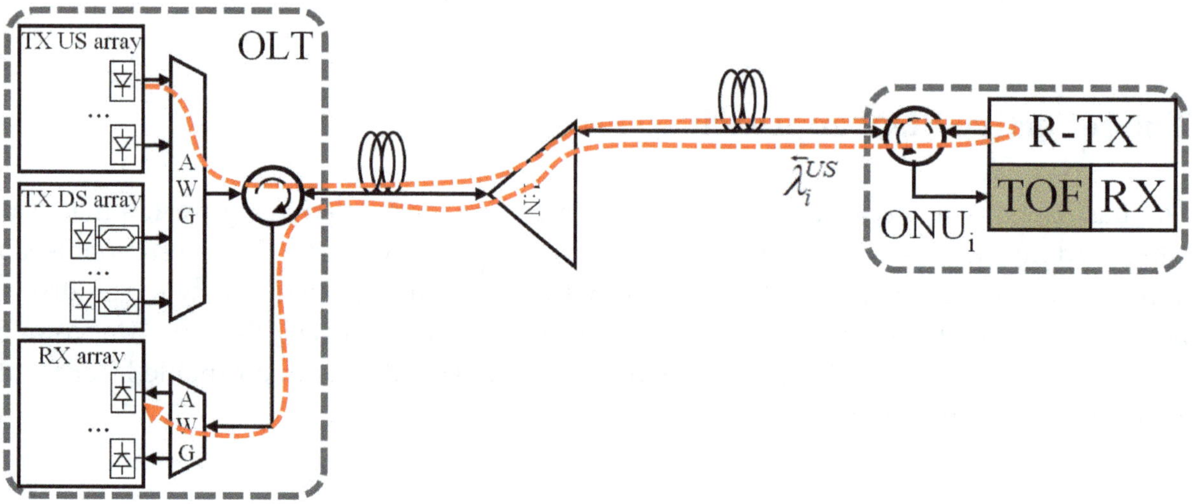

Figure 5. Scheme of a possible solution for Reflective-PON ONUs

2.1. Key components for the reflective ONU

Many variants of WDM reflective PON architectures can be found in literature, all around the common denominator of avoiding expensive tunable lasers at the ONU by means of using a reflective modulator for the upstream transmission that sends back a centrally-generated seed signal. One of the most common approaches adopted in order to obtain this effect at the ONU side is based on the use of Reflective Semiconductor Optical Amplifier (RSOA), which combine the amplification and modulation capabilities in a single device. They are composed by a high reflectivity coating on one facet, along with a curved waveguide and ultra-low reflectivity coating on the other facet, to produce a highly versatile reflective gain medium [17]. Today, few optical devices manufacturers propose commercial RSOA devices (e.g. as CIP, MEL and Kamelian), since they are just used in the reflective PON scenario and at a research level.

Figure 6 schematically depicts a RSOA in a typical low-cost TO-CAN package, while Figure 7 shows the electrical bandwidth measurement result of a typical commercial RSOA (butterfly-packaged solutions may have higher bandwidth, but their cost seems too high for application in the extremely cost-sensitive scenario of ONU). From this graph, it is possible to notice that the 3 dB electrical bandwidth of the device under test is about 500 MHz; anyway, the signal obtained by modulating the RSOA bias current I_b with a 1 Gbit/s OOK modulation (shown in the inset of Figure 7) is received without inter-symbol interference.

Since 2000, the interest of RSOA as upstream transmitters for WDM-PON applications based on reflective ONU has grown-up. The first results have been proposed in [18], where the uplink data stream was reflected and modulated via the RSOA at 1.25 Gbit/s, as then further developed and investigated in several later works (e.g. [19-22]).

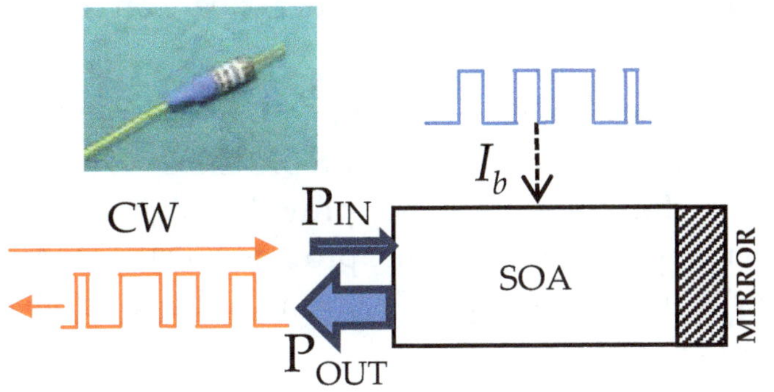

Figure 6. Schematic view of a RSOA

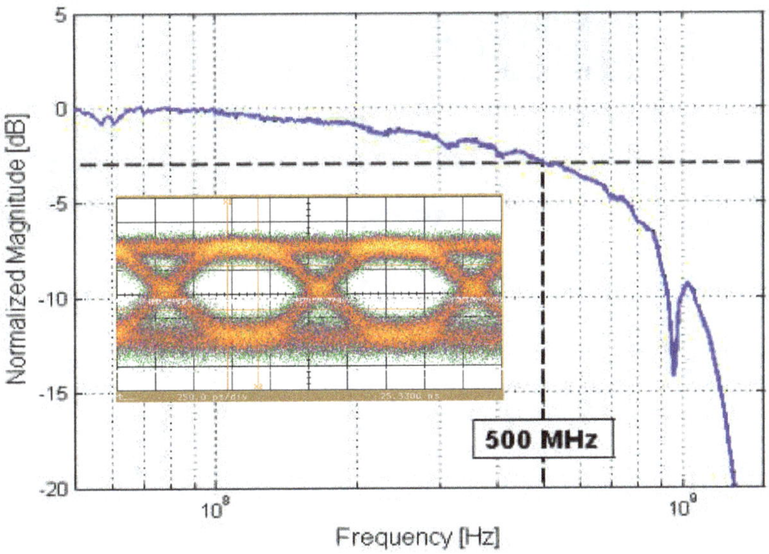

Figure 7. Characterization of a RSOA

In order to increase the upstream data rate up to 10 Gbit/s in such an architecture, using this low-bandwidth devices, different approaches have been proposed by research groups, like the use of optical filtering and electronic equalization [23-25] or the adoption of advanced modulation formats [26].

Another device that can be adopted as reflective transmitter at the ONU side is the Reflective Electro-Absorption Modulator (REAM). EAM are semiconductor devices usually made in the form of a waveguide with electrodes for applying an electric field in a direction perpendicular to the modulated light beam. Their principle of operation is based on a change in the absorption spectrum caused by the applied electric field, which changes the band-gap energy without involving the excitation of carriers by the electric field: the so called Franz-Keldysh effect [27]. Reflective EAM include an EAM and a mirror, as schematically depicted in Figure 8, and they can be used to reflect and modulate the incoming light by means of the applied electrical signal (V_b). Thanks to EAM, a modulation bandwidth close to 10 GHz can be achieved, as shown in

Figure 9, which makes these devices useful for optical fiber communication at 10 Gbit/s and above.

Differently from the RSOA, these devices do not amplify the light but, if coupled in cascade to a Semiconductor Optical Amplifier (SOA), they represent a very interesting solution for the ONU of a reflective PON architecture, since they provide a high-speed modulation capability, combined with the linear amplification of signal provided by the SOA. This solution has been proposed in several works, like for example [28, 29], in order to achieve a 10 Gbit/s upstream transmission in a WDM-PON scenario.

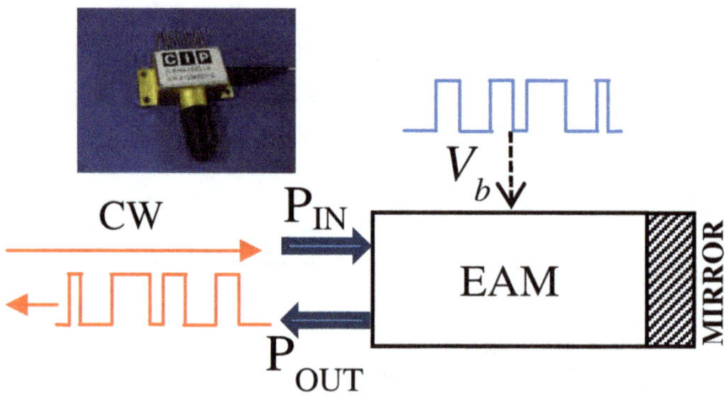

Figure 8. Schematic view of a REAM

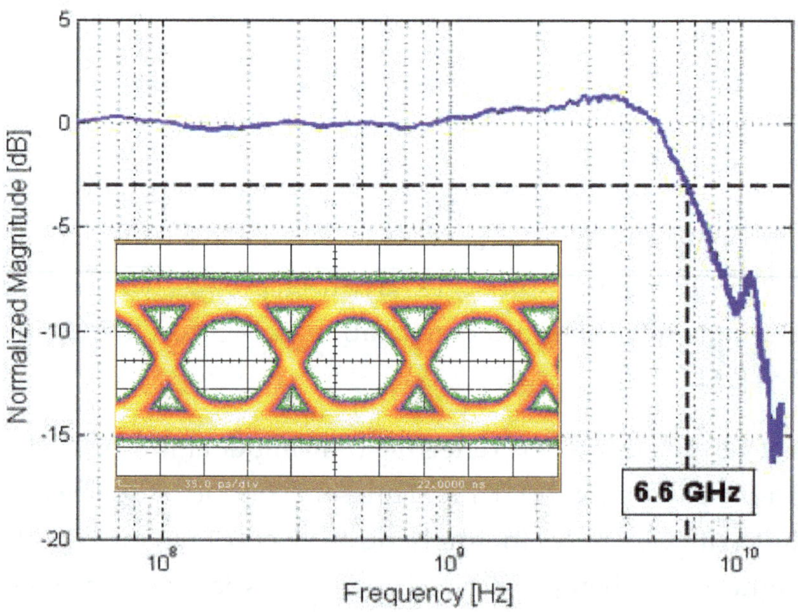

Figure 9. Characterization of a REAM

At the time of writing this Chapter, another solution for the reflective transmitter implementation is under analysis inside the FP7 EU STREP project titled "FABULOUS" [30]. It consists on a Mach-Zehnder based subsystem allowing polarization independent reflective carrier suppressed optical modulation for application in frequency division multiple access (FDMA) PON. The architecture of this subsystem, presented at first in [31], relies on using a Mach-Zehnder modulator (MZM) simultaneously in both directions within a polarization diversity loop made of a polarization beam splitter (PBS) looped on itself through the modulator. The MZM provides RF access to both the electrodes independently, allowing modulation to be efficient in both the forward and the backward directions. On one electrode of the MZM, RF drive power is applied in the forward direction while, on the other electrode, the modulation is applied in the backward direction (the same signal is applied to both sides). The two polarizations of the incoming optical signal are split through the PBS and sent to the MZM in counter-propagating directions (the MZM having polarization maintaining fibers on both the input and the output). After being modulated, they are recombined through the PBS and sent back towards the OLT, as schematically depicted in Figure 10.

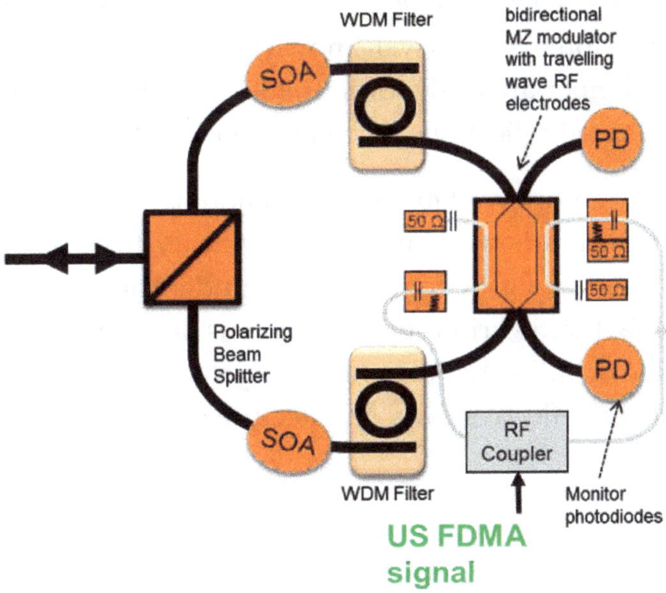

Figure 10. Architecture of R-MZM subsystem

2.2. Transmission issues in reflective PON

The reflective WDM-PON architectures proposed in literature in the last ten years seem to be completely incompatible with TWDM-PON for a set of different reasons such as:

- most of them require the use of AWGs in the ODN, while as previously mentioned most telecom operators believes that backward compatibility in PON is a must, and thus the splitter-based architecture should be maintained also in NG-PON2;
- the achievable ODN loss is limited to typically 20-25 dB, due to several spurious effects such as the Rayleigh back-scattering and the low received optical power;
- most of them use a dedicated wavelength for each ONU, and do not support burst-mode TDMA in the upstream.

In a reflective architecture like the one depicted in Figure 5, the upstream signal undergoes the ODN path loss twice (from the OLT to the ONU as CW seed signal and back to the OLT after the ONU reflection). The optical power at the OLT input is thus determined by the following formula:

$$S = \vec{P}_{CW} - 2L_{ODN} + G_{RONU} \tag{1}$$

where $\vec{P}_{CW}$ is the optical power of the CW signal at the input of the fiber, L_{ODN} is the ODN loss and G_{RONU} is the gain of the optical amplifier installed at the ONU. Since the TWDM-PON standard fixed the maximum value for the launched optical power at the ODN input to +11 dBm per wavelength and the gain of the optical amplifier at the ONU is typically of the order of $G_{RONU} = 20\ dB$, the optical power at the OLT input, for the lowest ODN class N1 ($L_{ODN} = 29\ dB$), is of the order of $S = -27\ dBm$, which is lower than the sensitivity of the standard direct-detection receivers. Moreover, it is well known that, in such an architecture, the spurious back reflections for a SMF fiber are of the order of 30-35 dB below the forward propagating signal, due to the concentrated reflections on components and the Rayleigh backscattering.

Rayleigh backscattering is an unavoidable phenomenon in optical fiber propagation. It is a fundamental loss mechanism arising from random density fluctuations frozen in the fiber during manufacturing. There is a growing interest in understanding Rayleigh backscattering since it can be a limiting factor in various optical systems. It must be taken into account in the performance computation of bidirectional lightwave system, especially in a wavelength-reuse system, like for example the reflective PON architecture depicted in Figure 11.

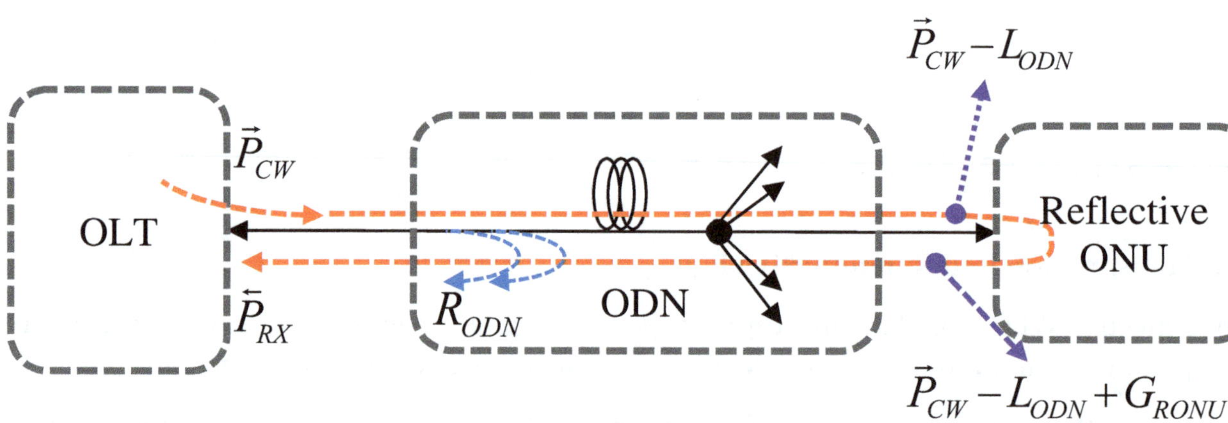

Figure 11. Rayleigh backscattering effect in reflective PON architecture at the OLT

In such a system, the received optical power is composed by the useful signal S, as specified above, and the interference signal I due to reflections, given by:

$$I = \vec{P}_{CW} - R_{ODN} \tag{2}$$

where R_{ODN} is the power of the reflections. Therefore, the signal to interference ratio is given by:

$$\left(\frac{S}{I}\right)_{dB} = -2L_{ODN} + G_{RONU} + R_{ODN} \tag{3}$$

For a standard direct-detection receiver, even if the best solutions proposed in literature to mitigate the Rayleigh backscattering effect are adopted, the signal to interference ratio should be greater than 5-10 dB. This sets an important limit to the maximum achievable ODN loss. Indeed, the maximum reachable ODN loss is lower than 25 dB, since the optical power of the spurious back reflections is about 30-35 dB below the forward propagating signal.

To improve the performance, one could in principle increase the G_{RONU}, but there are technological component issues that limit the maximum reachable gain of optical amplifiers; in addition, another issue can arise from the Rayleigh backscattering that is generated at the ONU side, as depicted in Figure 12 and explained in details in [32]. This means that, in order to satisfy the NG-PON2 requirements with a reflective PON architecture, direct-detection at the OLT side is out of question.

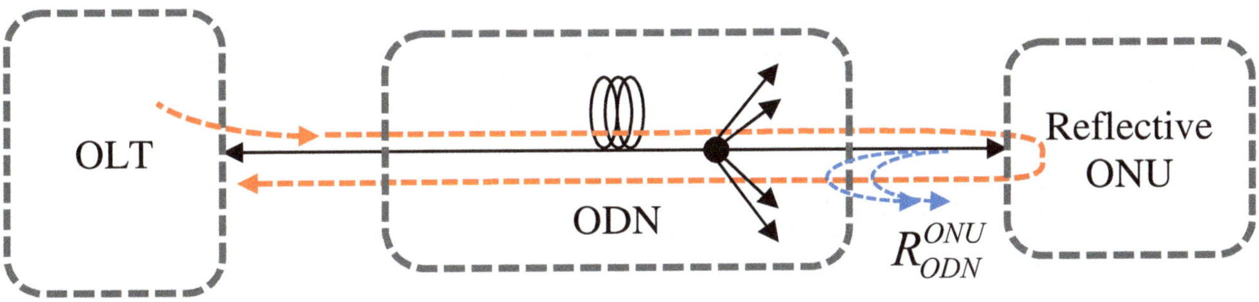

Figure 12. Rayleigh backscattering effect in reflective PON architecture at the ONU

2.3. Self-coherent detection in RPONs

Because of the aforementioned problems related to direct-detection, a coherent detection at the OLT seems to be a must in order to satisfy the NG-PON2 requirements, whether a reflective architecture approach is adopted.

The fundamental concept behind coherent detection is to use the product of the electrical fields of the modulated signal light (centred at f_S) and a CW local oscillator (centred at f_{LO}). This produces a frequency down-conversion of the signal to the frequency $f_{IF} = |f_S - f_{LO}|$, as schematically depicted in Figure 13.

A possible implementation of coherent detection in a reflective PON architecture is depicted in Figure 14: in this case, the CW light source placed at the OLT side is used both as a feed to be sent downstream to the reflective ONU and as a local oscillator for the OLT coherent

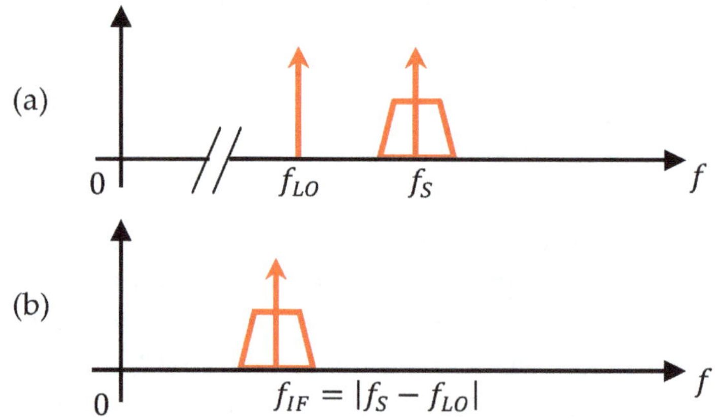

Figure 13. Spectra of (a) the optical signal and (b) the down-converted IF signal

receiver, executing a self-coherent detection. Therefore, after the optical-to-electrical conversion, the signal is down-converted to baseband.

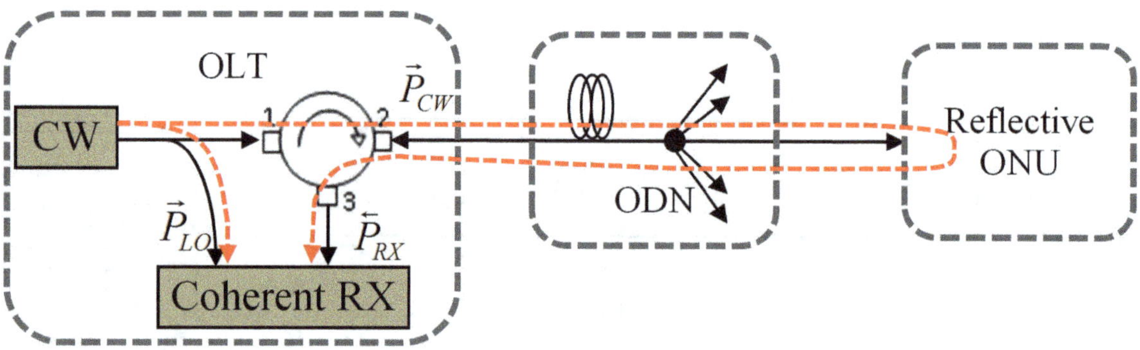

Figure 14. Scheme of a possible solution for self-coherent OLT in a reflective-PON architecture

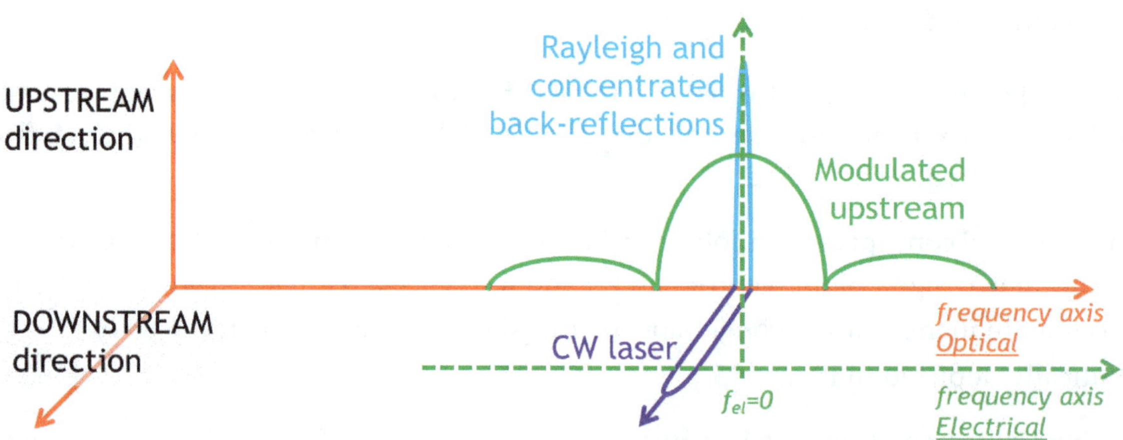

Figure 15. Spectral analysis of signal and reflections in reflective PON

As demonstrated by the results available in literature presented in the following, this setup supports higher ODN loss values with respect to the limit of reflective PONs based on direct-detection, in particular for the following reasons:

- much better sensitivity than direct-detection;
- much larger resilience to spurious back reflections.

Indeed, using a coherent receiver, the Rayleigh backscattering reflections appear as added close to the "electrical" DC, as depicted in Figure 15, thus they can be easily filtered out by electrical high-pass filters.

3. Latest research results

WDM reflective PON "variant", characterized by self-coherent detection at the OLT side, was initially proposed in [33], where a RSOA-based WDM PON architecture, employing a novel self-homodyne receiver and a novel signal processing technique to eliminate the penalty caused by the back-reflection of the seed light, is proposed. In addition, they successfully demonstrated a long-reach fiber-loopback system of over 100 km. These results indicate that the proposed architecture could provide an attractive solution to realize a long-reach WDM PON with high-split ratio.

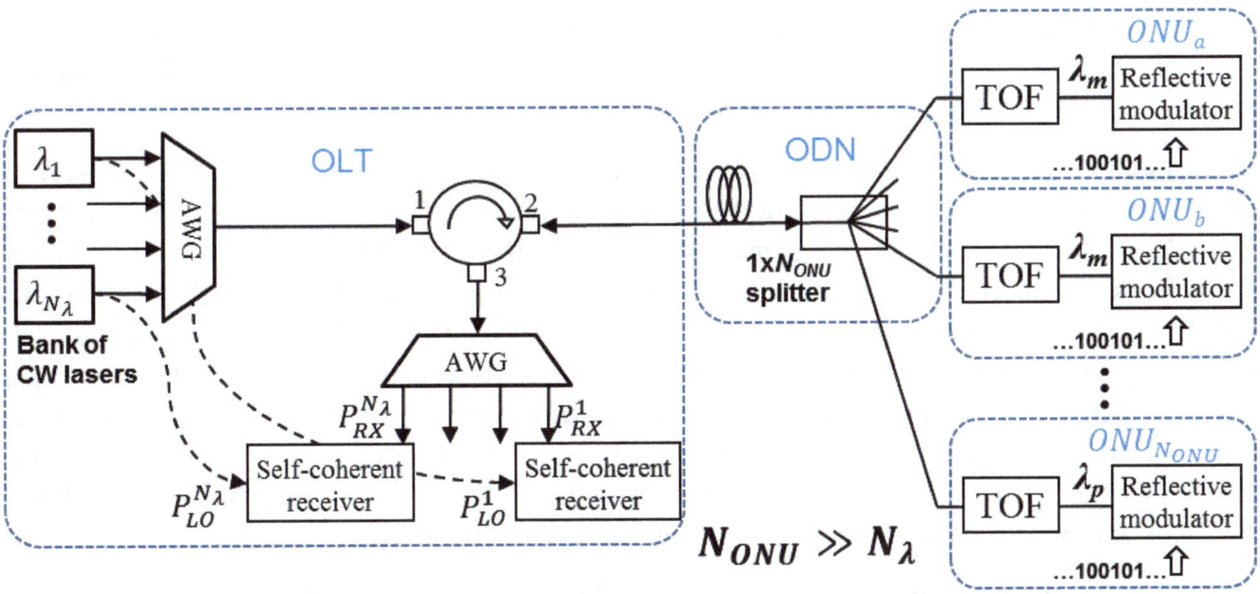

Figure 16. Scheme of self-coherent reflective PON architecture (only upstream transmission)

A possible implementation of the upstream transmission in a self-coherent reflective PON is shown in Figure 16; this architecture may offer the following advantages:

- the upstream wavelengths comb accuracy is completely set by the OLT and not by each individual ONU. In this solution, each ONU needs to tune its optical filter by locking it on one of the N_λ already existing wavelength;
- subsequent upgrades to dense-WDM (DWDM) seem more feasible when using a number of wavelengths N_λ significantly higher than 4, and possibly a narrower frequency spacing (such as 50 GHz), since it is possible to avoid tunable lasers at the ONU;
- for each upstream wavelength, the required CW laser and coherent receiver at the OLT are shared by several ONUs (of the order of N_{ONU} / N_λ) so that their cost may be more reasonably justified.

In addition to WDM, a second level of multiplexing is necessary in order to overcome the limit of a dedicated wavelength per user; this can be obtained sharing time, basing on Time Division Multiplexing (TDM) in the downstream and Time Division Multiple Access (TDMA) in the upstream, or frequency (FDM and FDMA) between users.

3.1. TDM-WDM results

Basing on the architecture depicted in Figure 16, the results presented in [34] experimentally demonstrate the simultaneous transmission of two independent TDMA ONUs with upstream bit rate equal to 2.5 Gbit/s, working in TDMA burst mode and with performance compatible with E2 XG-PON class, thus compatible with US TWDM-PON requirements. At the OLT, the signal demodulation is performed thanks to a self-coherent receiver and a custom burst-mode digital-signal processing technique. In particular, a conventional DSP solution based on Viterbi-Viterbi carrier-phase estimation and LMS adaptive equalization [35, 36] has been modified to work in burst mode operation, focusing on the alignment procedures on the received packets and on the convergence speed of the LMS algorithm [37, 38].

The reflective ONU of such architecture is emulated as depicted in Figure 17. The CW signal reaching the ONU is reflected and modulated by means of a REAM. The SOA placed in front of the REAM amplifies the optical signal twice, first on the feed CW downstream, and then on the reflected and modulated upstream signal. In order to emulate the ONU wavelength selection functionality, a TOF is placed between the SOA and the REAM. This TOF is also useful to partially filter out the ASE noise and, thanks to its approximately 4 dB insertion loss, to avoid excessive SOA saturation in the upstream direction. The SOA+REAM structure, even though not yet commercial, has been integrated in several research projects (such as in [40]). The 2.5 Gbit/s upstream signal is generated at the ONU by driving the REAM by pure NRZ signal, while the optical packets are generated by switching on and off the SOA.

Each packet contains a short dummy-pattern at the beginning and end of each burst, 127 bits of sync-pattern and 1000 bits payload (8B/10B coded), as shown in Figure 18. The dummy-pattern is useful for "absorbing" the rise and fall-time of the SOA acting as a gate, and also the transient effect of the coherent receiver AC-coupling. The sync-pattern is used for identifying

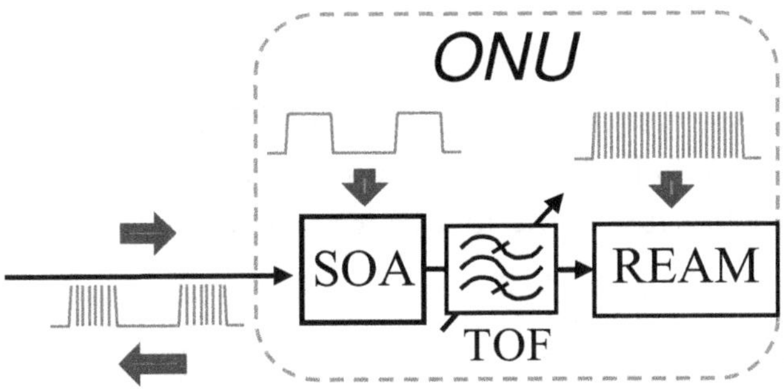

Figure 17. Reflective ONU implementation

the start of each packet and also for training the LMS algorithm, before switching to the LMS "decision-directed" mode, which should carry out the burst payload elaboration. This system emulates a TDMA transmission from two independent ONUs with 25 ns guard-time, where the ONU 1 is the reference ONU and the ONU 2 is the interfering ONU, reproducing the worst-case condition in terms of interference for such a system.

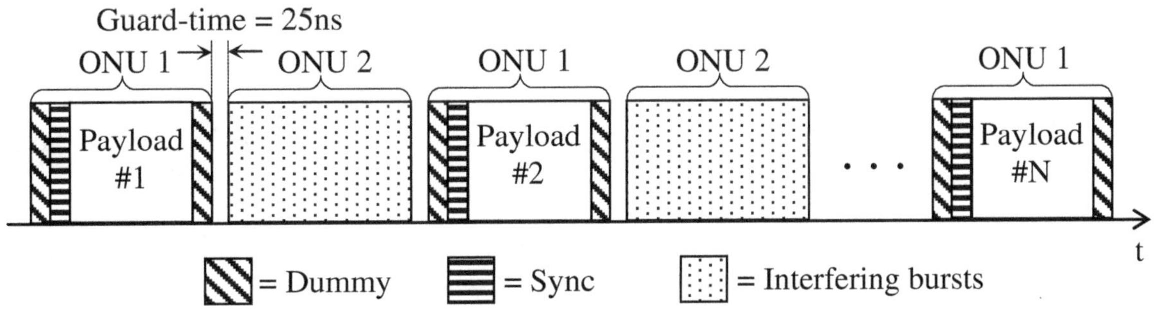

Figure 18. Time relations of the useful and interfering optical US packets for two ONUs

The authors of [34] expressed the performance of the system in terms of Bit Error Rate (BER) as a function of the ODN loss and the acceptable differential optical path loss (DOPL).

The results show that a BER<10^{-3} (a reasonable Forward Error Correcting code threshold for simple FECs, as those used in XG-PON) is obtained up to 35 dB of ODN loss and more than 15 dB of DOPL, as required by XG-PON class E2.

The same setup was also tested increasing the upstream bit rate and using commercial DFB lasers rather than ECL lasers at the OLT side [39]. The bit rate was set to 10 Gbit/s per wavelength (from the previous 2.5 Gbit/s rate), a bit rate currently under consideration in FSAN for the point-to-point WDM (PtP-WDM) for symmetric traffic, and for the TWDM-PON longer term evolution. Replacing the ECL lasers with DFB lasers would provide a significant cost reduction at the OLT. As reported in [39], in these working conditions, the FEC threshold is reached after 31 dB and for a DOPL higher than 15 dB, satisfying the requirement of N2 class.

3.2. FDM-WDM results

One of the main drawbacks of the more traditional TDMA-PON approach deployed nowadays, is that it does not scale well above 10 Gbit/s per wavelength in term of cost/complexity and power efficiency, mostly due to the fact that each ONU should work on the aggregated bit rate.

On the contrary, FDM/FDMA approach allows ONUs to operate at low DSP speeds (and so reduce their cost and power consumption); indeed, the speed at which the ONUs operates is equal to the maximum speed that the customers are allowed to communicate at (e.g. 1 Gbit/s) which is much smaller than the aggregated line rate (e.g. 20 Gbit/s).

The work presented in [41] proposes an innovative reflective PON approach for the upstream transmission using a special configuration based on FDMA, where each ONU is assigned a portion of the available electrical spectrum to perform an high spectral efficiency M-QAM modulation format as depicted in Figure 19.

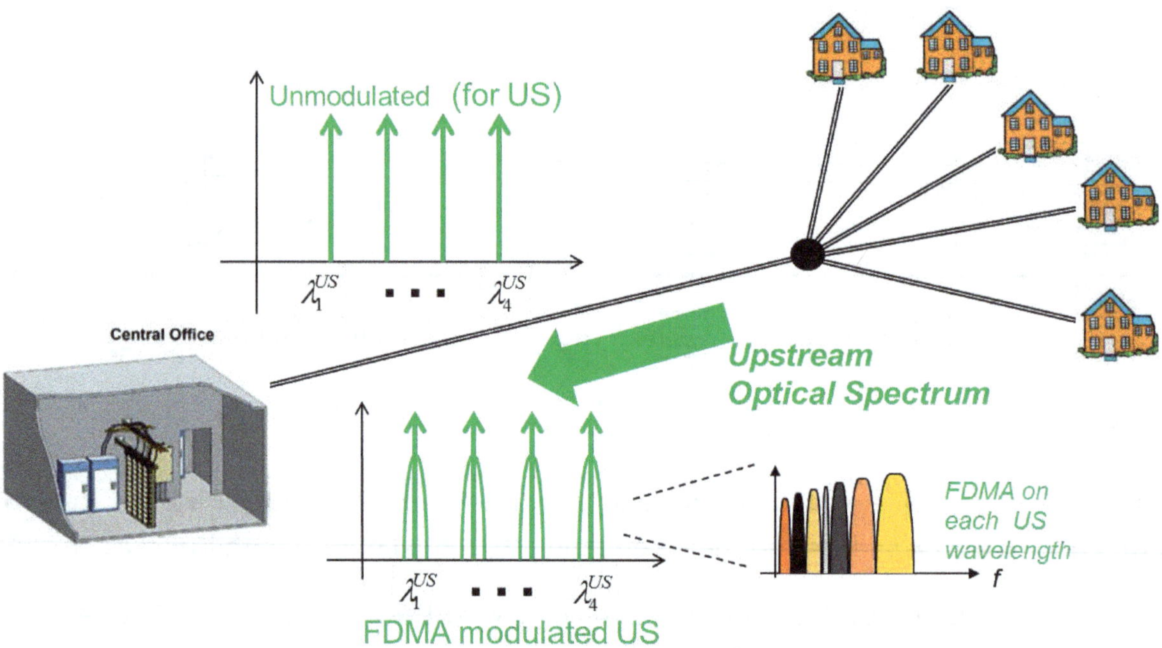

Figure 19. FDMA reflective PON architecture

The results presented in [41] demonstrate that this architecture targets an upstream capacity of 32 Gbit/s per optical carrier, outperforming NG-PON2 with a tenfold increase in the upstream capacity, with power budget and max reach overcoming ITU-T class N2, as defined in the XG-PON standard (up to 40 km and 31 dB ODN loss). An important feature of this approach consists in the fact that, due to the absence of optical sources at the customer premises, the ONU can be realized as a Photonic Integrated Circuit with an unprecedented level of integration. This holds true for self-coherent reflective PONs in general, but the highest level of integration is to date relevant to the architecture shown in [41], where the Mach

Zehnder Modulator, the two tunable filters (for multi-wavelength transmission) and the polarizing beam splitter that characterize the proposed ONU will be realized on silicon platform, the two SOA will be suitable for photonic integration and finally the electrical driver will be flip-chipped on top; for a detailed description see [42, 43].

4. Conclusions

In this Chapter, we have given a general description of the self-coherent reflective PON architecture as a possible technological approach to the NG-PON2 requirements. We have shown that, thanks to reflective ONU and the self-coherent detection at the OLT side, the performance required by NG-PON2 are satisfied even without the need for a tunable laser at the ONU side. If properly integrated, the reflective ONU structure can be less expensive than the TWDM-PON solution in term of CAPEX, and have a much simpler wavelength control. The only significant increase in cost in the proposed solution is due to the presence of one coherent receiver per each upstream wavelength (a cost to be shared among all the users with the same wavelength). While the cost of coherent receivers used today for long-haul is likely still too high for an OLT, it can be observed an enormous effort in the long-haul technical community to make small-form factor and low-power coherent receivers possible, so that it is reasonable to envision a sharp decrease in coherent receiver costs that may allow the use of coherent detection at OLT in the near term.

Author details

S. Straullu[1*], S. Abrate[1] and R. Gaudino[2]

*Address all correspondence to: straullu@ismb.it

1 Istituto Superiore Mario Boella, Torino, Italy

2 Politecnico di Torino, Torino, Italy

References

[1] E. Harstead, et al., "Technologies for NGPON2: Why I think 40G TDM PON (XLG-PON) is the clear winner", in Proc. Opt. Fiber Commun. Conf./Nat. Fiber Opt. Eng. Conf. Workshop, Mar. 2012

[2] X. Hu, et al., "Energy-efficient WDM-OFDM-PON employing shared OFDM modulation modules in optical line terminal", Opt. Exp., vol. 20, no. 7, pp. 8071–8077, Mar. 2012.

[3] W. Wei et al., "Optical orthogonal frequency division multiple access networking for the future internet", IEEE/OSA J. Opt. Commun. Netw., vol. 1, no. 2, pp. 236–246, Jul. 2009.

[4] H. K. Lee et al., "A WDM-PON with an 80 Gb/s capacity based on wavelength-locked Fabry-Perot laser diode", Opt. Exp., vol. 18, no. 17, pp. 18077–18085, Aug. 2010.

[5] K. Grobe et al., "PON in adolescence: from TDMA to WDM-PON", IEEE Commun. Mag., vol. 46, no. 1, pp. 26–34, Jan. 2008.

[6] Y. Luo at al., "Time-and Wavelength-Division Multiplexed Passive Optical Network (TWDM-PON) for Next-Generation PON Stage 2 (NG-PON2)", IEEE J. LIGHTWAVE TECH. 31 no. 4, 587-593 (2013).

[7] M. Zhu et al.,"Upstream multi-wavelength shared PON with wavelength-tunable self-seeding Fabry-Perot laser diode," in Proc. Opt. Fiber Commun. Conf./Nat. Fiber Opt. Eng. Conf., Mar. 2011, pp. 1–3.

[8] D. Liang et al., "Low threshold electrically-pumped hybrid silicon microring lasers," IEEE J. Sel. Topics Quantum Electron., vol. 17, no. 6, pp. 1528–1533, Nov./Dec. 2011.

[9] A. Kato et al., "Tunable optical filter with cascaded waveguide Fabry-Pérot resonators featuring liquid crystal cladding," Photon. Technol. Lett., vol. 24, no. 4, pp. 282–284, Feb. 2012.

[10] R. Murano et al., "Tunable 2.5 Gb/s receiver for wavelength-agile DWDM-PON," presented at the presented at the Opt. Fiber Commun. Conf./Nat. Fiber Opt. Eng. Conf., 2008, Post Deadline Paper.

[11] G. W. Yoffe et al., "Widely-tunable 30 mW laser source with sub-500 kHz linewidth using DFB array," in Proc. IEEE Lasers Electro-Optics Soc., Nov. 2008, pp. 892–893.

[12] D. Ton et al., "2.5-Gb/s modulated widely tunable laser using an electroabsorption modulated DFB array and MEMS selection," IEEE Photon. Technol. Lett., vol. 16, no. 6, pp. 1573–1575, Jun. 2004.

[13] Y. Liu et al., "Uncooled DBR laser directly modulated at 3.125 Gb/s as athermal transmitter for low-cost WDM systems," IEEE Photon. Technol. Lett., vol. 17, no. 10, pp. 2026–2028, Oct. 2005.

[14] P. Ossieur et al.,"Demonstration of a 32 512 Split, 100 km Reach, 2 32 10 Gb/s hybrid DWDM-TDMA PON using tunable external cavity lasers in the ONUs," J. Lightw. Technol., vol. 29, no. 24, pp. 3705–3718, Dec. 2011.

[15] F. Wei et al. "Tunable external cavity diode laser with a PLZT electrooptic ceramic deflector," IEEE Photon. Technol. Lett., vol. 23, no. 5, pp. 296–298, Mar. 2011.

[16] J. Zheng et al., "Optically tunable ring external-cavity laser," in Proc. IEEE Photon. Conf., Oct. 2011, pp. 644–645.

[17] http://www.freepatentsonline.com/7046435.html

[18] P. Healey et. al., "Spectral slicing WDM-PON using wavelength-seeded reflective SOAs", Electronics Letters, 37, 19, 1181 – 1182 (2001).

[19] H.C. Shin et al., "Reflective SOAs optimized for 1.25 Gbit/s WDM-PONs", Lasers and Electro-Optics Society (LEOS), Vol. 1 (2004).

[20] L. Wooram et al., "Bidirectional WDM-PON based on gain-saturated reflective semiconductor optical amplifiers", Photonics Technology Letters, IEEE, 17, 11, 2460 – 2462 (2005).

[21] C. Arellano et al, "Bidirectional single fiber transmission based on a RSOA ONU for FTTH using FSK-IM modulation formats", Optical Fiber Communication Conference, 2005, OFC/NFOEC, Vol. 3 (2005).

[22] E. Wong et al., "Low-cost WDM passive optical network with directly-modulated self-seeding reflective SOA", Electronics Letters, 42, 5, 299 – 301 (2006).

[23] I. Papagiannakis et al., "Investigation of 10-Gb/s RSOA-Based Upstream Transmission in WDM-PONs Utilizing Optical Filtering and Electronic Equalization", Photonics Technology Letters, IEEE, 20, 24, 2168 – 2170 (2008).

[24] K.Y. Cho et al., "10-Gb/s Operation of RSOA for WDM PON", Photonics Technology Letters, IEEE, 20, 18, 1533 – 1535 (2008).ù

[25] G. Qi et al., "Low-Bandwidth RSOA Using Partial Response Equalization", Photonics Technology Letters, IEEE, 23, 20, 1442 – 1444 (2011).

[26] T. Duong et al., "Experimental demonstration of 10 Gbit/s upstream transmission by remote modulation of 1 GHz RSOA using Adaptively Modulated Optical OFDM for WDM-PON single fiber architecture", ECOC 2008.

[27] http://users.df.uba.ar/bragas/Web roberto/Papers/Ralph Franz Keldish.pdf

[28] E.K. MacHale et al., "Extended-Reach PON Employing 10Gb/s Integrated Reflective EAM-SOA", ECOC 2008.

[29] D. Smith et al., "Colourless 10Gb/s reflective SOA-EAM with low polarization sensitivity for long-reach DWDM-PON networks", ECOC 2009.

[30] http://www.fabulous-project.eu/

[31] B. Charbonnier et al., "Reflective polarisation independent Mach-Zenhder modulator for FDMA/OFDMA PON", Electronics Letters, 46, 25, 1682 – 1683 (2010).

[32] C. Arellano et al., "Reflections and Multiple Rayleigh Backscattering in WDM Single-Fiber Loopback Access Networks", IEEE J. LIGHTWAVE TECH. 27 no. 1, 12-18 (2009).

[33] S.P. Jung et al., "Demonstration of RSOA-based WDM PON employing self-homodyne receiver with high reflection tolerance", Optical Fiber Communication Conference, OFC/NFOEC (2009).

[34] S. Straullu et al., "Compatibility between coherent reflective burst-mode PON and TWDM-PON physical layers", OPEX 22, Issue 1, pp. 9-14 (2014), http://dx.doi.org/10.1364/OE.22.000009

[35] K. Y. Cho et al., "10-Gb/s operation of RSOA for WDM PON," IEEE PHOTON. TECHNOL. LETTER, 20, 18, 1533 – 1535 (2008).

[36] S. J. Savory, "Digital filters for coherent optical receivers," OPEX 16, 2, 804-817 (2008).

[37] F. Vacondio et al., "Experimental demonstration of a PDM QPSK real-time burst mode coherent receiver in a packet switched network," ECOC, Tu.3.A.1, (2012).

[38] F. Vacondio, et al. "Flexible TDMA Access Optical Networks Enabled by Burst-Mode Software Defined Coherent Transponders", ECOC, We.1.F.2, London, UK, (2013).

[39] S. Straullu et al., "TWDM-PON-compatible 10 Gbps Burst-mode coherent reflective ONU achieving 31 dB ODN loss using DFB lasers", ECOC, Cannes, France (2014).

[40] A. Naughton et al., "Optimisation of SOA-REAMs for hybrid DWDM-TDMA PON applications", Optics Express, 19, 26 (2011).

[41] S. Straullu et al., "Reflective FDMA-PON with 32 Gbps upstream capacity per wavelength and more than 32 dB ODN loss", ECOC, Cannes, France (2014).

[42] S. Abrate et. al., "FDMA-PON architecture according to the FABULOUS European Project" Proc. SPIE 8645, Broadband Access Communication Technologies VII, 864504 (2013).

[43] S. Menezo et al., "Reflective silicon Mach-Zehnder modulator with Faraday rotator mirror effect for self-coherent transmission", OFC/NFOEC 2013.

5

Writing of Long Period Fiber Gratings using CO_2 Laser Radiation

João M.P. Coelho, Catarina Silva, Marta Nespereira,
Manuel Abreu and José Rebordão

1. Introduction

The development of optical fiber gratings (OFGs) had made significant advances both in terms of research and development of optical communications and sensors. OFGs are intrinsic devices that allow modulate the properties of light propagation within the fiber. Grating structures are comparatively simple and in its most basic form, consist on a periodic modulation of the properties of an optical fiber (usually the refraction index of the core). Its application as a sensing element is advantageous because of the intrinsic characteristics of the fiber sensors, such as remote sensing, electromagnetic immunity, weight and compactness, and capability for real time sensing and low cost [1].

Among OFGs, long period fiber gratings (LPFGs) are one of the most important fiber-based sensors. They were first presented by Vengsarkar and co-workers in 1996 [1] as band-rejection filters. Since then, LPFG technology has been receiving continuously growing attention from scientific community. Due to their spectral characteristics, LPFGs have found many applications in both optical communications and sensing systems. In the optical communications field, they have been demonstrated as gain equalizers [1], dispersion compensators [2], optical switches [3], components in wavelength division multiplexing (WDM) systems [4], band rejection filters and mode converters [5]. For optical sensing applications, LPFGs have been implemented as a temperature [6], strain [7] and refractive index sensor [8-10]. As element sensor a LPFG exhibits high sensitivity to the refractive index (RI) of the material surrounding the cladding surface. Other strengths are their low insertion losses, low back-reflection, polarization independence, relatively simple fabrication, and remote sensing easily multiplexed.

These gratings devices operate in transmission mode and are manufactured with periods typically in the range from 100 μm to 1000 μm [6]. Their large modulation period promotes the coupling of the light from the fundamental core mode to co-propagating cladding modes in a single-mode fiber [1]. The light coupled to the cladding decays quickly due to the absorption and scattering by the coating over the cladding.

A commonly-used optical fiber typically consists of a core and a cladding. In Figure 1 it is schematized the coupling mode that occurs in a LPFGs inscribed in single-mode fiber (SMF). As a result, the transmission spectrum of a LPFG has a series of resonant loss peaks (attenuation bands) centered at discrete wavelengths. A typical (theoretical) example is shown in Figure 2.

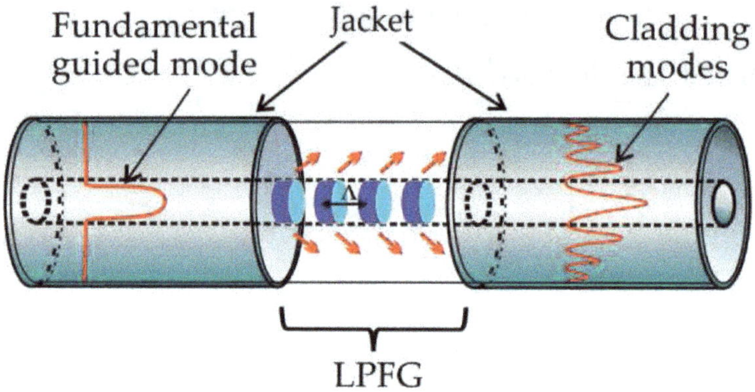

Figure 1. Schematic diagram of a mode coupling in a long period grating [11].

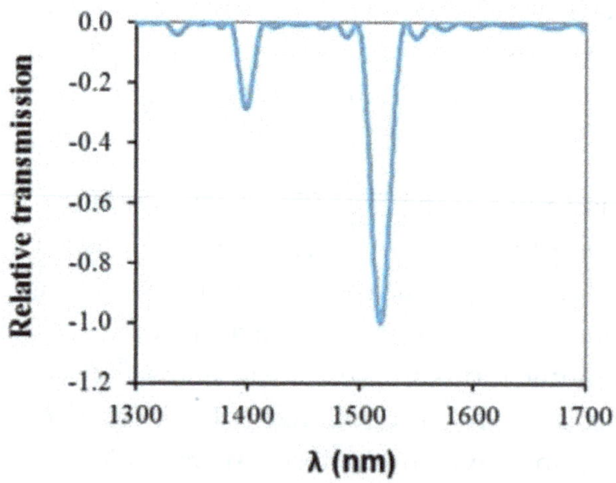

Figure 2. Theoretical example of the spectrum of a 500 μm-LPFG inscribed in a Corning SMF-28 fiber.

The resonant wavelength at which coupling takes place satisfies the phase matching condition

$$\lambda_{res}^{m} = \left(n_{eff,co} - n_{eff,cl}^{m}\right)\Lambda \tag{1}$$

in which $n_{eff,co}$ and $n_{eff,cl}$ are the effective refractive indices of the fundamental core mode and the m_{th} cladding mode, respectively; λ_{res} is the center wavelength of the transmission resonance; and Λ is the period of refractive index modulation [12].

Both resonant wavelength and attenuation amplitude of LPFGs are sensitive to a several physical parameters: temperature, strain, external refractive index, fiber dimensions, grating pitch, etc. These physical parameters affect the coupling between the core and cladding modes, which could lead to both amplitude and wavelength shift of the attenuation bands in the transmission spectrum [13]. The measurement of these spectral parameters in response to the environment surrounding the grating region is the basis of sensing with these devices.

In particular, LPFGs exhibits high sensitivity to changes in the RI of the medium surrounding the fiber due to the dependence of the phase matching condition upon the effective refractive index of the cladding modes. This characteristic makes these structures very attractive for sensing applications and several configurations, as well as applications, of LPFG devices for the measurement of physical and chemical quantities have been studied [6-13].

As mentioned before, LPFGs are created by inducing a periodic refractive index modulation (typically 10^{-4} [14-19]) in the core of an optical fiber with period lengths of several hundred micrometers. This can be made by permanent modification of the refractive index of the optical fiber's core or by physical deformation of the fiber.

Since the first demonstration of these devices by writing the grating with ultraviolet (UV) laser light through an amplitude mask in 1996 [1], several methods have been developed to create and improve the quality of the LPFGs. The conventional UV writing method is based on the exposure of photosensitive optical fibers to UV light through an amplitude mask, phase mask or by interferometry [8-9]. In germanium-doped (Ge-doped) silica fibers, UV light changes the refractive index of the core fiber, being this effect related with the formation of Ge-associated defects [20]. However, this method has some inherent limitations. It requires complex and time-consuming processes, including annealing and hydrogen loading for making the fibers photosensitive, and different amplitude masks when different dimensions are required. Also, the masks need replacement after prolonged usage and the required laser equipment is expensive.

There are, however, many non-photochemical methods available for gratings inscription. These include ion beam implantation [21], applying mechanical pressure [22], electric-arc discharge [23, 24] and irradiation by femtosecond laser pulses [25] or CO_2 laser beam [26, 27]. Among these methods, the latter is particularly flexible, as it can be applied to different types of fibers and the writing process can be computer-controlled to fabricate complicated gratings profiles without using amplitude masks. Furthermore, the use of infrared radiation as showed that the resulting interaction mechanisms are more efficient and allow creating devices with particular characteristics.

Taking this in consideration, this chapter addresses the application of CO_2 laser radiation in writing LPFGs and the physical principles involved in the process. A special emphasis will be given to the modulation of the refractive index resultant from the interaction between the mid-infrared (MIR) radiation (emitted by these lasers) and a conventional Ge-doped SMF.

In section 2 it will be described the fabrication process for applying the MIR radiation, starting with a review on the use CO_2 lasers in the creation of LPFGs. Experimental work is presented as well as the physical principles that may be responsible to induce the periodic refractive index modulation in the fiber's core.

Section 3 will address the subject of simulating the thermal mechanical processes involved in the process. Analytical and numerical models will be analysed and compared. In particular, a 3D finite element method (FEM) model will be presented, including the temperature dependence of the fiber's main parameters.

In section 4, it is presented a practical example of writing LPFGs on a Ge-doped fiber using a CO_2 laser. A comparison between calculated data and experimental data is made, and future work towards a full understanding of the physical processes is foreseen.

2. CO_2 laser induced LPFG

The use of CO_2 lasers to write LPFGs has emerged as an important alternative. Compared with other LPFG fabrication methods, this irradiation technique provides many advantages, including high thermal stability, more flexibility, lower insertion loss, lower cost. This section addresses the application of the CO_2 laser irradiation in creation of LPFG and the physical principles involved in this process.

The first results of the application of the 10.6 μm wavelength radiation emitted by CO_2 lasers for the fabrication of LPFG in a conventional fiber were published in 1998 [26-28]. Since then, different experimental methodologies have been described. The most common is the point-to-point technique using a static asymmetrical irradiation with a CO_2 laser emitting in a specific mode (continuous wave, CW, or pulsed) and a lens focusing the beam on the fiber. Akiyama [28] used continuous wave emission, while Davis [26] applied laser pulses with powers of about 0.5W.

In Figure 3 it is presented a schematic representation of a typical LPFG fabrication system based on the point-to-point technique employing a CO_2 laser. The optical fiber with its buffer stripped is placed on a motorized translation stage. In order to keep the fiber straight during the writing process, a small weight is applied at the end of the fiber and the laser beam is focused on the fiber. In this technique the periodicity of the LPFG writing is accomplished by moving the fiber along its axial direction via a computer controlled translation stage, which also controls the CO_2 laser beam emission. A broadband source and an optical spectrum analyzer (OSA) are employed to monitor the evolution of the spectrum during the laser irradiation. This method has the advantage of requiring a simpler setup, although the irradiation occurs on just one of the sides of the fiber. Also, the translation stage movement can generate vibrations that may be transmitted to the fiber, affecting the quality and repeatability of the LPFG. This problem can be solved using a system where the beam delivery system is moving instead of the fiber [29] or if the fiber is maintained static and a two-dimensional galvanometric mirrors system scans the beam over its surface [29-30]. Some authors reported hybrid methods, combining point-by-point and scanning [31,32]. In this chapter we will consider the method presented by Alves *et al.* [32] that combines a translation stage to move the fiber synchronized with a one-dimensional scan over a cylindrical lens.

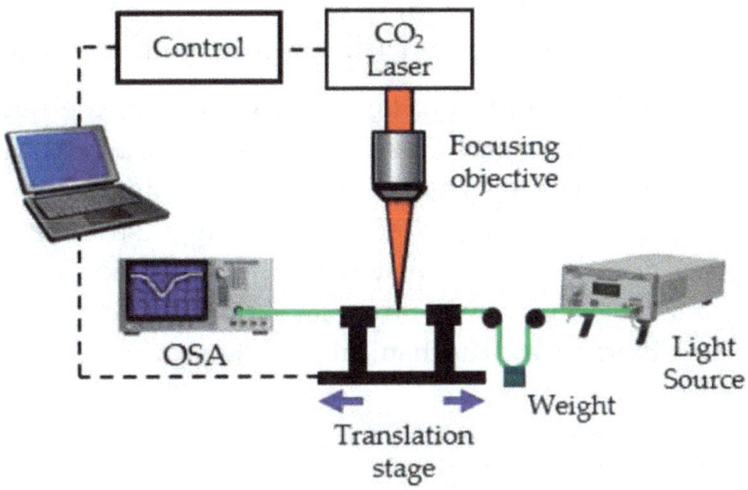

Figure 3. Schematic diagram of a typical LPFG fabrication system based on the point-to-point technique using a CO_2 laser.

Since the silica glass has strong absorption around the wavelength of the CO_2 laser radiation, the beam intensity is gradually attenuated along the incident direction, resulting in asymmetric RI modulation. Such distribution could cause coupling of the core mode to both the symmetric cladding modes and the asymmetric cladding modes [33]. As a result, high fiber grating birefringence and high polarization-dependent loss can become inevitable [30].

In order to solve the birefringence problem, different methods have been proposed. Single–side and symmetric exposures to the laser radiation were compared by Oh *et al.* [34], demonstrating that the polarization-dependent loss of the single-side exposure (1.85 dB at 1534 nm) could be significantly reduced to 0.21 dB by applying the second method. Nevertheless, due to its simplicity, the single-size exposure is the most common applied methodology. Grubsky [35] proposed a simple fabrication method for obtaining high-quality LPFGs with a CO_2 laser using a reflector to make the fiber's exposure axially uniform. Zhu *et al.* [36] used a high frequency CO_2 laser system to write LPFGs with uniform RI modulation by introducing twist strains to the fiber and then exposing the two sides of the fiber to the laser radiation. Experimental results showed that twisted LPFGs (T-LPFGs) exhibited clear spectra, low insertion losses, and low polarization-dependent losses when compared with those created by exposure to a single-side CO_2 laser beam [37].

Gratings inscription was also achieved through the use physical deformations (or geometrical deformation). Wang *et al.* [38] proposed a new method for writing an asymmetric LPFG by means of carving periodic grooves on one side of a fiber with a focused CO_2 laser beam. The periodic grooves do not cause a large insertion loss because these grooves are totally confined within the cladding and have no influence on the light transmission in the fiber's core. The grooves enhance the efficiency of grating fabrication and introduce unique optical properties (extremely high strain sensitivity). In 2006, Su *et al.* [39] demonstrated the possibility of producing long-period grating in multimode fibers by deforming the geometry of the fiber periodically with a focused CO_2 laser beam and applied them to strain measurements.

Besides conventional single-mode fibers (SMFs), CO_2 laser have been used to write LPFGs in other types of fiber, such as boron doped SMF [40,41], and photonic crystal fibers (PCFs) [42,44].

There have been many studies focused on the understanding of the physical mechanisms involved in the CO_2 laser writing process for different kinds of fibers. Most of the existing works consider that the main mechanisms responsible for creating a refractive-index change in the CO_2 laser irradiation-induced LPFGs are residual stress relaxation processes [26,29,35,40-44]. In these processes, heat created by the absorption of laser energy in the material play an important role and, as will be explained next, modelling the writing process requires considering both thermal and mechanical processes.

3. Thermo-mechanical modelling

As mentioned before, the main effect behind LPFG fabrication using CO_2 lasers is heating, where the refraction index change is achieved by irradiating a fiber submitted to a tensile stress. The high absorption of the glass material to the MIR radiation emitted by these lasers leads to an excess of energy due to the excitation of the lattice which is transformed into heat, increasing the material's temperature from its surface to its bulk by heat conduction. This effect depends on the irradiation time and on the thermal diffusivity of the material, it is localized and periodically induced in the fiber's length, being responsible for the creation of the gratings.

Considering the temperature, *T*, changing with time, *t*, (transient regime) due to the action of a heat source *Q*(*r*,*t*), the resulting energy balance leads to the heat conduction equation:

$$\left[\frac{\partial\rho}{\partial t}+\rho\nabla\cdot\vec{u}\right]\int C_p\,dT+\rho C_p\left(\frac{\partial T}{\partial t}+\vec{u}\cdot\nabla T\right)-\nabla\cdot K\nabla T=q(T)+Q(r,t) \tag{2}$$

where *r* represents the geometric coordinates (depending on the geometry) and being ρ the density, $\vec{u}$ the velocity vector, C_p the specific heat and *K* the thermal conductivity. The convective and radiative heat flux is represented by [45]:

$$q(T)=h(T_{ext}-T)+\sigma_B\,\varepsilon\left(T_{amb}{}^4-T^4\right) \tag{3}$$

being T_{ext} the external temperature, T_{amb} the ambient temperature, *h* the heat transfer coefficient, σ_B the Stefan-Boltzmann constant and ε the surface emissivity.

If enough energy is applied, differences in the thermal expansion coefficients and viscosity of core and cladding lead to residual thermal stresses and draw-induced residual stresses, and refractive index change results from frozen-in viscoelasticity [46]. The analysis of these effects is complex and highly dependent on the physical characteristics of the different materials composing the optical fiber. For simplifying the subject, we will consider silica-based single mode fibers since they are at the base of most LPFGs manufactured using CO_2 lasers. Also, from the different irradiation methodologies explained in the previous section, we will consider the coordinate referential illustrated in Figure 4. The main interfaces between regions of interest in the fiber, illustrated in Figure 4(b), are represented by a point in the upper surface (relative to laser incidence), two points in the cladding/core interface and one point in the bottom interface. In order to simplify the plots regarding calculated data in section 4, and since early work demonstrated low variation in temperature between the cladding/core interfaces [47-49], we use the core's middle point instead.

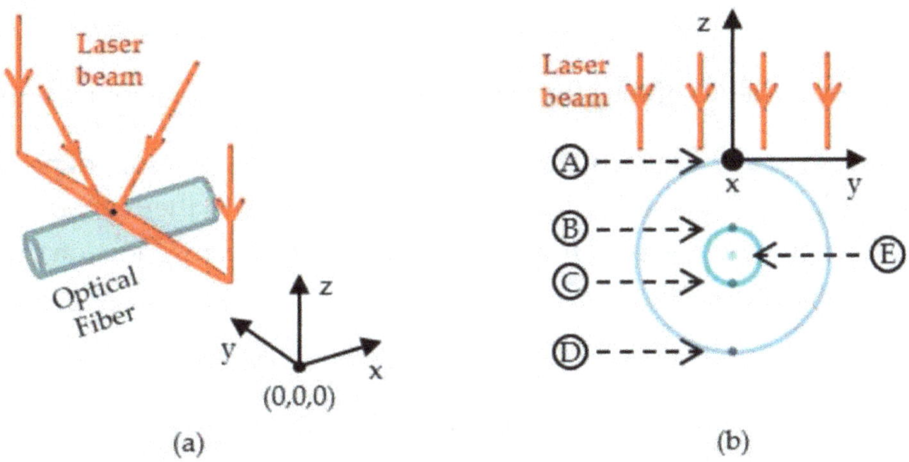

Figure 4. (a) Schematic of coordinates used in this work and (b) optical fiber cross-section indicating the considered referential, the interfaces between the different regions, and points of interest: A – irradiated surface, B – core/cladding interface (upper), C – core/cladding interface (lower), D – bottom surface and E – middle point. The origin of the reference system is in the middle of the laser line.

3.1. Laser heating modelling

When considering an homogeneous isotropic material, the condition of mass conservation, and introducing the thermal diffusivity k, given by $K/(\rho C_p)$, Equation (2) can be simplified to:

$$\left(\frac{\partial T}{\partial t}+\vec{u}\cdot\nabla T\right)-k\nabla\cdot\nabla T-q(T)=\frac{Q(r,t)}{\rho C_p} \tag{4}$$

For a laser beam incident on a surface and propagating in the z-direction:

$$Q(r,t)=a_T(1-R)I(r)\varphi(t) \tag{5}$$

where a_T is the attenuation coefficient of the material, R its reflectance and $I(r)$ the irradiance. For continuous wave emission with a duration t_{on}:

$$\varphi(t)=\begin{cases}0, \text{if } t\leq 0 \vee t>t_{on} \\ 1, \quad \text{if } 0<t\leq t_{on}\end{cases} \tag{6}$$

Accordingly with the irradiation geometry schematized in Figure 4, and considering that the laser beam has an elliptical Gaussian distribution at the surface being irradiated, then [50]:

$$I(r)=I(x,y,z)=\frac{2a_T P}{\pi r_x r_y}\exp\left[-2\left(\frac{x^2}{r_x^{\ 2}}+\frac{y^2}{r_y^{\ 2}}\right)-a_T|z|\right] \tag{7}$$

where r_x and r_y are the values of the ellipse's semi-minor and semi-major axis, respectively.

If one neglects radiative and convective losses and considers temperature dependent material properties, then it is possible to obtain analytical expressions to the temperature. Typically [47,50,51], the heat equation is solved numerically using the Green's function method and the temperature can be obtained [49]:

$$\begin{aligned}T(x,y,z,t)=\frac{(1-R)P}{4\pi K r_x r_y}\int_0^{\sqrt{4kt}}\Psi(x,y,s)\cdot\Bigg[&\exp(a_T|z|)erfc\left(\frac{a_T s}{2}+\frac{|z|}{s}\right)+\\ &+\exp(-a_T|z|)\cdot\sqrt{a^2+b^2}\cdot erfc\left(\frac{a_T s}{2}-\frac{|z|}{s}\right)\Bigg]ds\end{aligned} \tag{8}$$

with

$$\Psi(x,y,s)=\frac{a_T s}{\dfrac{s^2}{(r_x r_y)}+1}\cdot\exp\left[\frac{x^2}{r_x^2+s^2}-\frac{y^2}{r_y^2+s^2}+\frac{(a_T s)^2}{4}\right] \tag{9}$$

From (8) an on-axis approximation can be used [51] and the temperature rise can be approximated through simple analytical expressions:

$$\Delta T(0,0,\infty)=\frac{(1-R)P}{2\sqrt{\pi r_x r_y}K} \tag{10}$$

for the steady state condition ($t >> r_x^2\, r_y^2 / 4k$) and

$$\Delta T(0,0,t) = \frac{(1-R)P}{\pi^{3/2} K \sqrt{r_x\, r_y}} \tan^{-1}\left(\sqrt{\frac{4kt}{r_x\, r_y}}\right) \quad (11)$$

under transient conditions. Although Yang *et al.* [51] proved that these equations can be used to study the thermal transport in CO_2 laser irradiated fused silica, when modelling the effects in writing LPFGs, the full 3D temperature distribution is required and a numerical approach is mandatory [48]. Using a dedicated Finite Element Method (FEM) program, although computationally demanding, allows not only considering all the physical phenomena, but also including the temperature dependence of the materials and even the simulation of consecutive periods. If two laser shots are considered, one centred in (0,0,0) and at t=0 s, and the other at $x=\Delta x$ and $t=t_2=t_{on}+\Delta t$. In the latter case ($t=t_2$), equations (6) and (7) are altered accordingly to

$$\varphi(t') = \begin{cases} 0, \text{ if } t' \leq 0 \vee t' > t_{on} \\ 1, \quad \text{if } 0 < t' \leq t_{on} \end{cases}, \text{ with } t' = t - (t_{on} + \Delta t) \quad (12)$$

$$I(r) = I(x,y,z) = \frac{2a_T P}{\pi r_x r_y} \exp\left[-2\left(\frac{(x-\Delta x)^2}{r_x^2} + \frac{y^2}{r_y^2}\right)^2 - a_T |z|\right] \quad (13)$$

3.2. Residual elastic stresses modelling

Residual elastic stresses are considered those that are frozen into the fiber [52] and have an important impact on the production of LPFG since they affect the refractive index of an optical fiber. When using the considered MIR radiation, two categories of residual elastic stresses must be considered: thermal and draw-induced stresses.

3.2.1. Residual thermal stresses

Residual thermal stresses appear as the optical fiber is cooled down from high temperatures and regions with different thermal expansion coefficients contract differently in time. Dopants introduced increase the differences in viscosity and thermal expansion coefficients.

A solution for the resulting residual thermal stresses of an initially unstressed axisymmetric cylinder heated at a given temperature, T, can be obtained using the radial coordinate r [52,53]

$$\sigma_x = \frac{E}{1-\nu}\left[\frac{2\nu}{r_c^2}\int_{r=0}^{r_c} \alpha T r dr - \alpha T\right] \quad (14)$$

$$\sigma_r = \frac{E}{1-\nu}\left[\frac{1}{r_c^2}\int_{r=0}^{r_c} \alpha T r dr - \frac{1}{r^2}\int_{r=0}^{r} \alpha T r dr\right] \tag{15}$$

$$\sigma_\theta = \frac{E}{1-\nu}\left[\frac{1}{r_c^2}\int_{r=0}^{r_c} \alpha T r dr + \frac{1}{r^2}\int_{r=0}^{r} \alpha T r dr - \alpha T\right] \tag{16}$$

being r_c the radius (cladding or core), E the Young's modulus and υ the Poisson's ratio. This solution is valid when the elastic properties can be considered constant, which doesn't applies for optical fibers. Thus, and similarly to the approximated solution for heating, given by equations (8) and (9), not taking in consideration the temperature dependence of the different parameters can lead to non-accurate results. Again, the use of numerical methods, with particular focus on FEM, are appropriated and give the opportunity to combine both heating and thermal stresses models. In this case, by solving equation (4), the thermally-induced residual stresses, σ_T, can be obtained considering the constitutive equations for a linear isotropic thermoelastic material and the stress tensor obtained.

3.2.2. *Draw-induced stresses*

Residual stress effects on the refractive indices of fibers were reported for the first time by Hibido *et al.* in 1987 [54,55] regarding undoped silica-core single mode fibers. Considering a fiber with core and cladding, having different viscosities due to different dopants concentrations, during the draw process the higher viscosity glass will solidify first and support the draw tension. The low-viscosity glass solidifies conforming with the elastically stretched high-viscosity glass. Then, when the draw force is released at room temperature, the high-viscosity glass cannot contract due to the already solidified low-viscosity glass. The resulting residual axial elastic stresses can be expresses as

$$\sigma_{x,1} = F\left(\frac{E_1}{A_1E_1 + A_2E_2}\right) \tag{17}$$

and

$$\sigma_{x,2} = \frac{F}{A_2}\left(\frac{A_1E_1}{A_1E_1 + A_2E_2}\right) \tag{18}$$

being F the draw tension, A the cross-section area and E the Young's module for the considered regions. The indexes 1 and 2 in these equations represent the regions of low-viscosity and high-viscosity glasses, respectively.

Similarly to these stresses, frozen-in inelastic strains were also found if a fiber is rapidly cooled to room temperature while under tension (as in the considered case) [46,52]. Using the equivalent elastooptic relations [52], the isotropic perturbation on the refractive, Δn, index is

$$\Delta n = -\frac{n^3}{6}\chi\left(T_F\right)p\sigma \tag{19}$$

In the later expression, n is the nominal refractive index, $\chi(T_F)$ is the relaxation compressibility at the fictive temperature, T_F, p is the orientation average elastooptic coefficient, and σ the overall residual stresses (in MPa) in the fiber's axial direction. Accordingly with Yablon [52], stresses in the other directions can be neglected.

Besides stress-related refractive index change, localized heating can induce micro-deformation of the fiber and changes in the glass structure. The latter is likely to occur in the core for which the fictive temperature (the glass structure doesn't change below the fictive temperature) is lower [56,57]. As reported for a Ge-doped core (e.g. the fictive temperature ranges from 1150 K to 1500 K.

4. A practical example

To illustrate the application of the theory and also correlate it with experimental data, we will consider a common example of LPFG writing using NIR radiation: a standard single-mode fiber, SMF-28 [58], consisting of a core of 3.5 mol% Ge-doped SiO_2, is irradiated by a CW 10.6 μm wavelength CO_2 laser. For fused silica, $n \approx 1.45$ (in the near-infrared), $\chi(T_F) \approx 0.0568$ GPa^{-1} and $p \approx 0.22$, which allows simplifying equation (19):

$$\Delta n \approx -6.35\times 10^{-6}\,\sigma \tag{20}$$

Also, in this case, equations (17) and (18) represent, respectively, the residual axial elastic stresses at the core (low-viscosity glass) and the cladding (high-viscosity glass) [46].

The particular conditions used in both simulation and experimental works, as well as the obtained results, will be detailed next.

4.1. Simulation

The simulation of the writing process was made implementing a 3D FEM model using the COMSOL Multiphysics program. Whenever relevant, the material's dependence with temperature was considered and the proper geometry and FEM parameters defined.

Besides COMSOL, two other programs were used: Matcad and a simulation tool developed in MatLab by Baptista [60] based on the three layer model developed by Erdogran [14,60]. The latter was used to apply the refractive index data obtained in the FEM model to simulate the transmission spectrum of a LPFG. Matcad was used to solve equations (8) and (11) and compare the temperature data with that obtained through the FEM model.

4.1.1. Fiber's characteristics

When light interacts with matter, one of the main parameters is the absorption coefficient, a_T. As it can be deduced from the formulae of section 3, it plays a major role in the process of heating the fiber. Besides its dependence with the wavelength of the light, it varies with temperature. This variation is important and for the 10.59 μm CO_2 laser wavelength (λ_1), within 298 K–2,073 K temperature range can be obtained by [61]:

$$a_T(T) = \frac{4\pi}{\lambda_1}\left[1.82\times10^{-2} - 10.1\times10^{-5}(T-273.15)\right] \tag{21}$$

For the thermal conductivity, heat capacity, density and emissivity, the temperature dependence was modelled using native COMSOL functions for a Corning fused silica glass (7940) [48]. The doping effect on most of the parameters was disregarded mainly because the Ge concentration in the fiber's core is very low [62]. However, for the Young's modulus and Poisson's ratio (Figure 5), the function behaviour was extrapolated [63]. Also, both the heat transfer coefficient and reflectivity were considered constant and equals to 418.68 W m^{-2} K^{-1} [45] and 0.15 [51], respectively.

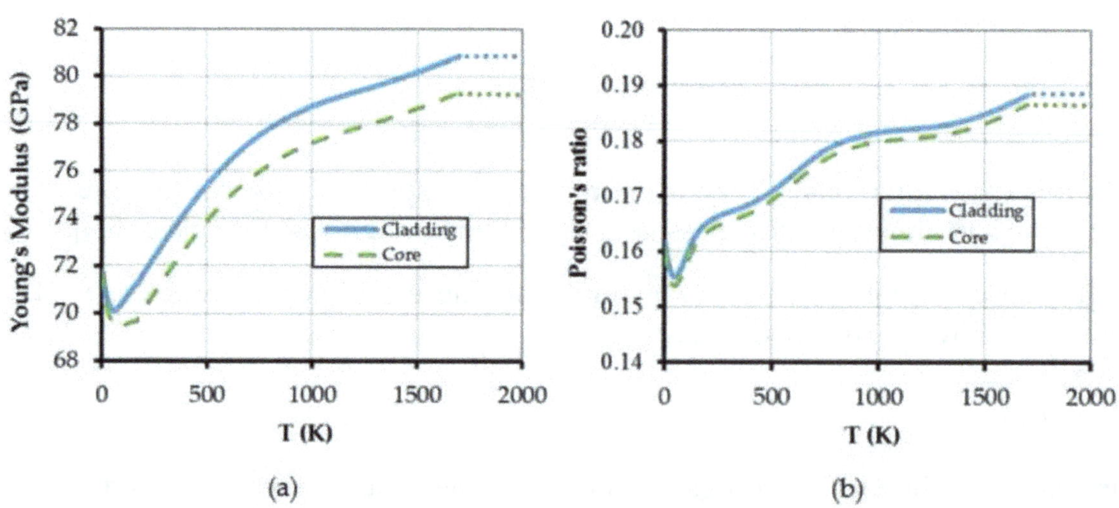

Figure 5. Variation of (a) Young's module and (b) Poisson's ratio with temperature for both fused silica (from COMSOL materials library) cladding and Ge-doped fused silica (extrapolated) core glasses.

4.1.2. Implementation

The physical problem was mathematically solved using the FEM model implemented using the COMSOL Multiphysics 3.5 program to create the transient heat conduction and (mechanical) stress-strain models under the conditions of this study. In order to introduce some of the complexity of stress-related issues regarding the processing of the optical fibers, the residual axial elastic stresses were implemented considering Equations (9) and (10) and the total resulting stress was obtained adding the thermally-induced residual stresses obtained with the program.

As illustrated in Figure 4, the implemented geometry consists of a set of (concentric) cylinders with radius of curvatures accordingly with the characteristics of the core and cladding of the fiber previously described. To avoid the influence of the external boundaries on the irradiated and analysed zones, the overall length for the geometry was set as 13 mm. However, to reduce the computational load and loosen the mesh dimensions in the volumes not affected heat source, the cylinders were implemented as three separate sets; the central one, where the laser incidence will be simulated, has a 1.7mm length. The outer set of cylinders are asymmetric since the second laser shot will be simulated just in the positive x-direction.

Table 1 presents the 3D geometry data and the mesh statistics and Figure 6 shows the implemented mesh, with particular focus on the central irradiation zone. Both outer boundary surfaces are defined as thermally isolated, being one of them fixed. The ambient temperature was considered to be 295 K and equal to the external temperature, T_{ext} in Equation (3).

	Central geometry		**External geometries**			
	Cladding	**Core**	**Cladding**		**Core**	
Geometry			Left	Right	Left	Right
Length (mm)	1.7		5.15	6.15	5.15	6.15
Radius (μm)	62.5	4.1	62.5		4.1	
Mesh (tetrahedral)						
# elements	26,122	1,795	26,262	26,825	1,710	1,795
min. quality	0.0495	0.1659	0.2287	0.2284	0.2004	0.1956
volume ratio	8.47×10^{-4}	0.1010	0.0028	0.0036	0.4906	0.3593

Table 1. Geometry data and mesh statistics.

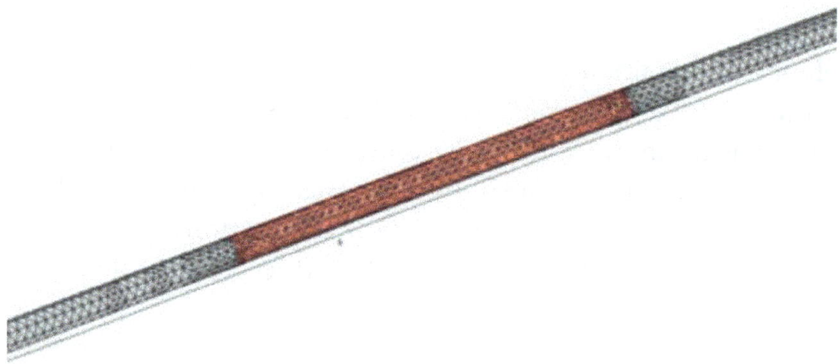

Figure 6. Image of the mesh implemented at the central zone of the FEM geometry. The coloured region corresponds to the central geometry with finer mesh (Table 1).

4.2. Experimental methodologies

The implemented irradiation methodology combines a translation stage to move the fiber synchronized with a one-dimensional scan over a cylindrical lens [32]. Figure 7 shows a schematic of the setup and Figure 8 a picture of its implementation. The light source is a Synrad 48-2 CO_2 laser emitting a 3.5 mm diameter CW laser beam with a wavelength of 10.6 μm. Two mirrors direct the beam towards the focusing lens, a ULO Optics ZnSe cylindrical lens (50 mm focal length) which produces a 0.15 mm x 1.75 mm (measured using the knife-edge method [64]) elliptical spot on the fiber with its longer axis perpendicular to the fiber's axis. One of the mirrors is a galvanometric mirror (Cambridge Technology 6860) which allows a scan over the lens (and, consequently, over the surface of the fiber).

A linear translation stage moves axially the fiber so the periodic refractive index change is accomplished. During the process, one of its ends is fixed and a weight (typically ~50 g ± 0.5 g) hangs on the other, thus creating a tension. Thus, this system acts like a XY writing system. The advantage of the translation stage is its long range with micrometric precision (10 cm with a repeatability of 1 μm). However, it has the disadvantage of having relative low speed when compared with a galvanometric system. Combining the two systems, we have the benefit of having a long X-axis range with a fast Y-axis range and, by moving the laser spot very fast over the fiber (Y-axis), lower interaction times can be achieved. Also, since the laser is not always over the fiber, it's possible to have the laser emitting continuously, which prevents the transients during the laser start-up and allows easier control of its power. Uncertainties regarding the irradiation data are: power, ±0.5 W; duration, ±1 ms.

To monitor the LPFG fabrication process, a broad band light source (Thorlabs S5FC1005S) and an OSA (Agilent 86140B) were used. The irradiated zones were analysed using an optical microscope (Zeiss AxioScope A1) with a maximum amplification of 1,000×. The LPFG period is limited by the laser spot size and by the translation stage minimum movement. In the considered setup, since the translation stage can have 1 μm steps, the laser spot size (150 μm) constitutes the major limitation.

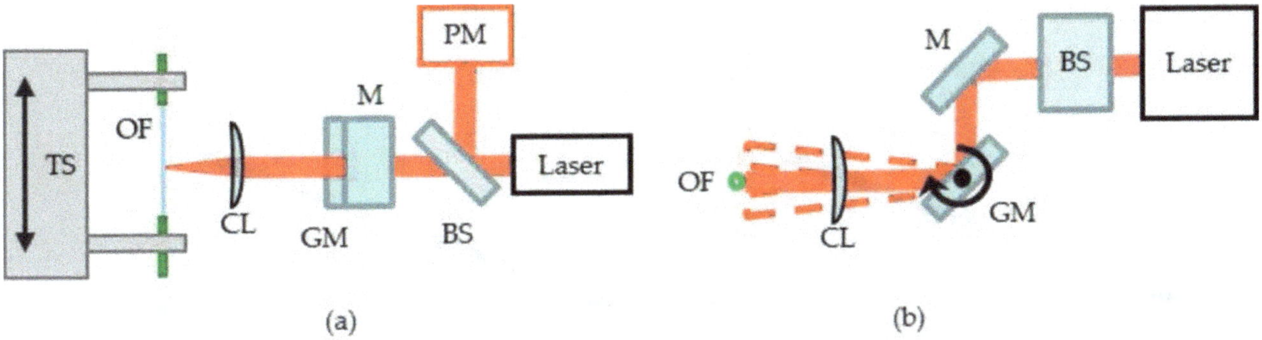

Figure 7. (a) Top and (b) lateral schematic views of the optical setup. BS – beam splitter; CL-cylindrical lens; GM – galvanometric mirror; M – mirror; OF – optical fiber; PM – power meter; TS – translation stage.

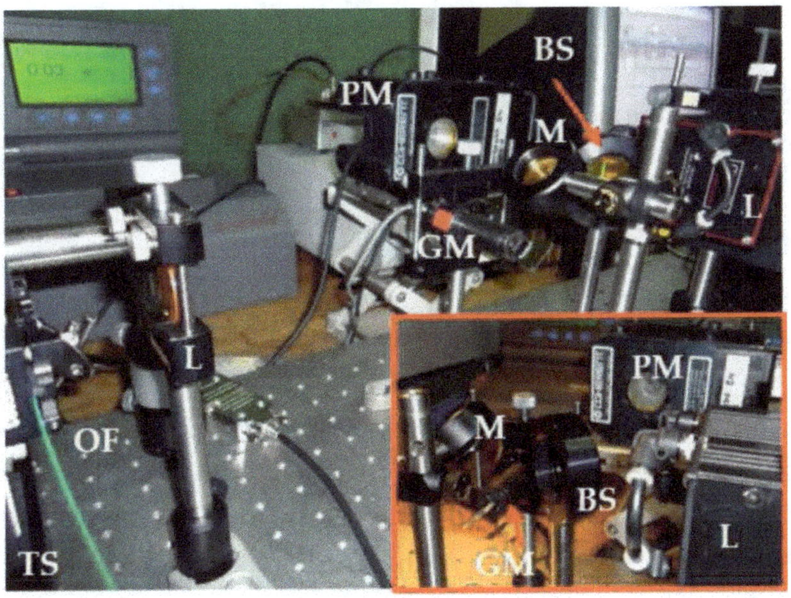

Figure 8. Picture of the setup implemented. The inset shows a detail view of the laser output area. BS – beam splitter; CL – cylindrical lens; GM – galvanometric mirror; L – laser; M – mirror; OF – optical fiber; PM – power meter; TS – translation stage.

4.3. Results and analysis

An example of the temperature distribution resulting from considering 6 W laser power, irradiation duration of 0.6 s and 47 g weight (F=0.461 N) is shown in Figure 9(a). The result of simulating a second irradiation (equivalent to have a LPFG period), 1 s after the first, at a distance of 500 μm, is shown in Figure 9(b).

Figure 10 shows the equivalent plots of temperature with time for the irradiated front, core (middle) and the back surfaces (Figure 4), at x=y=0 m and x=500 μm, y=0 m. These plots show that the (spatial) proximity between shots raises the temperature even when they are not under direct irradiation. As the distance reduces (shorter LPFG periods), this secondary heating increases (Figure 11). This is particularly important when the second shot is applied because

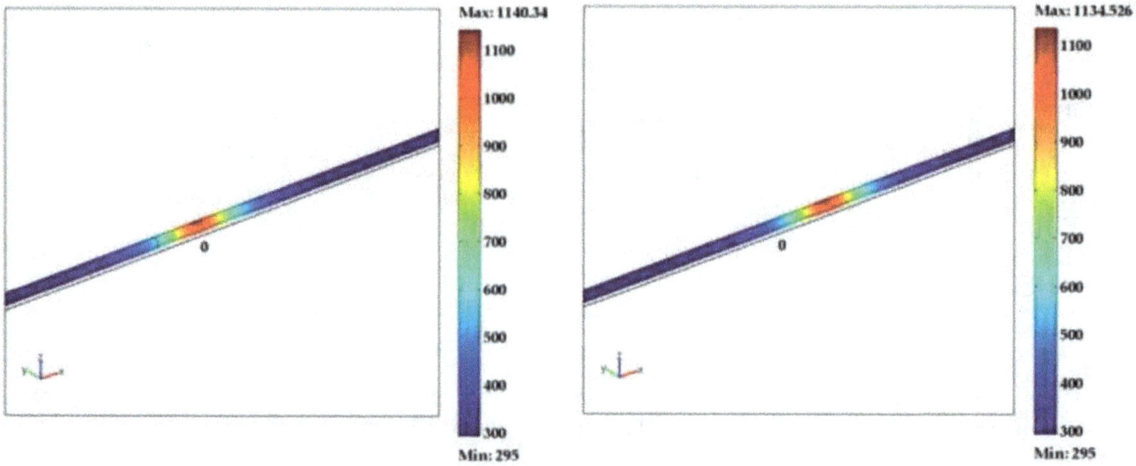

(a) (b)

Figure 9. Temperature distribution in the implemented 3D geometry for the laser irradiation of an optical fiber (P=6 W; t_{on}=0.6 s; F=0.461 N; Δt=1 s; Δx=500 μm), at (a) t=0.6 s and (b) t=2.2 s. Colour bar values are in K.

of the possibility of annealing, which could (totally or partially) relieve the previously induced internal stresses.

Similarly, as we analyse the temperature distribution at the fiber's axial direction plotted in Figure 12(a) (for the same conditions of Figure 10), the superposition of thermally affected areas in both shots underlines one critical aspect when writing LPFGs: the influence of the grating's period on the pitch width, and consequent "softening" of the spatial refractive index gradient. As it can be seen in Figure 12(b), as the period decreases, the interception's x-coordinate values decreases and the temperature at that point increases.

The practical consequence of this behaviour it's not clear at this time and further research is necessary, but these phenomena can be responsible by the reduction of the success rate in producing LPFGs with shorter periods found in other studies [65].

The importance of using the FEM simulation instead of the analytical equation (11) or the approximated integral equation (8) is observed in Figure 13 for the temperature at the core (middle). In fact, disregarding radiative and convective losses, adding to the important variation on the parameters values with temperature, deviates the solution by a significant amount (about 200 K at its maximum values, between each solution).

As expected, as the laser power increases, the temperature also increases. From the data obtained using the FEM model and plotted in Figure 14, a 3rd order polynomial can be used to relate these two parameters. Considering

$$T(P) = a_3 P^3 + a_2 P^2 + a_1 P + a_0 \tag{22}$$

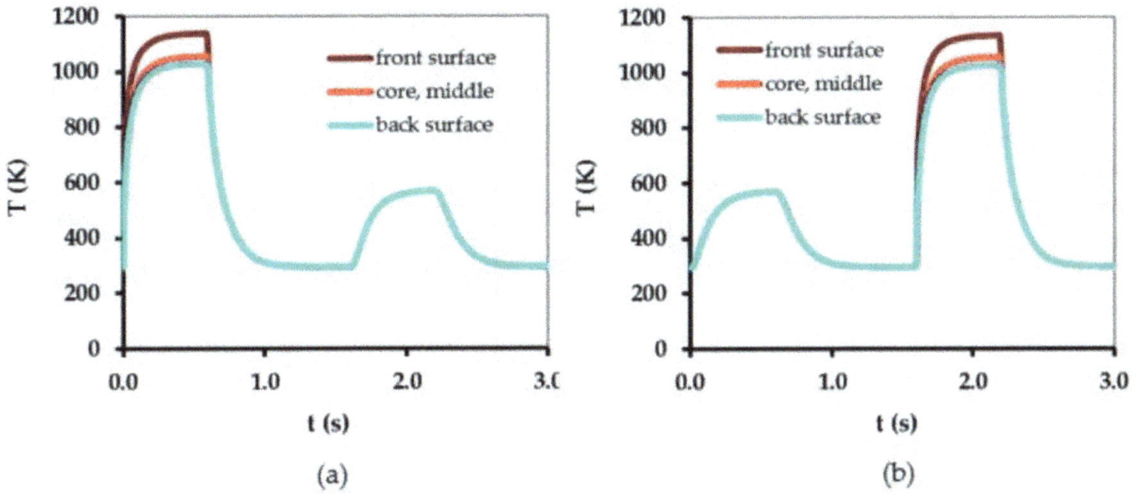

Figure 10. Plots of the temperature evolution during laser irradiation and cooling at (a) x=0 m and (b) x=Δx. (P=6 W; t_{on}=0.6 s; F=0.461 N; Δt=1 s; Δx=500 μm).

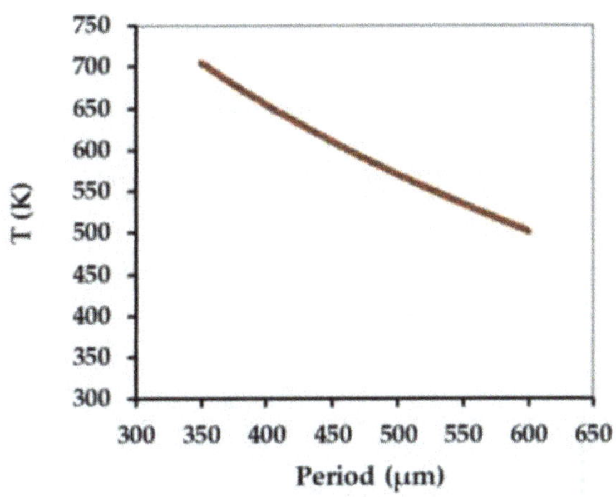

Figure 11. Maximum temperature at x=0 m when the laser is irradiating at $x=\Delta x$ for different values of Δx (LPFG period). (P=6 W; t_{on}=0.6 s; Δt=1 s; F=0.461 N).

(with the laser power, P, in W, and the temperature, T, in K), Table 2 gives the equivalent parameters for the different zones of the fiber.

For the considered example, at t=0.6 s, axial residual thermal stresses values along z-axis were determined as having a maximum of about −0.8 MPa. Axial elastic stresses act in the opposite direction and were calculated as being $\sigma_{x,1}$=35 MPa and $\sigma_{x,2}$=0.153 MPa. This results in refractive index changes (difference between final and initial values) for the core and cladding of the order of -2×10^{-4} and 4×10^{-6}, respectively. Figure 15 shows the calculated (maximum) changes for different laser power. Under the considered conditions, it is clearly observed that for the cladding, a well-defined step occurs between 4.5 W and 5 W. Regarding the core, although its refractive index shows minor variations, it is possible to observe the beginning of the contribution of thermal stresses around 5 W.

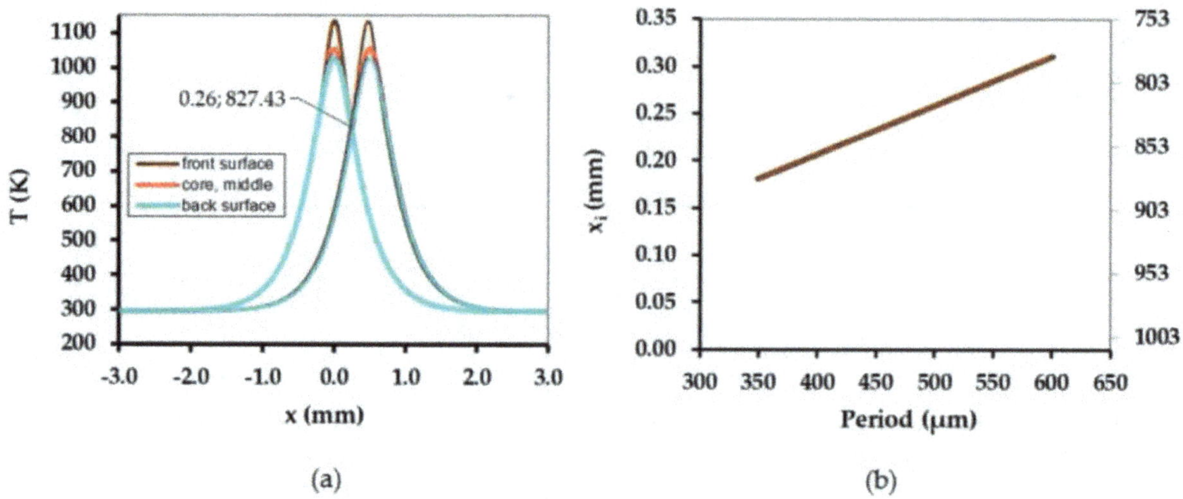

Figure 12. (a) Plots of the temperature distribution at the fiber's axial direction simulated for t=0.6 s and t=2.1 s, with Δx=500 μm (the label shows the position, in mm, and the temperature, in K, of the interception between plots), and (b) the variation on the interception point for different values of Δx (LPFG period). (P=6 W; t_{on}=0.6 s; Δt=1 s; F=0.461 N).

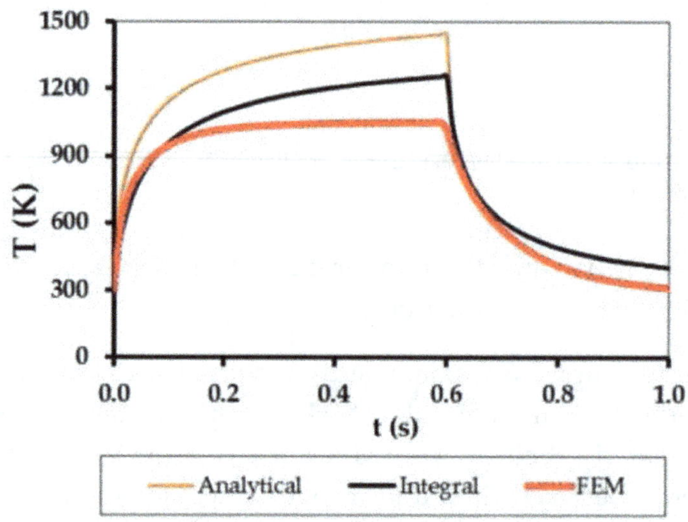

Figure 13. Plots of the temperature evolution during laser irradiation and cooling obtained using equations (11), analytical, and (8), integral, and the FEM model. (P=6 W; t_{on}=0.6 s; F=0.461 N).

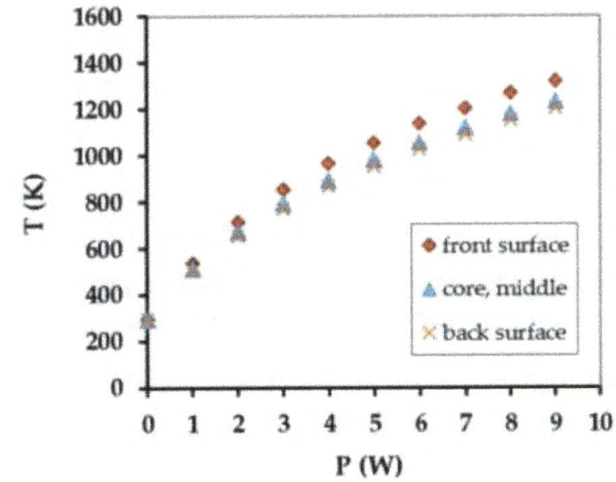

Figure 14. Maximum temperature calculated for different laser powers, at x=0 m. (P=6 W; t_{on}=0.6 s; F=0.461 N).

	a_3	a_2	a_1	a_0
Front surface	1.21	-26.64	256.4	295
Core, middle	1.06	-22.91	224.7	302
Back surface	0.97	-21.18	212.8	302

Table 2. Coefficients for equation (22), depending on the analysis point being considered (Figure 4).

Another parameter to take in consideration is the draw force due to the weight applied on the fiber. Figure 16 shows the behaviour of the refractive index change with the applied force (and corresponding weight), considering P=6 W and t_{on}=0.6 s. This parameter affects mainly the core and a linear relation (with F in N) can be obtained:

$$\Delta n_{\text{core}} = \left(-4.9388\,F + 0.0166\right) \times 10^{-4} \tag{23}$$

A microscope photograph of an example of an irradiated fiber under the experimental conditions considered in this work is shown in Figure 17. The imaged zone comprises one 500 μm period of a 25 mm grating irradiated with 6 W, for a duration of 0.6 s (each pulse) and subjected to a weight of 47 g. It is also visible a (small) micrometric deformation of the fiber in each irradiated region. Figure 18 shows the spectral transmission of the written LPFG, comparing the experimental data with the simulated spectrum, obtained using the refractive index changes calculated by the FEM model and using the simulation tool developed by Baptista [59].

In spite the relative spectral transmission data agreement, the experimental work demonstrated that different operational parameters can influence the resulting LPFG. In particular, the way the weight is positioned and laser power fluctuations can easily change the final result.

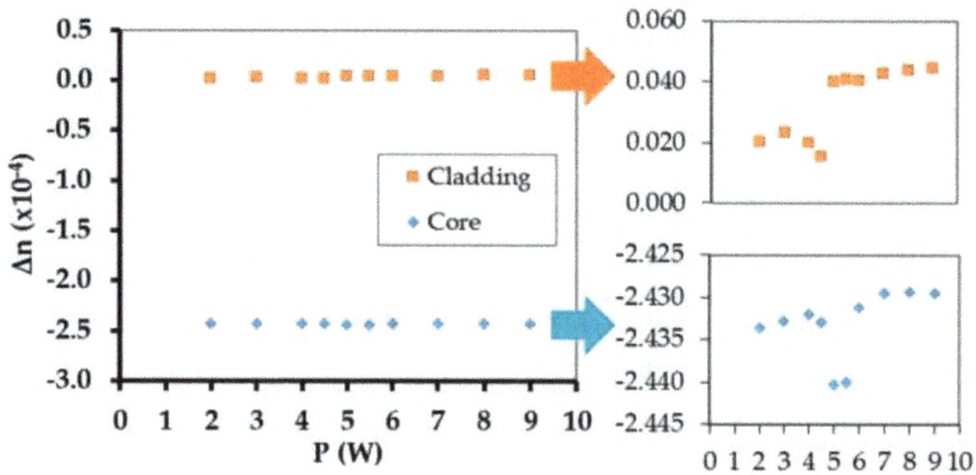

Figure 15. Calculated (maximum) refractive index change at the core and cladding for different applied laser powers. (t_{on}=0.6 s; F=0.461 N).

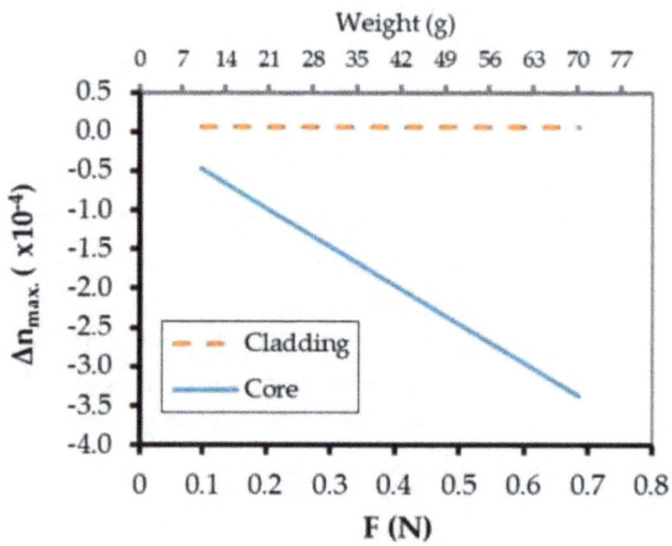

Figure 16. Calculated refractive index change (maximum change for core and cladding) for different applied draw tensions. (P=6 W; t_{on}=0.6 s; F=0.461 N).

In general, the process involves monitoring the transmission spectrum and iterative action on the length of the LPFG so a well-defined resonance is obtained. In our setup, the feedback on the emitted laser power reduces the problem, but not completely.

Although theoretically we could simulate considering different laser powers and weights, experimentally it was observed that using lower laser powers (<5 W) no LPFGs were obtained. Also, using higher laser powers (>8 W) or higher weights (typically >60 g, F > 0.6 N) tapering occurs, a phenomena not included in our model.

The values obtained by the model are in agreement with those estimated by other authors for the refractive index modulations necessary for achieving a fiber optic grating. Temperatures

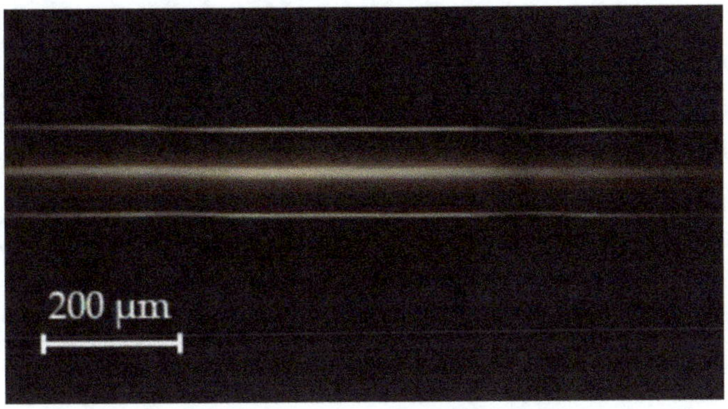

Figure 17. Picture showing two irradiated zones from a 25 mm LPFG with 500 μm period written on a SMF-28 optical fiber. ($P \approx 6$ W; t_{on}=0.6 s; Δt=1 s; $F \approx 0.5$ N)

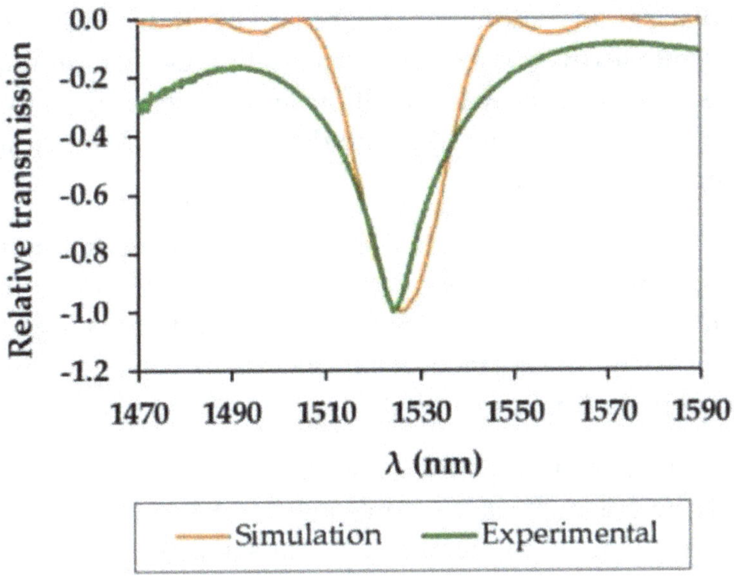

Figure 18. Experimentally obtained and simulated relative normalized spectral transmission under the same conditions considered in Figure 17.

calculated are similar to those obtained by other authors for arc-induced LPFG (e.g. in the range 1,100 K–1,400 K according to [56]) and the refractive index changes are within the overall range mentioned in other works [46,52,56]. Also, the behaviour of the refractive index change as the applied drawing force increases complies with recent experimental indications that the refractive index of the core decreases while the opposite occurs in the cladding, and that this change occurs primarily in the core [66,67].

Nevertheless, a complete model of the complex physical phenomena involved, in particular regarding the refractive index change dependence on stress, requires further research. It's expected that future work will focus on experimental measurements of temperature, stresses and refractive index changes induced by the MIR laser radiation. Also, although published works can contribute in assessing the validity of the results, the influence of specific characteristics of the fibers is a well-recognized issue. In particular, the effect of pre-existing stresses (typically from the fiber manufacture or preparation), differences in the materials, or other unaccounted phenomena can influence the performance of the FEM model when compared with real data. Similarly, the impact of the several approximations considered (e.g., transverse stresses are neglected), unaccounted phenomena like eventual changes on the glass polarizability and using standard material data must be analysed in detail, as well as the influence of the experimental data uncertainties on the model.

5. Conclusions

In Summary, the CO_2 laser irradiation technique is a highly efficient, low cost and versatile technique to write high-quality LPFGs in different types of optical fibers, such as conventional single mode fibers, polarization-maintaining fibers, and photonic crystal fibers. This technique offers a number of advantages over other fabrication techniques. It eliminates the need of using a mask as well as the need for pre-hydrogenation of the fiber and consequent post-thermal annealing to stabilize the gratings. The LPFGs induced by CO_2 laser exhibits unique grating properties, such as high thermal stability.

Although simplifications can lead to analytical equations, FEM modelling allows more realistic simulations of the physical processes involved in the writing of LPFGs using MIR radiation. The 3D model presented simulates the writing of one period and allows the analysis of both thermal and stress data. All the main practical parameters are considered as inputs and thermal dependence of the material's data is included.

The model performance was evaluated by considering a practical example of writing LPFGs on a Ge-doped fiber. Different analysis were presented and it was demonstrated that refractive index changes predicted by the FEM model led to transmission spectra with resonance peaks similar to those obtained experimentally. So, although additional work should be performed to further validate the analysis done (mainly regarding the characterization of stresses acting in the optical fiber and experimentally measuring refractive index changes), the FEM results are in accordance with literature and experimental data.

Acknowledgements

This work was partially supported by FEDER funding through the Programa Operacional Factores de Competitividade – COMPETE and by national funding by the FCT – Portuguese Fundação para a Ciência e Tecnologia through the project PTDC/FIS/119027/2010. The authors gratefully acknowledge José Luis Santos, Orlando Frazão, Pedro Jorge and Paulo Caldas from INESC-Porto for their advices and crucial contributions. A special thanks to David Castro Alves, Fernando Monteiro and António Oliveira for their support to the experimental activities described in this chapter.

Author details

João M.P. Coelho[1,2], Catarina Silva[1], Marta Nespereira[1], Manuel Abreu[1] and José Rebordão[1]

1 Laboratório de Óptica, Lasers e Sistemas, Faculdade de Ciências, Universidade de Lisboa, Pólo do Lumiar, Estrada do Paço do Lumiar, Lisboa, Portugal

2 Instituto de Biofísica e Engenharia Biomédica, Faculdade de Ciências, Universidade de Lisboa, Campo Grande, Lisboa, Portugal

References

[1] Vengsarkar A, Lemaire P, Judkins J, Bhatia V, Sipe J. Long-Period Fiber-Grating-Based Gain Equalizers. Optics Letters 1996; 21(5):335-338.

[2] Das M, Thyagarajan K. Dispersion compensation in transmission using uniform long period fiber gratings. Optics Communications 2001; 190:159–163.

[3] Eggleton BJ, Slushe RE, Judkins JB, Stark JB, Vengsarkar AM. All-optical switching in long-period fiber gratings. Optical Letters 1997; 22(12): 883-885.

[4] Zhu C, Lu Y, Lacquet BM, Swart PL, Spammer SJ. Wavelength-tunable add/drop multiplexer for DWDM using long period gratings and fiber stretchers. Optics Communications 2002; 208(9): 337-344.

[5] Vengsarkar AM, Lemaire PJ, Judkins JB, Bhatia B, Erdogan T, Sipe JE. Long-period fiber gratings as band-rejection filters. In Optical Fiber Communication Conference, OFC95, San Diego, CA, Feb. 1995, PD4-2, 1995.

[6] Khaliq S, James S, Tatam R. Enhanced sensitivity fibre optic long period grating temperature sensor. Measurement Science and Technology 2002; 13(5):792–795.

[7] Patrick H, Chang C, Vohra S. Long Period Fiber Gratings for Structural Bend Sensing. Electronics Letters 1998;34(18):1773–1775.

[8] James S, Tatam R. Optical fibre long-period grating sensors: Characteristics and application. Measurement Science and Technology 2003;14:R49-R61.

[9] Bhatia V, Vengsarkar A. Optical fiber long-period grating sensors. Optics Letters 1996;21(9): 692-694.

[10] Patrick H, Kersey A, Bucholtz F. Analysis of the response of long-period fiber gratings to external index of refraction. Journal of Lightwave Technology 1998;16(9): 1606–1612.

[11] Silva C, Coelho J, Caldas P, Jorge P. Fibre Sensing System Based on Long-Period Gratings for Monitoring Aqueous Environments. In Yasin, M., Harun S, Arof H (Eds.) Fiber Optic Sensors. Ridjeka: InTech 2012. p317–342.

[12] Erdogan T. Fiber grating spectra. Journal of Lightwave Technology 1997;15(8): 1277-1294.

[13] Bhatia V. Applications of long-period gratings to single and multi-parameter sensing. Optics Express 1999;4(11): 457-466.

[14] Erdogan T. Cladding-mode Resonances in Short-and Long-period Fiber Grating Filters. Journal of the Optical Society of America A 1997;14(8): 1760-1773.

[15] Costa RZV, Kamikawachi RC, Muller M, Fabris JL. Thermal characteristics of long-period gratings 266 nm UV-point-by-point induced. Optics Communications 2009;282(5): 816-823.

[16] Zhang Y, Chen X, Wang Y, Cooper K, Wang A. Microgap Multicavity Fabry-Pérot Biosensor. Journal of Lightwave Technology 2007;25(7): 1797-1804.

[17] Lee CE, Gibler WN, Atkins RA, Taylor HF. In-Line Fiber Fabry-Pérot-Interferometer with High-Reflectance Internal Mirrors. Journal of Lightwave Technology 1992;10(10): 1376-1379.

[18] Nespereira M, Silva C, Coelho J, Rebordão J. Nanosecond Laser Micropatterning of Optical Fibers. In: Costa M. (ed.) International Conference on Applications of Optics and Photonics: proceedings of SPIE on CD-ROM: 8001, 3-7 May 2011, Braga, Portugal. Bellingham: SPIE 2011.

[19] Kawasaki BS, Hill KO, Johnson DC, Fujii Y. Narrow-band Bragg reflectors in optical fibers. Optical Letters 1978;3(2): 66–68.

[20] Douay M, Xie WX, Taunay T, Bernage P, Niay P, Cordier P, Poumellec B, Dong L, JF Bayron, Poignant H, Delevaque E. Densification involved in the UV-based photosensitivity of silica glasses and optical fibers. Journal of Lightwave Technology 1997;15(8):1329-1342.

[21] Fujimaki M, Nishihara Y, Ohki Y, Brebner JL, Roorda S. Ion-implantation-induced densification in silica-based glass for fabrication of optical fiber gratings. Journal of Applied Physics 2000;88(10): 5534-5537.

[22] Cardenas-Sevilla GA, Monzon-Hernandez D, Torres-Gomez I, Martinez-Rios A. Mechanically induced long-period fiber gratings on tapered fibers. Optics Communications 2009;282(14):2823-2826.

[23] Rego G, Okhotnikov O, Dianov E, Sulimov V. High-temperature stability of long period fiber gratings produced by using an electric-arc. Journal of Lightwave Technology 2001;19(10): 1574-1579.

[24] Humbert G, Malki A. High performance bandpass filters based on electric arc-induced π-shifted long-period fibre gratings. Electronics Letters 2003;39(21): 1506-1505.

[25] Nikogosyan D N. Long-period Gratings in a standard telecom fibre fabricated by high-intensity fentosecond UV and near-UV laser pulses. Measurement Science and Technology 2006;17(5): 960-967.

[26] Davis DD, Gaylord TK, Glytis EN, Kosinski SG, Mettler SC, Vengsarkar AM. Long Period Fibre Grating Fabrication with Focused CO_2 Laser Pulses. Electronics Letters 1998;34(3): 302-303.

[27] Kim BH, Ahn T, Kim DY, Lee BH,Chung Y, Un-Chul Paek, Won-Taek Han. Effect of CO_2 laser irradiation on the refractive-index change in optical fibers. Applied Optics 2002;41(19): 3809-3815.

[28] Akiyama M, Nishide K, Shima K, Wada A, Yamauchi R. A Novel Long-period Fiber Grating Using Periodically Releases Residual Stress of Pure-silica Core Fiber. In: Optical Fiber Communication Conference (OFC), San José, CA, pp. 276-277, Feb. 1998. [Techn. Dig. Conf. Opt. Fiber Commun., 276, 1998].

[29] Wang Y. Review of Long Period Fiber Gratings Written by CO_2 Laser. Journal of Applied Physics 2010;108(8): 081101.

[30] Rao YJ, Zhu T, Ran ZL, Wang YP, Jiang J, Hu AZ. Novel Long-period Fiber Gratings Written by High-frequency CO_2 Laser Pulses and Applications in Optical Fiber Communication. Optics Communications 2004;229(1-6): 209–221.

[31] Chan H, Perez E, Alhassen F, Tomov I, Lee H. Ultra-Compact Long-Period Fiber Grating and Grating Pair Fabrication using a Modulation-Scanned CO_2 Laser. In: Conference on Optical Fiber Communication and the National Fiber Optic Engineers Conference (OFC/NFOEC), Anaheim, CA, pp. 1-3, Mar. 2007. [Proc. of Nation Fiber Optic Engineers Conference, Optical Society of America, paper JWA5, 2007].

[32] Alves DC, Coelho J, Nespereira M, Monteiro F, Abreu M, Rebordão JM. Automation methodology for the development of LPFG using CO_2 laser radiation. In: Costa M. (ed.) 8th Iberoamerican Optics Meeting and 11th Latin American Meeting on Optics, Lasers, and Applications: proceedings of SPIE 8785, 18 November 2013: 6 pages.

[33] Ryu HS, Park YW, Oh ST, Chung YJ, Kim DY. Effect of asymmetric stress relaxation on the polarization-dependent transmission characteristics of a CO_2 laser-written long-period fiber grating. Optical Letters 2003;28(3): 155–157.

[34] Oh ST, Han WT, Paek UC, Chung Y. Azimuthally Symmetric Long-period Fiber Gratings Fabricated with CO_2 Laser. Microwave and Optical Technology Letters 2004;41(3): 188-190.

[35] Grubsky V, Feinberg J. Fabrication of axially symmetric long-period gratings with a carbon dioxide laser. IEEE Photonics Technology Letters 2006;18(21): 2296–2298.

[36] Zhu T, Chiang K, Rao Y, Shi C, Song Y, Liu M. Characterization of long period fiber gratings written by CO_2 laser in twisted single mode fibers. Journal of Lightwave Technology 2009. 27(21): 4863–4869.

[37] Shang RB. Fabrication of twisted long period fiber gratings with high frequency CO_2 laser pulses and its bend sensing. Journal of Optics 2013;15(7): 075402.

[38] Wang YP, Wang DN, W. jin, Rao YJ, Peng GD. Asymmetric long period fiber gratings fabricated by use of CO_2 laser to carve periodic grooves on the optical fiber. Applied Physics Letters 2006;89: 151105.

[39] Su L, KS Chiang, Lu C. CO_2-laser-induced long-period gratings in graded-index multimode fibers for sensor applications. *IEEE* Photonics Technology Letters 2006;18(1): 190–192.

[40] Kim CS, Han Y, Lee BH, Han WT, Paek UC, Chung Y. Induction of the refractive index changes in B-doped optical fibers through relaxation of the mechanical stress. Optics Communications 2000;185(4-6): 337-342.

[41] Kim BH, Park Y, Ahn TJ, Lee BH, Chung Y, Paek UC, Han WT. Residual stress relaxation in the core of optical fiber by CO_2 laser radiation. Optical Letters 2001;26(21): 1657-1659.

[42] Kakarantzas G, Birks TA, Russell PS. Structural long-period gratings in photonic crystal fibers. Optical Letters 2002;27(12): 1013–1015.

[43] Lee HW, Liu Y, ChiangKS. Writing of long-period gratings in conventional and photonic-crystal polarization-maintaining fibers by CO_2-laser pulses. IEEE Photonics Technology Letters 2008;20(2):132–134.

[44] Liu Y, Chiang KS. Recent development on CO_2-laser written long-period fiber gratings. In: V, Ming-Jun Li, Ping Shum, Ian H. White, Xingkun Wu, (eds.) Passive Components and Fiber-based Devices, proceedings of SPIE 7134: pp. 713-437,. Hangzhou, China October 26, 2008.

[45] Grellier AJC, Zayer NK, Pannell CN. Heat transfer modelling in CO_2 laser processing of optical fibers. Optics Communications 1998;152(4-6): 324-328.

[46] Yablon AD, Yan MF, Wisk P, DiMarcello FV, Fleming JW, Reed WA, Monberg EM, DiGiovanni DJ, Jasapara J. Refractive Index Perturbations in Optical Fibers Resulting from Frozen-in Viscoelasticity. Applied Physics Letters 2004;84(1):19-21.

[47] Coelho J, Nespereira M, Silva C, Rebordão J. Modeling refractive index change in writing long-period fiber gratings using mid-infrared laser radiation. Photonic Sensors 2013;3(1): 67-73, 2013.

[48] Coelho J, Nespereira M, Silva C, Rebordão J. 3D Finite Element Model for Writing Long-Period Fiber Gratings by CO_2 Laser Radiation. Sensors 2013;13(8): 10333-10347.

[49] Coelho J, Nespereira M, Silva C, Pereira D, Rebordão J, Advances in optical fiber laser micromachining for sensors development, In: Sulaiman Wadi Harun and Ham-

zah Arof (Eds.) Current Developments in Optical Fiber Technology. Ridjeka: Intech; 2013. p375-401.

[50] Coelho J, Abreu MA, Carvalho-Rodrigues F. Modelling the spot shape influence on high-speed transmission lap welding of thermoplastics films. Journal of Optics and Lasers in Engineering 2008;46(1): 55-61.

[51] Yang S, Matthews M, Elhadj S, Draggoo, V, Bisson S. Thermal transport in CO_2 laser irradiated fused silica: in situ measurements and analysis. Journal of Applied Physics 2009;106: 103-106.

[52] Yablon AD. Optical and Mechanical Effects of Frozen-in Stresses and Strains in Optical Fibers. IEEE Journal of Selected Topics in Quantum Electronics 2004;10(2): 2004.

[53] Timoshenko SP, Goodier JN. Theory of elasticity, 2nd Ed.; New York: McGraw-Hill; 1970. p409-410.

[54] Hibino Y, Hanawa F., Abe T, Shibata S. Residual stress effects of refractive indices in undoped silica-core single-mode fibers. Applied Physics Letters 1987;50(22): p1565-1566.

[55] Hibino Y, Edahiro T, Horiguchi T, Azuma Y, Shibata N. Evaluation of residual stress and viscosity in SiO_2-core/F-SiO_2 clad single mode fibers from Briouin gain spectra. Journal of Applied Physics 1989;66(9): 4049-4052.

[56] Rego, GM. Arc-induced long-period fibre gratings. Fabrication and their application in communications and sensing. Ph.D. dissertation, Dept. Elect. Comp. Eng., Univ. of Porto, Porto, Portugal, 2006.

[57] Lancry, M, Réginier, E, Poumellec, B. Fictive temperature in silica-based glasses and its application to optical fiber manufacturing. Progress in Material Sciences 2012;57: 63-94.

[58] Corning® SMF-28 optical fiber product information. New York: Corning Inc.: 2002.

[59] Baptista, FDV Simulação do Comportamento Espectral de Redes de Período Longo em Fibra Óptica. Ms. dissertation, Centro de Ciências Exactas e da Engenharia, Univ. of Madeira, Funchal, Portugal, 2009.

[60] Erdogan, T. Cladding-mode resonances in short-and long-period fiber grating filters: errata. Journal of the Optical Society of America A 2000;17: p2113.

[61] McLachlan A, Meyer F. Temperature dependence of the extinction coefficient of fused silica for CO_2 laser wavelengths. Applied Optics 1987;26(9):p1728–1731.

[62] André P, Rocha A, Domingues F, Facão M. Thermal effects in optical fibers. In Marco Aurélio dos Santos Bernardes (Ed.) Developments in Heat Transfer; Ridjeka: Intech; 2011. p1–20.

[63] Clowes J, Syngellakis S, Zervas M. Pressure sensitivity of side-hole optical fiber sensors. IEEE Photonics Technology Letters 2009;10(6): 857–859.

[64] Siegman AE, Sasnett MW, Johnston TF. Choice of clip level for beam width measurements using knife-edge techniques. IEEE Journal of Quantum Electronics 1991;27(4): 1098–1104.

[65] Nespereira M, Alves, DC, Monteiro, F, Abreu, M, Coelho J, Rebordão JM. Repeatability analysis on LPFGs written by a CO_2 laser. In: Costa M. (ed.) II International Conference on Applications of Optics and Photonics: proceedings of SPIE 9286, 26-30 May 2014, Aveiro, Portugal. Bellingham: SPIE 2014.

[66] Li Y, Wei T, Montoya JA, Saini SV, Lan X, Tang X, Dong J, Xiao H. Measurement of CO_2-laser-irradiation-induced refractive index modulation in single-mode fiber toward long-period fiber grating design and fabrication. Applied Optics 2008;47(29): 5296-5304.

[67] Hutsel MR, Gaylord TK. Residual-stress relaxation and densification in CO_2-laser-induced long-period fiber gratings. Applied Optics 2012;51(25): 6179-6187.

6

Brillouin Scattering in Optical Fibers and its Application to Distributed Sensors

Weiwen Zou, Xin Long and Jianping Chen

1. Introduction

Brillouin based distributed optical fiber sensors have been studied for more than two decades because they have incomparable abilities over the pointed or multiplexed fiber-optic sensors based on fiber Bragg grating and/or inline Fabry-Perot resonator. They originated from the intrinsic fiber-optic nonlinearity in optical fibers, i.e. Brillouin scattering, and have many distinguished advantages, such as high accuracy due to the frequency revolved interrogation, multiple sensitivities of measurands (strain, temperature etc.), no dead zones of sensing location due to the distributed sensing ability, and immunity to the electro-magnetic interference. Nowadays, they have been thought as great potentials in industrial applications to smart materials and smart structures.

This chapter introduces the basic principle and recent advances of Brillouin scattering in optical fibers. The working mechanism, different interrogation techniques, difficulty or challenge of the sensing ability, and recent breakthroughs of Brillouin based distributed optical fiber sensors are demonstrated, respectively.

2. Brillouin scattering in optical fibers

2.1. Principle

Light scattering phenomena in optical fibers occur regardless of how intense the incident optical power is. They can be basically categorized into two groups, i.e. spontaneous scattering and stimulated scattering[1]. Spontaneous scattering refers to the process under conditions such that the material properties are unaffected by the presence of the incident optical fields.

For input optical fields of sufficient intensities spontaneous scattering becomes quite intense and stimulated scattering starts. The nature of the stimulated scattering process grossly modifies the optical properties of the material system and vice versa. Spontaneous and stimulated scattering in optical fibers are composed of Rayleigh, Raman, and Brillouin scattering processes. Each scattering process is always present in optical fibers since no fiber is free from microscopic defects or thermal fluctuations which originate the three processes. For a monochromatic incident lightwave of frequency f_0 at $\lambda_{0=}$~1550 nm (telecom wavelength), three processes are schematically described by the spectrum of the scattered light as shown in Fig. 1. The components, whose frequency is beyond f_0, correspond to anti-Stokes while those below f_0 correspond to Stokes.

Brillouin scattering is a "photon-phonon" interaction as annihilation of a pump photon creates a Stokes photon and a phonon simultaneously. The created phonon is the vibrational modes of atoms, also called a propagation density wave or an acoustic phonon/wave. In a silica-based optical fiber, Brillouin Stokes wave propagates dominantly backward [2] although very partially forward[3]. The frequency (~9-11 GHz) of Stokes photon at ~1550-nm wavelength is in quantity dramatically different from or smaller by three orders of magnitude than Raman scattering (see Fig. 1) and is dominantly down-shifted due to Doppler shift associated with the forward movement of created acoustic phonons. In a polymer optical fiber, the frequency is ~2-3 GHz due to the different phonon property[4].

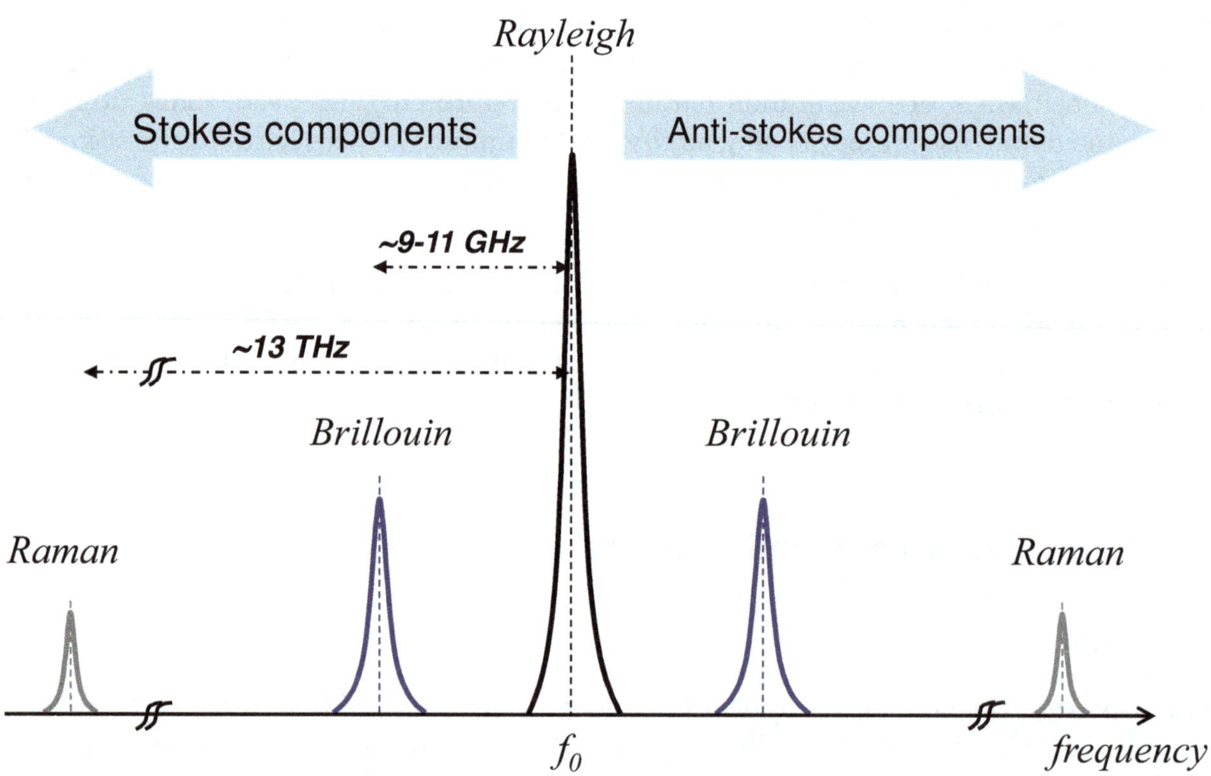

Figure 1. Schematic spectrum of scattered light resulting from three scattering processes in optical fibers.

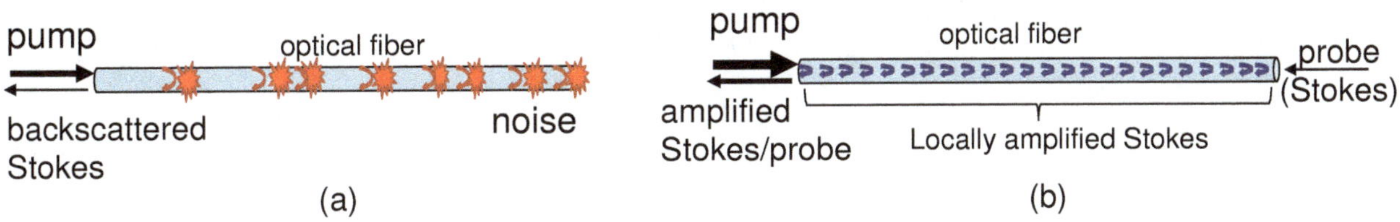

Figure 2. Comparison of (a) spontaneous Brillouin scattering (SpBS) and (b) stimulated Brillouin scattering (SBS) in optical fibers.

Figure 2 illustrates the difference between spontaneous Brillouin scattering (SpBS) and stimulated Brillouin scattering (SBS) in optical fibers. In principle, the SpBS (see Fig. 2(a)) is started from a noise fluctuation and influences the pump wave (E_p); the SBS (see Fig. 2(b)) occurs when the pump power for SpBS is beyond the so-called Brillouin threshold value (P_{th}) or when two coherent waves with a frequency difference equivalent to the phonon's frequency are counter-propagated. Brillouin scattering dynamics in optical fibers are generally governed by the following coupling equations [5, 6]:

$$\left(\frac{1}{v_g}\frac{\partial}{\partial t}+\frac{\partial}{\partial z}\right)E_p=-\frac{\alpha}{2}E_p+i\kappa_1\rho E_s, \tag{1}$$

$$\left(\frac{1}{v_g}\frac{\partial}{\partial t}-\frac{\partial}{\partial z}\right)E_s=-\frac{\alpha}{2}E_s+i\kappa_1\rho^* E_s, \tag{2}$$

$$\left(+\frac{\partial}{\partial t}+\frac{\Gamma_B}{2}+2\pi i v_B\right)\rho=i\kappa_2 E_p E_s^*+N, \tag{3}$$

where E_p and E_s stand for the normalized slowly-varying fields of pump and Stokes (or probe) waves, respectively; and $P_p=|E_p|^2$ and $P_s=|E_s|^2$ correspond to their optical powers; ρ denotes the acoustic (or phonon) field in terms of the material density distribution; N represents the random fluctuation or white noise in position and time [5]; v_g is the group light velocity in the fiber; α is the fiber's propagation loss; Γ_B is the damping rate of the acoustic wave, which equals to the reciprocal of the phonon's lifetime ($1/\Gamma_B=\tau_\rho=\sim 10$ ns) and is related to the acoustic linewidth ($\Delta\nu_{B=}\Gamma_B/\pi$) [7]; κ_1 and κ_2 are the coupling coefficients among E_p, E_s, and ρ. If SpBS is considered, it is reasonable to assume E_s is sufficiently small so that the second term of N dominates in the right side of Eq. (3). In contrast, the first term of $i\kappa_2 E_p E_s^*$ dominates for SBS.

Taken into account the SBS power transfer between P_p and P_s under the assistance of the acoustic wave and the so-called acousto-optic effect, Eqs. (1-3) can be rewritten as

$$\left(\frac{1}{v_g}\frac{\partial}{\partial t}+\frac{\partial}{\partial z}+\alpha\right)P_p=-g(\nu)P_pP_s, \tag{4}$$

$$\left(\frac{1}{v_g}\frac{\partial}{\partial t}-\frac{\partial}{\partial z}+\alpha\right)P_s=+g(\nu)P_pP_s, \tag{5}$$

where the sign difference between the right hands of Eq. (4) and Eq. (5) means that the pump power is reduced or depleted but the probe (Stokes) power is increased or amplified.

g(ν) is called Brillouin gain spectrum (BGS), the key phraseology to represent Brillouin scattering in optical fibers. It denotes the spectral details of the light amplification from strong pump wave to weak counter-propagating probe/Stokes wave in SBS or those of the noise-initialized scattered phonons in SpBS. The BGS is generally expressed by [8-10]

$$g(\nu)=\sum_l g^{(l)}(\nu), \tag{6}$$

$$g^{(l)}(\nu)=\left\{g_{B0}\cdot\frac{\nu_B\cdot\Delta\nu_B}{\nu_{ac}^{(l)}\cdot\Delta\nu_{ac}^{(l)}}\right\}\cdot\left\{\frac{(\Delta\nu_{ac}^{(l)}/2)^2}{[\nu-(f_0-\nu_{ac}^{(l)})]^2+(\Delta\nu_{ac}^{(l)}/2)^2}\right\}\cdot\left\{\frac{1}{A_{ao}^{(l)}}\right\}, \tag{7}$$

where Eq. (6) means that the BGS is the summation of all the longitudinal acoustic modes' gain spectra and Eq. (7) corresponds to the *l*th-order one assigned the subscript of "*l*". In Eq. (7), $\Delta\nu_{ac}^{(l)}$ is the *l*th-order linewidth or full width at half magnitude (FWHM) which can be assumed to be approximately the same for all the acoustic modes; $\nu_{ac}^{(l)}$ is the effective acoustic velocity; $A_{ao}^{(l)}$ (in μm²) is the so-called acousto-optic effective area of the *l*th-order one. $\nu_{ac}^{(l)}$ and $A_{ao}^{(l)}$ are qualitatively different among all acoustic modes.

There are two basic methods to theoretically and numerically analyze the BGS in optical fibers [8, 11]. One method [11] is based on Bessel or modified Bessel functions for optical fibers with regular geometric and dopant distribution, such as step-index optical fibers. The other one [8] is called two-dimensional finite-element-method (2D-FEM) modal analysis of BGS for optical fibers with complicated or arbitrary distribution. The 2D-FEM modal analysis has been used to study a Panda-type polarization-maintaining optical fiber (PMF) [8], a SMF with arbitrary residual stress [12], a *w*-shaped triple-layer fiber [13], or optical fibers with non-uniform optical/acoustic profiles such as solid or microsctrucuted photonic crystal fibers (PCF) [14-21].

The contribution of the fundamental acoustic mode to the entire BGS is basically dominant, which has a Lorentzian feature as schematically depicted in Fig. 3(a). Besides, it modulates the refractive index of optical fiber and changes the group velocity of optical fields in a profile shown in the inset of Fig. 3(a), which has been adopted for Brillouin slow or fast light[22, 23].

There are three basic parameters of Brillouin frequency shift (BFS, ν_B), Brillouin gain peak (g_{B0}), and Brillouin linwidth ($\Delta\nu_B$) in the main-peak BGS (i.e. the fundamental acoustic mode or l=1 in Eq. (7)). ν_B is defined as

$$\nu_B \equiv \nu_{ac}^{(1)} = \frac{2}{\lambda_0} \cdot n_{eff} \cdot V_a, \tag{8}$$

where λ_0 is the light wavelength (λ_0=c/f_0 with c the light speed in vacuum), n_{eff} is the effective refractive index of the fiber, and V_a is the effective acoustic velocity of the fundamental acoustic mode. n_{eff} and V_a in Eq. (8) as well as $A_{ao}{}^{(l)}$ in Eq. (7) are all determined by the respective waveguide structures of the optical modes (n_0 and n_1) and those of the longitudinal acoustic modes (V_{l1}, V_{l0}), relative to the silica dopant materials and distributions in the cross section [24]. Figure 3(b) illustrates a simple example of step-index single-mode optical fiber (SMF), even for which the BGS comprises of several (typically, four) longitudinal acoustic modes due to the different contrast of optical and acoustic waveguides [8, 11]. The measured BGS of the SMF and a high-delta nonlinear optical fiber at 1550nm are depicted in Fig. 3(c)[13]. It is worth noting that ν_B in optical fibers suffers strong influence from the residual elastic and inelastic strains induced by different draw tensions during fiber fabrication[12].

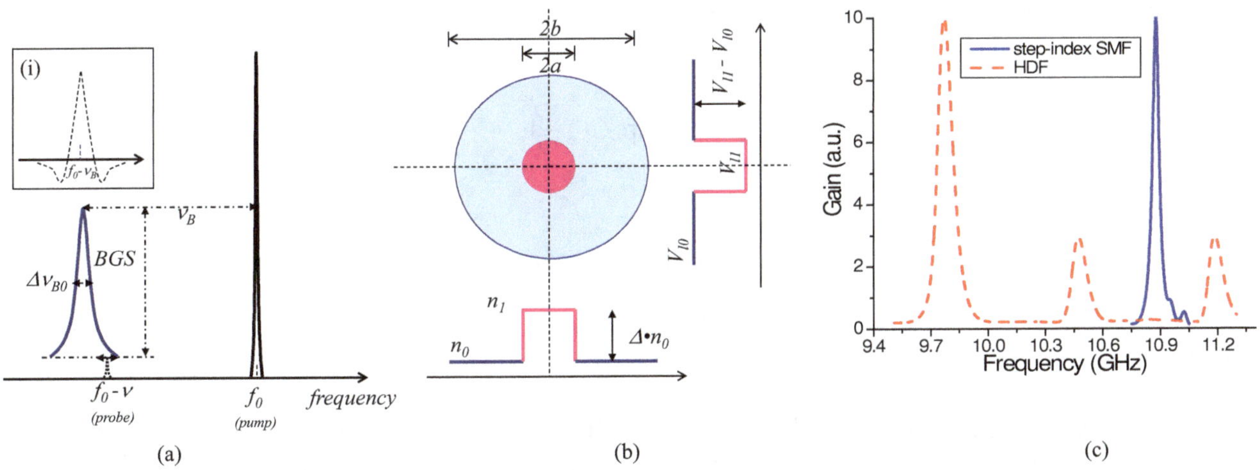

Figure 3. (a) Brillouin gain spectrum (BGS) in optical fibers. The inset "i" denotes the change of group velocity of optical fields. (b) Cross section of a step-index SMF. Δ=(n_1-n_0)/n_0 is the relative index difference between the core (n_1) and the cladding (n_0). (c) Measured BGS of a step-index SMF (solid curve) and of a 17.0-mol% high-delta fiber (dashed curve). (Fig. 3(c) after Ref. [13]; © 2008 OSA.)

g_{B0} in Eq. (7) is determined by

$$g_{B0} = \frac{4\pi n_{eff}{}^8 p_{12}{}^2}{\lambda_0{}^3 \rho_0 c \nu_B \Delta\nu_{B0}}, \tag{9}$$

where ρ_0 is the density of silica glass (~2202 kg/m^3) and p_{12} the photo-elastic constant (~0.271). In most silica-based fibers, the peak gain value of g_{B0} lies in the range of 1.5~3× 10^{-11} m/W [25].

$\Delta\nu_{B0}$ in silica optical fibers with a typical value of 30~40 MHz is characteristic of SpBS. However, in the SBS process, it was theoretically proved that $\Delta\nu_B$ strongly depends on the pump power, which is expressed as follows [5, 26]:

$$\Delta \nu_B = \Delta \nu_{B0} \sqrt{\frac{\ln 2}{G_s}}, \tag{10}$$

where G_s is the single pass gain experienced by the weak probe wave from the strong pump wave, defined by

$$G_s = \frac{g_{B0} P_{pump} L_{eff}}{K \cdot A_{\mathrm{eff}}^{ao}}, \tag{11}$$

where K (=1~2) is a polarization factor (=2 for a complete polarization scrambling process), L_{eff}=[1-exp(-αL)]/α is the effective length of the fiber with α the optical loss (m^{-1}) and L the fiber length, P_{pump} the pump power, and $A_{\mathrm{eff}}^{\mathrm{ao}}=A_{ao}^{(1)}$ is the acoustic-optic effective area of the fundamental acoustic mode. It is noted that the exponential (G_e) or logarithmic (G_{dB}, *in dB*) gain of the weak probe power (P_{probe}) are presented by

$$G_e = \frac{\Delta P_{probe}}{P_{probe}} = \exp(G_s) - 1, \tag{12}$$

$$G_{dB} = 10\log_{10}(G_e) \approx 4.342 G_s. \tag{13}$$

Eq. (13) is valid when G_e>>1.

From Eq. (10), one could estimate that the linewidth goes gradually to zero for very high gain, which can be obtained by either increasing the pump power or the interaction length of the fiber (see Eq. (11)). A zero linewidth corresponds to acoustic oscillation with an infinite time. However, pump depletion always occurs when the single pass gain (G_s) increases, which results in a limited effective single pass gain and in turn leads to a finite linewidth instead of zero one.

The experimental characterization of the phenomenon of the Brillouin linwidth's narrowing in three SMFs are depicted in Fig. 4 [27]. When the pump power is intensified to a high value of above ~24 dBm (~250 mW), the Brillouin main-peak linewidth of 180-m-long SMF becomes increasing. This is because the pump power depletes much faster than its contribution to the single-pass gain G_s since the probe wave experiences an amplification of more than G_{dB}=20 dB

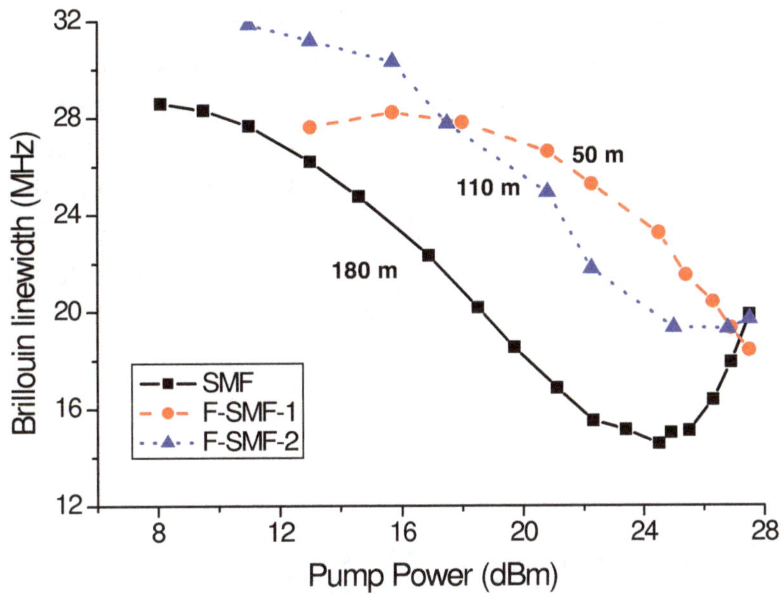

Figure 4. Measured Brillouin linewidth in three SMF varying with increase of Brillouin pump power. F-SMF denotes a SMF with pure-silica core and F-doped-silica cladding. (After Ref. [27]; © 2008 OSA.)

for a greater pump power than ~24 dBm. From the other point of view, during the pump-probe-based BGS measurement, which will be described later, the Brillouin probe wave with a down-shifted frequency of just ν_B feels more significantly the depletion of the pump power than the one with a downshifted frequency of a finite offset from ν_B. More details of the effect of pump depletion in SBS will be demonstrated in **Section 2.3**.

From Eq. (11), one can derivate the so-called pump threshold value of SBS originating from SpBS (also called Brillouin generator). It is given by

$$P_{th} = \kappa \frac{A_{\text{eff}}^{ao}}{L_{eff}} \frac{K}{g_{B0}}, \tag{14}$$

where κ is a numerical factor (=~21) [28] that may change in terms of the fiber length [29]. If a long-enough SMF is considered, α=0.2 dB/km (or 0.046 /km) meaning L_{eff}=21.7 km. $A_{\text{eff}}^{\text{ao}} \approx A_{\text{eff}}$=100 μm^2 and g_{B0}=2× 10^{-11} m/W. For a perfectly-linearized pump wave, $P_{\text{th}} \approx$ 3.2 mW; for a completely polarization-scrambled pump wave, $P_{\text{th}} \approx$ 4.6 mW.

2.2. Experimental characterization

The experimental characterization of BGS in optical fibers can be implemented by two individual ways that depend on which principle of SpBS or SBS is based on. The SpBS based configuration is illustrated in Fig. 5(a). A pump wave is amplified by an erbium doped fiber amplifier (EDFA) and its polarization is optimized by a polarization controller (PC), scrambled by a polarization scrambler (PS) or switched by a polarization switcher (PSW). It is launched through an optical coupler or circulator into the fiber under test (FUT). A weak Stokes wave with downshifted frequency around ν_B is backscattered towards the coupler/circulator and can be detected by three different schemes, which are depicted in Fig. 5(b). First, an optical filter (Etalon, Fabry-Perot filter or FBG) is inserted before a photo-detector (PD) so as to eliminate the influence of Rayleigh scattering on Stokes wave and spectrally analyzed by an electrical spectrum analyzer (ESA) [2, 30]. Second, Stokes wave is optically mixed with a part of the pump wave (serving as an optical oscillator) and then detected or heterodyne-detected by a high-speed PD or a high-speed balanced PD, which is directed to the ESA [31]. Third, heterodyne detection can be carried out at an intermediate frequency (IF) range by tuning the frequency of the optical oscillator or further using of a local microwave (RF) oscillator before ESA [32].

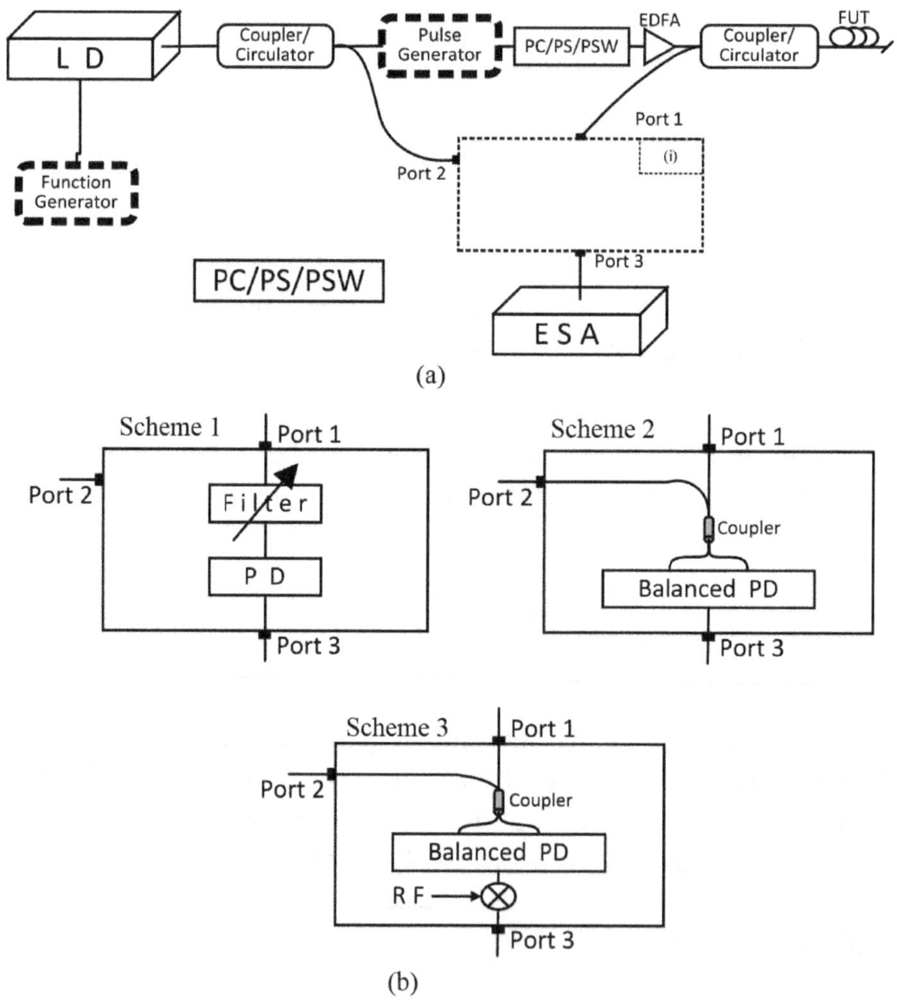

Figure 5. Experimental setup of BGS measurement based on spontaneous Brillouin scattering (SpBS). (a) Basic configuration. LD: laser diode; PC: polarization controller; PS: polarization scrambler; PSW: polarization switcher; EDFA: erbium-doped fiber amplifier; ESA: electrical spectrum analyzer. The dashed boxes of Function Generator or Pulse generator are used for distributed SpBS measurement. (b) Three different methods to detect the weak Stokes wave, corresponding to the dotted box "i" in (a) with two optical ports (port 1 and port 2) and one electrical port (port 3).

The pump-probe-based experimental configuration, depicted in Fig. 6(a), is more attractive to investigate the BGS in optical fibers (especially with very short length) since SBS process occurs and high Brillouin gain can be utilized. Two light waves from a laser unit are the optical sources: the one with larger optical frequency (f_0) serves as SBS pump and the other with lower optical frequency (f_0-ν) works as SBS probe wave. They are launched into the opposite ends of FUT so as to ensure their counter-propagation and generate intense SBS interaction. The pump wave transfers intense energy to the probe wave, called Brillouin gain; in contrast, the probe wave absorbs energy from the pump wave, called Brillouin loss. The magnitude of both Brillouin gain and loss depends on the frequency offset (ν) between the pump and probe waves, determined by the Lorentzian feature of BGS (see Eq. (7)). Subsequently, the BGS can be characterized by monitoring the power of the probe wave at PD1 (i.e. Brillouin gain) [8] or that of the pump wave at PD2 (i.e. Brillouin loss) [33] as a function of ν (provided by the laser unit), respectively. Recently, a scheme based on a combination of Brillouin gain and loss was newly proposed to enhance the signal to noise ratio (SNR) of the BGS measurement [34]. It can be realized by periodic switching of the pump and probe wave and detected at either PD1 or PD2. Besides, the simultaneous detection of PD1 and PD2 followed by a subtraction may work equally.

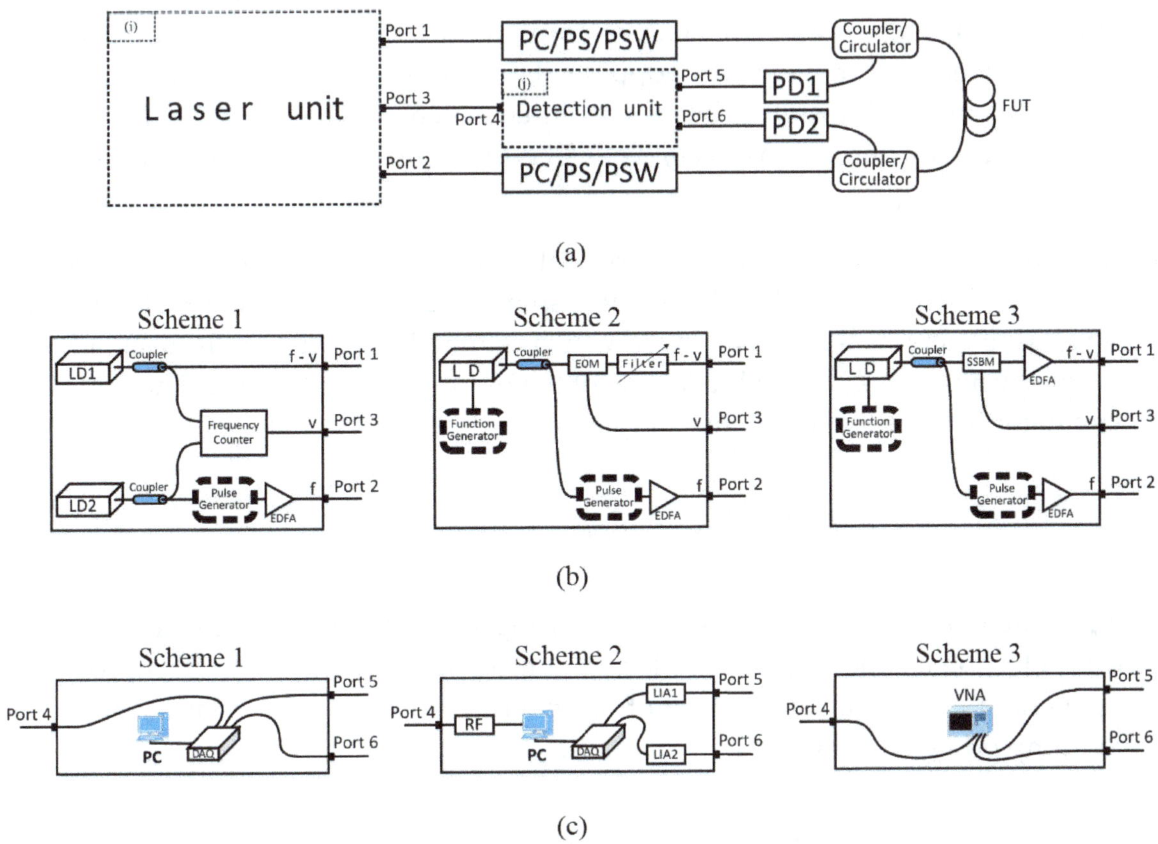

Figure 6. Experimental setup of BGS measurement based on stimulated Brillouin scattering (SBS). (a) Basic configuration. (b) Three different methods of laser unit (dotted box "i" in (a)) with two optical ports and one electrical port. The dashed boxes of Function Generator or Pulse generator are used for distributed SBS measurement. (c) Three schemes of detection unit shown in dotted box "*j*" in (a).

The laser unit shown in Fig. 6(a) comprises three ports (two corresponding to optical fields and the other to the revealed value of ν or a microwave/RF input), which has three different schemes as illustrated in Fig. 6(b). First, one can utilize two individual lasers under frequency/phase locking and frequency countering [33]. Second, one laser is divided into two parts. One part is amplified by an EDFA working as pump wave; the second part serving as probe wave is modulated by an electro-optic intensity modulator (EOM) to generate two sidebands working as probe wave[25]. An optical filter is inserted before launching into the FUT or laid after PD1 so as to cut off the influence of the frequency-upshifted sideband. Third, the second part can be also modulated by a single-sideband modulator [35] to get well-suppressed frequency-downshifted sideband directly serving as probe wave. There are also three schemes, depicted in Fig. 6(c), to realize the detection unit shown in Fig. 6(a). The first and simple scheme is related to the first scheme of the laser unit. A personal computer with a multi-channel data acquisition card (DAQ) can catch the value of ν and record the data of PD1 and/or PD2 so as to pick up the Brillouin signal (gain and/or loss) as a function of ν. The second and third schemes in Fig. 6(c) can be used for either the second and/or third laser unit in Fig. 6(b), respectively. For instance, a high-cost vector network analyzer provides a frequency-tuned RF signal to modulators and simultaneously detect the Brillouin signal [36]. Alternatively, the RF signal can be achieved from a microwave synthesizer and the data of PD1 and/or PD2 can be picked up by a DAQ with or without a lock-in amplifier (LIA). It is notable that the use of LIA for detection unit requires an intensity-chopping of the pump wave by an additional EOM [10] or periodic switching of upshifted or downshifted sideband at the SSBM [34], which is advantageous for characterization of very weak BGS or a short-length FUT due to its high SNR and accuracy[12, 34, 37].

Figure 7(a) depicts a high-accuracy experimental setup of pump-probe SBS-based BGS characterization by use of SSBM and LIA for a short-length FUT [37]. An EDFA is inserted after the SSBM to increase the probe power, which is aimed to reduce the impact of Rayleigh scattering or splicing/crack induced reflection of the pump wave in the FUT. The optical lights after a circulator include the following components:

$$P_{tot} = P^{0}_{probe} + \delta P_{probe} + P^{R}_{pump}, \tag{15}$$

where $P_{\mathrm{probe}}{}^{0}$ denotes the probe power experiencing no Brillouin amplification, $\Delta P_{\mathrm{probe}}$ the amplified probe power, and $P_{\mathrm{pump}}{}^{\mathrm{R}}$ the reflected pump power. Thanks to the lock-in detection, the component of $P_{\mathrm{probe}}{}^{0}$ is effectively cut off by the LIA since it has no relationship with the chopped pump power given by

$$P_{pump}(t) = P^{0}_{pump} \cdot \cos(2\pi f_{ch} t), \tag{16}$$

where f_{ch} is the chopping frequency. Simply assuming that there is no depletion for pump power and no optical propagation loss for either probe or pump light, $\Delta P_{\mathrm{probe}}$ can be expressed by

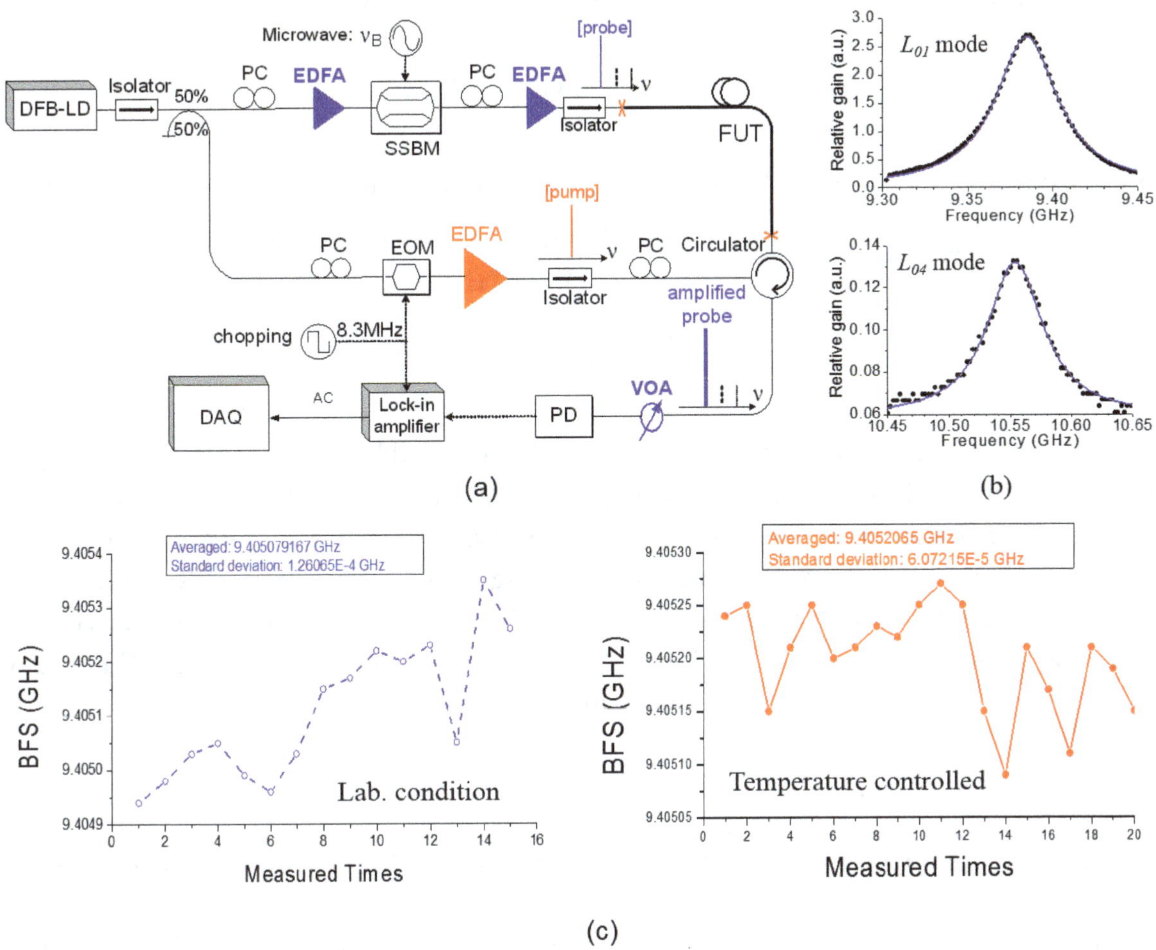

Figure 7. High accuracy pump-probe-based BGS characterization. (a) Experimental setup. (b) Characterized BGS. (c) Measurement accuracy. ((a) and (b) after Ref. [37]; © 2007 OSA.)

$$\delta P_{probe} = g(\nu)\cdot P^0_{probe}(f_0-\nu)\cdot P_{pump}, \tag{17}$$

If R denotes the reflectivity of pump power arising from both Rayleigh scattering and some reflection points, the reflected pump power can be expressed:

$$P^R_{pump} = [P^0_{pump} - P_{loss}]\cdot R, \tag{18}$$

where P_{loss} is the so-called Brillouin loss of pump power during Brillouin interaction which is approximately equal to ΔP_{probe} as

$$P_{loss} = \delta P_{probe} = g(\nu)\cdot P^0_{probe}\cdot P^0_{pump}\cdot \cos(2\pi f_{ch}t), \tag{19}$$

The demodulated electric amplitude via a LIA at f_{ch} is given by

$$P_{de} = \left[P^0_{probe} \cdot g(\nu) \cdot (1-R) + R \right] \cdot P^0_{pump}, \tag{20}$$

where the part of $\left[P^0_{probe} \cdot g(\nu) \cdot (1-R) \right] \cdot P^0_{pump}$ is the signal to be detected by the LIA and the rest part of $R \cdot P^0_{pump}$ is the noise level. From it, one can deduce the signal-to-noise ratio (SNR):

$$SNR = \frac{P^0_{probe} \cdot g(\nu) \cdot (1-R)}{R}, \tag{21}$$

which is independent on the pump level, but just determined by the probe power of $P_{probe}{}^0$ and the reflection rate of R as well as the BGS of $g(\nu)$. In other words, an increase of probe power can drastically enhance SNR and also improve the system accuracy. As an example, Figure 7(b) shows a characterized fundamental-order or higher-order resonance BGS in a *w*-shaped high-delta fiber with fluorine inner cladding (F-HDF) [37]. The measurement system has a high accuracy of 0.13-MHz standard deviation at laboratory condition or 0.05 MHz for well-temperature-controlled condition, as illustrated in Fig. 7(c).

2.3. Pump depletion effect

As mentioned above (see Fig. 4), pump depletion effect influences the linewidth of pump-probe-based BGS. Early in 2000 [38], it was first observed that the spectrum broadening and hole burning occurs in a SBS generator (i.e. noise-started spontaneous Brillouin scattering). The reason was thought as the waveguide interaction among different angular components of the pump and backscattered Stokes signals. Besides, during the application of SBS-based amplifier, two coherent optical waves with precise frequency difference equal to Brillouin frequency shift are launched into optical fibers; then a frequency-scanned weak signal could suffer non-uniform amplification if the two waves' powers are too high [39, 40]. In SBS-based distributed fiber optical sensor, which will be introduced in **Section 3**, two coherent waves (pulse and/or continuous wave (CW)) are injected into the two opposite ends of the sensing fiber. Recently, it was found that pump depletion of pump-probe-based system configuration could induce a significant measurement error of the local Brillouin frequency shift in the far end of the probe (Stokes) wave [41].

Assuming that CW probe wave, $P_s(0)$, is injected at the near end of the fiber (z=0) while CW pump wave, $P_p(l)$, is launched at the far end of the fiber (z=l with l the fiber length). Considering the steady-state condition and neglecting the transmission loss of the fiber, the coupling equations of Eq. (4) and Eq. (5), describing the SBS interaction, can be modified to the dimensionless equations [42]:

$$\frac{dQ_p}{dx} = kQ_pQ_s, \tag{22}$$

$$\frac{dQ_s}{dx} = kQ_pQ_s, \tag{23}$$

where $Q_p=P_p/P_s(0)$ and $Q_s=P_s/P_s(0)$ represent the normalized pump and probe waves with respect to the injected probe wave of $P_s(0)$, $x=z/l$ is the normalized position, and $k=G\bullet l\bullet P_s(0)$ is the normalized Brillouin gain with $G=g(\nu)/A_{\text{eff}}$. Figure 8(a) illustrates the normalized BGS at different positions of an arbitrary-length SMF, which are numerically calculated according to Eqs. (22) and (23). It is found that BGS gradually gets broadened, saturated, and hole-burned when the position moves from the far end ($z=l$) towards the near end of the fiber ($z=0$). We define the BGS saturation as the critical condition of spectral hole burning phenomenon.

Further introduce the injected power ratio between pump and probe waves, defined as $\gamma=Q_p(x=1)=P_p(l)/P_s(0)$. Since $dQ_p/dx=dQ_s/dx$, the difference between Q_p and Q_s maintains a constant (A), i.e. $A=Q_p-Q_s$, which is determined by k and γ. Consequently, the analytical solutions to Eqs. (22) and (23) are derived as

$$Q_p = \frac{A(A+1)e^{-kAx}}{(A+1)e^{-kAx}-1}, \tag{24}$$

$$Q_s = \frac{A}{(A+1)e^{-kAx}-1}. \tag{25}$$

The critical condition of the spectral hole burning phenomenon can be theoretically expressed by:

$$\frac{dQ_s}{d\nu} = 0. \tag{26}$$

By numerically solving Eqs. (24)-(26), one can interpret the critical condition by two different ways: (1) the critical position x_c for the fixed pump and probe power; (2) the critical powers for a specific position of the fiber. Figure 8(b) depicts the calculated relation of x_c to k and γ, which indicates that x_c moves towards the fiber far end when k and γ reach higher values (i.e. the fiber gets longer or the injected powers are stronger). It means that the pump depletion gets worse since much longer segments in the fiber suffer spectral hole burning. It is notable to address that the physical nature of the critical powers is essentially the same as that of the critical position because they can be also deduced by the contour (i.e. k-γ curve) at a fixed position x in Fig. 8(b):

$$P_s(0) = k / GL, \tag{27}$$

$$P_p(L) = \gamma k / GL. \tag{28}$$

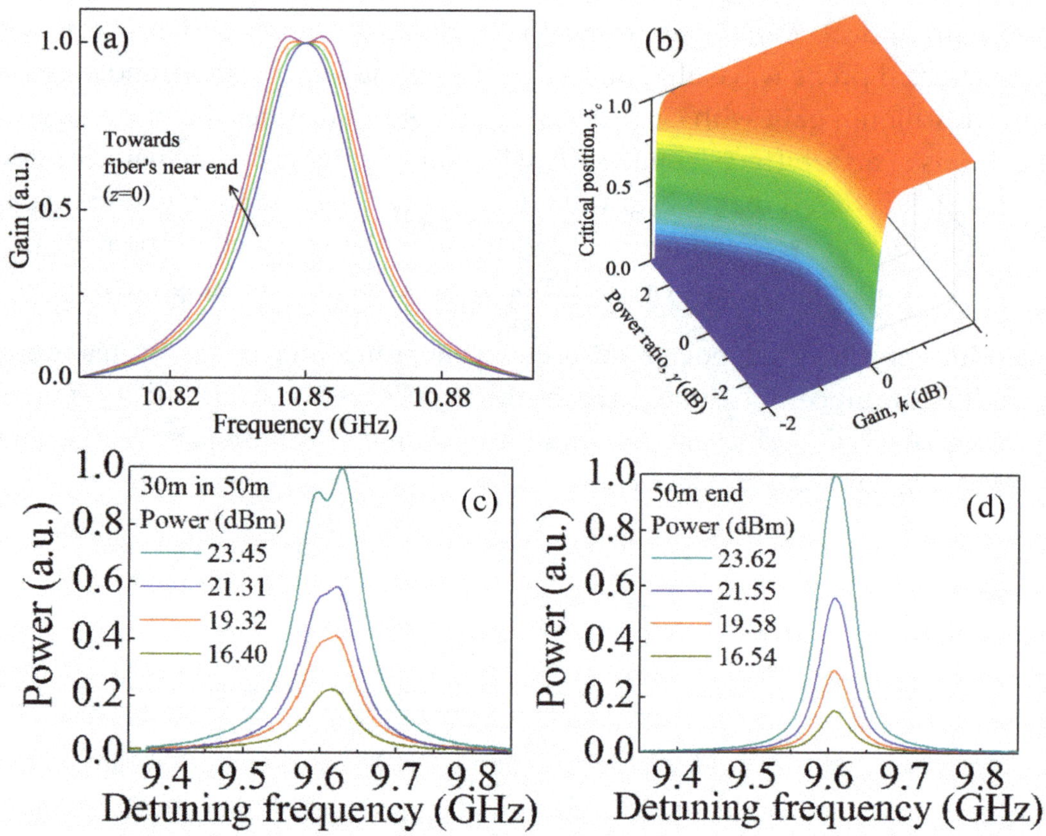

Figure 8. (a) Simulated BGS at different positions in the fiber. (b) Critical position x_c, determined by the normalized gain k and injected power ratio γ. The critical positions divide the fiber into two parts with and without the spectral hole burning phenomenon. Measured BGS in the middle (c) or at the end (d) of a 50-m-long fiber. (after Ref. [42]; © 2014 JJAP.)

Figure 8(c) and 8(d) illustrates the measured BGS under different pump power for two different positions (in the middle and at the far end, respectively) of a 50-m-long dispersion compensated fiber (DCF). The probe power is fixed at 9.8 dBm. At the far end, the BGS [see Fig. 8(d)] rises with the power increased but always preserves the Lorentz shape. While in the middle, the experimental result [see Fig. 8(c)] is in a qualitative accordance with the numerical analysis [see Fig. 8(a)]. The Brillouin gain keeps rising with the increase of optical power, while the peak at the local Brillouin frequency shift seems to be saturated gradually and a hollow starts appearing when it reaches ~20 dBm, which is just the spectral hole burning phenomenon. The hollow in the BGS may introduce great errors to pump-probe-based Brillouin distributed sensors since it disables the peak-searching of the Brillouin frequency shift.

The pump power leading to the BGS saturation is approximately characterized as the critical pump power (for instance, 21.3 dBm at z=30 m). The measured critical powers for two positions (z=20 m or 30 m) of 50-m-long DCF are depicted in Fig. 9, where the simulated critical powers are compared. It illustrates that the critical pump powers approximately measured for several probe powers (open symbols) have very similar trend as the theoretical analysis (curves). It is clear that the position of bigger z requires greater critical powers. This is because the pump depletion is weaker and the distortion of the BGS is less serious if the position is much closer to the fiber far end.

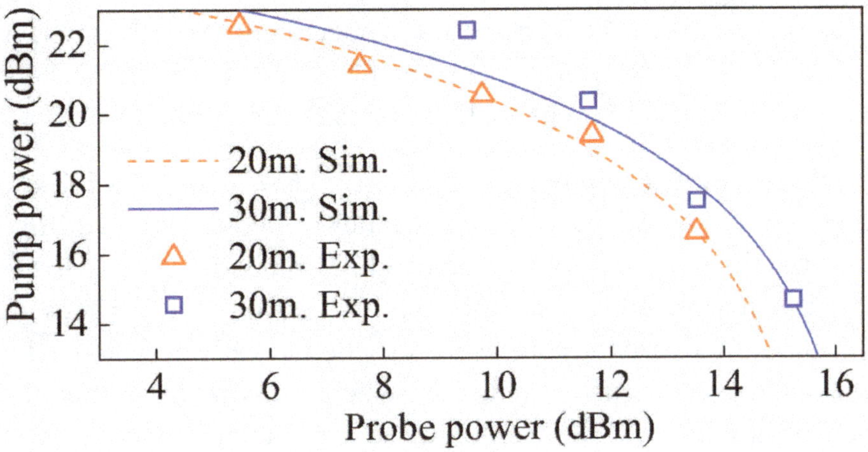

Figure 9. Critical powers for two different positions in a 50-m-long DCF. Solid and dashed curves, simulation; open symbols: experiment. (after Ref. [42]; © 2014 JJAP.)

3. Brillouin–based distributed sensors

3.1. Sensing of measurands

The first report of Brillouin based distributed optical fiber sensors [43] was based on the same principle as that of optical time domain reflectometry (OTDR) or Raman based OTDR (ROTDR) technique as a non-destructive attenuation measurement technique for optical fibers. In that proposal [33], SBS process was performed by injecting an optical pulse source and a continuous-wave (CW) light into two ends of FUT. When the frequency difference of the pulse pump and CW probe is tuned offset around ν_B of the FUT, the CW probe power experiences Brillouin gain from the pulse light through SBS process. Similarly like the case of OTDR, the SBS distributed measurement could measure attenuation distribution along the fiber having no break from an interrogated optical power as a function of time, but it has much higher signal-to-noise ratio (more than ~10 dB) than OTDR due to SBS high gain. Later, Horiguchi and co-researchers found that this non-destructive can be extended into **a frequency-resolved technique** because ν_B of optical fibers has linear dependence on measurands of strain and temperature as follows [44, 45]:

$$\nu_B - \nu_{B0} = A \cdot \delta\varepsilon + B \cdot \delta T, \tag{29}$$

where ν_{B0} is measured at room temperature (25°C) and in the "loose state" as a reference point, $\Delta\varepsilon$ the applied strain and ΔT the temperature change. The "loose state" means that the FUT is laid freely in order to avoid any artificial disturbances. *A (or* $C\nu\varepsilon$) is the strain coefficient in a unit of MHz/με and *B (or* $C_T\varepsilon$) is the temperature coefficient in a unit of MHz/°C. Figure 10 illustrates the characterized strain or temperature dependence in a standard SMF under the experimental setup of Fig. 7(a), where the BGS always moves towards higher ν_B and its gain reduces or increases when $\Delta\varepsilon$ or ΔT is increased, respectively. At 1550 nm, A=0.04~0.05 MHz/με and B=1.0~1.2 MHz/°C, which depends on the fiber's structure and jackets. **Note that Eq. (29) is the basic sensing mechanism of Brillouin-based distributed sensors.**

The nowadays telecom optical fibers (ITU-T G.651, G.652, G.653, and G.655) mostly have GeO_2-doped fiber cores [46] and pure-silica (or other-doped-silica) cladding. Naturally, the GeO_2 doping induces the reduction of the longitudinal acoustic velocity in GeO_2-doped core V_{l1} with respect to that in pure-silica cladding V_{l2} (i.e. $V_{l1} < V_{l2}$) [24, 47]. It provides a waveguide of longitudinal acoustic modes in the core region as schematically depicted in Fig. 3(b). A recent study further proves that the acoustic modes sense better confinement than the optical modes in a GeO_2-doped optical fiber [13]. The enhanced confinement results in the existence of multiple L_{0l} acoustic modes in a single-mode optical fiber (SMF) [8], and also leads to that the first-order L_{01} acoustic mode among all L_{0l} modes is best confined in the core and even better confined than the fundamental LP_{01} optical mode [13]. Furthermore, the enhanced confinement shows that the effective acoustic velocity of L_{01} mode (V_a) is close to V_{l1} (i.e. $V_a \approx V_{l1}$), the longitudinal acoustic velocity in the core. Therefore, the change of the L_{01} mode's effective acoustic velocity V_a is dominantly due to the change of the core's acoustic velocity V_{l1} but negligibly (less than 1%) due to that of the cladding's acoustic velocity V_{l2}, even though the core's acoustic velocity V_{l1} and the cladding's acoustic velocity V_{l2} vary equally [12].

The longitudinal acoustic velocity V_{l1} in the GeO_2-doped core (approximately, the L_{01} mode's effective acoustic velocity V_a) is determined by the Young's modulus (E_1) and the density (ρ_1) [48]:

$$V_a \approx V_{l1} = \sqrt{E_1 / \rho_1}, \tag{30}$$

For convenience, we introduce a normalized strain coefficient ($A'=A/\nu_{B0}$, in a unit of 10^{-6}/με) and a normalized temperature coefficient ($B'=B/\nu_{B0}$, in a unit of 10^{-6}/°C). The normalized strain coefficients include three respective factors[49]:

$$A' \equiv \frac{A}{\nu_{B0}} = A'_{neff} + A'_{\rho} + A'_{E}, \tag{31}$$

$$B' \equiv \frac{B}{\nu_{B0}} = B'_{neff} + B'_{\rho} + B'_{E}. \tag{32}$$

Each three parts in the right sides of Eq. (31) and Eq. (32) are determined by relative change rates in n_{eff}, E_1, and ϱ_1 due to the applied strain $\Delta\varepsilon$ or the temperature change ΔT. A'_{neff} and B'_{neff} are determined by the elasto-optic and thermo-optic effects; A'_{ρ} and B'_{ρ} are subject to the strain-induced distortion and the thermal expansion; A'_E and B'_E are decided by the strain-induced second-order nonlinearity of Young's modulus and the thermal-induced second-order nonlinearity of Young's modulus.

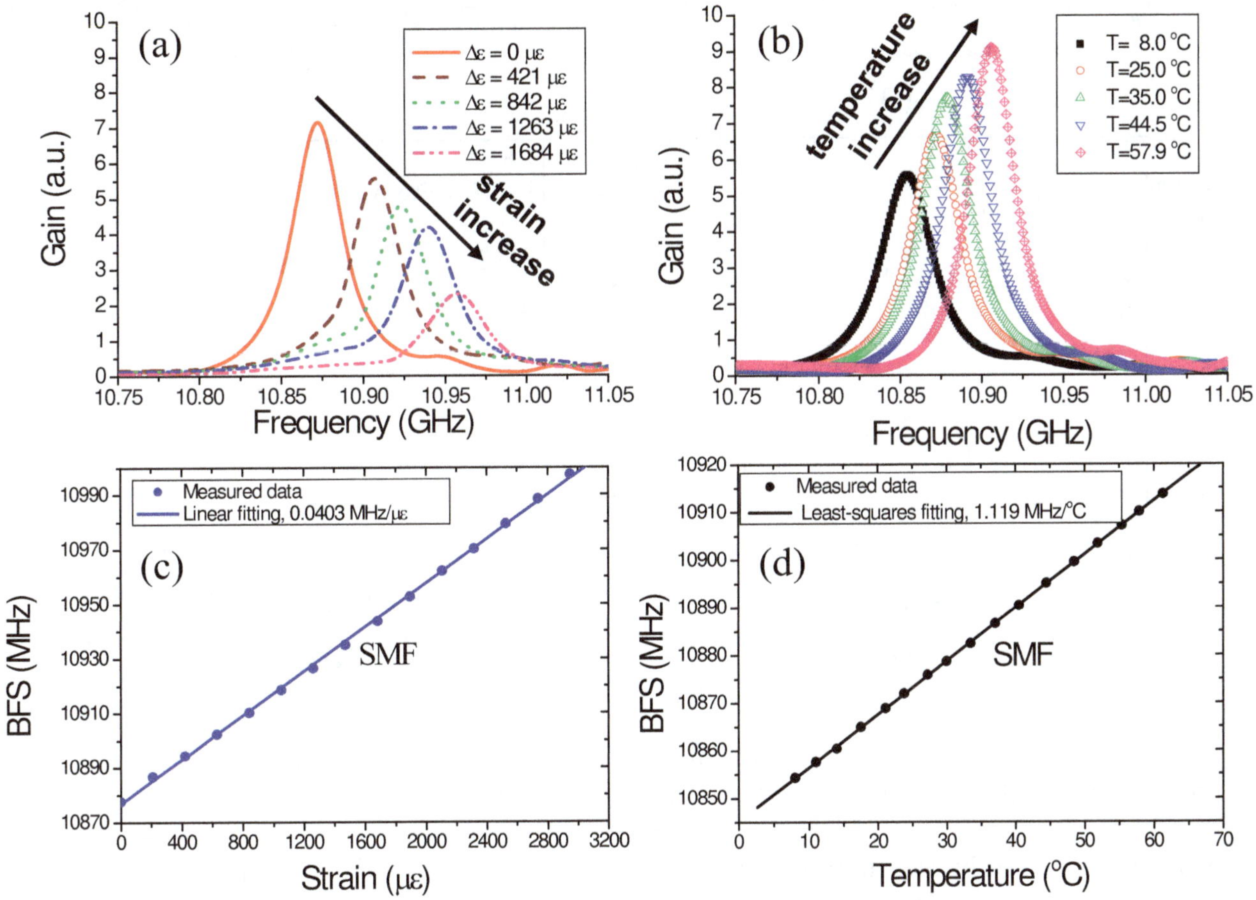

Figure 10. (a) Stain and (b) temperature dependences of BGS in SMF; (c) Strain and (d) temperature dependences of Brillouin frequency shift ν_B in SMF.

Strict experimental characterization on a series of optical fibers with different GeO_2 concentration is depicted in Fig. 11 [49]. The BFS has linear dependence on the GeO_2 concentration in the fiber's core (i.e.-87.3 MHz/mol%), which corresponds to ν_{B0} change of-87.3 MHz regarding 1-mol% increase of GeO_2 concentration in the core (i.e. an incremental Δ of 0.1 %). It specifies the previously reported values [25, 50, 51]. Besides, the frequency spacing between neighbouring acoustic modes increases by orders when the GeO_2 concentration is enhanced, for example, ~50-60 MHz for Fiber-A (SMF, 3.65 mol%) versus ~700-720 MHz for Fiber-C (HNF, 17.0 mol%).

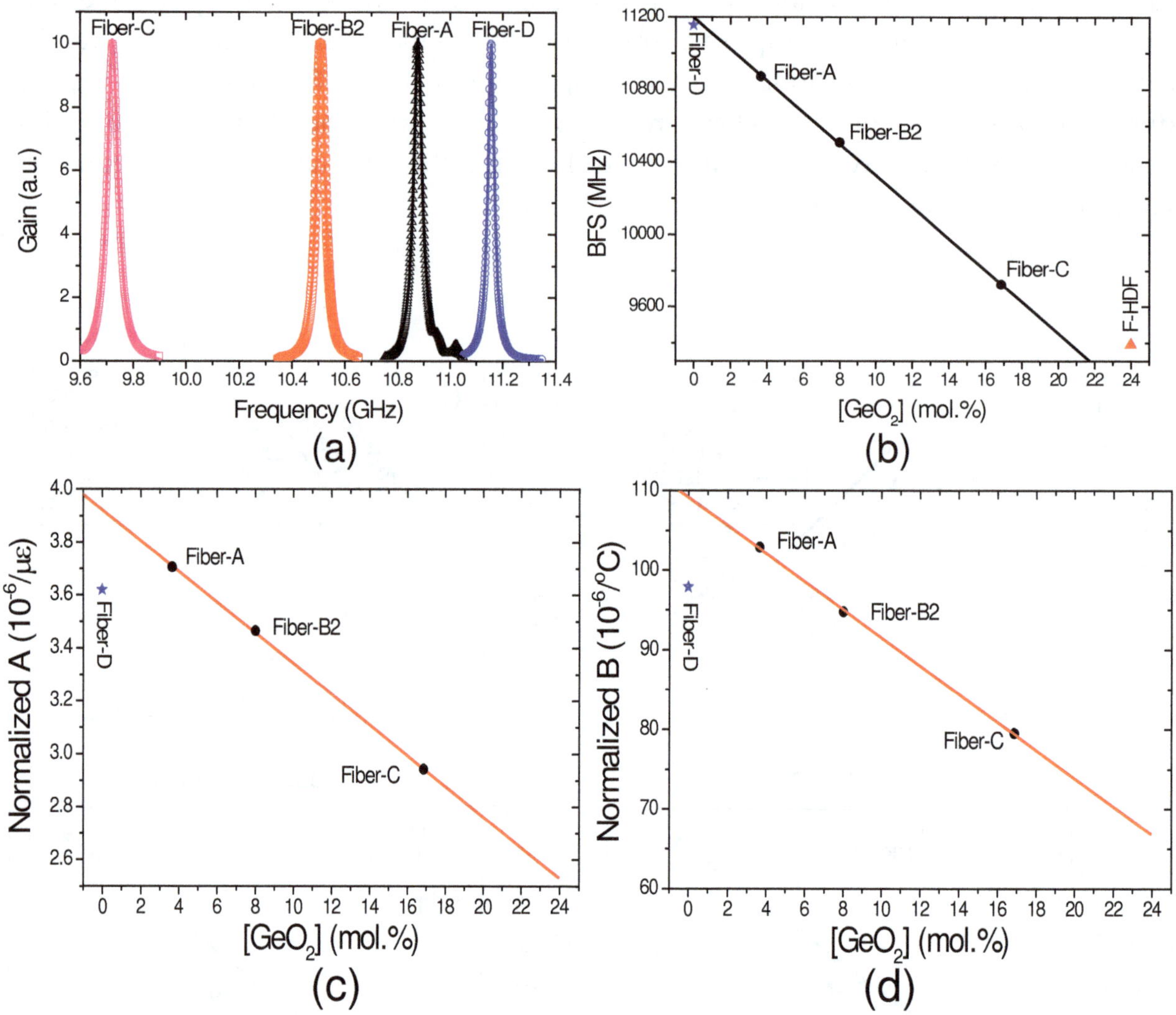

Figure 11. (a) BGS, (b) BFS, (c) normalized strain coefficient, and (d) normalized temperature coefficients in silica optical fibers with different GeO_2 concentration. (After Ref. [49]. © 2008 OSA/IEEE.)

The normalized strain and temperature coefficients defined in Eq. (31) and Eq. (32) were characterized by repeating the BGS measurement under different applied strain and temperature change. It shows a linear dependence of A' or B' on GeO_2 concentration with slope of-1.48 %/mol% or-1.61 %/mol%, which denote that A' and B' are relatively decreased by-1.48 % and-1.61 % for an incremental Δ of 0.1 %. The theoretical study further indicates that both the strain and temperature dependences in Eq. (29) are dominantly (~92%) responsible from the strain-induced and thermal-induced second-order nonlinearities of Young's modulus, that is, Eq. (31) and Eq. (32) [49].

3.2. Sensing of location

Besides the sensing of measurands (see Eq. (29)), the mapping of spontaneous or stimulated Brillouin scattering process (not just non-destructive attenuation measurement [43]) is another key issue to realize distributed optical fiber sensing [52-54]. Two different mapping ways, as schematically illustrated in Fig. 12, were proposed. One is to repeat the localized BGS in scanned positions along the FUT; the other is to repeat the Brillouin interaction under different frequency offset.

There are three different mapping or position-interrogation techniques, including time domain [33, 52-56], frequency domain [57, 58], and correlation domain [59-61]. Regarding the injection ways of optical fields, there are two opposite groups, i.e. analysis versus reflectometry. The analysis is two-end injection based on SBS; while the reflectometry is one-end injection based on SpBS. Comparably, the analysis has much higher SNR than the reflectometry. Note that there is an additional method between analysis and reflectometry, called one-end analysis [55, 62, 63]. Its only difference from the traditional (two-end) analysis is the one-end injection and its SBS process occurs between the forward pump and the backward probe wave that is reflected at the far end of FUT.

The basic principle of time-domain sensing technique is the "time-of-flight" phenomenon in FUT. For two-end or one-end analysis, named Brillouin optical time domain analysis (BOTDA) [33, 52, 54], one of pump and probe waves is pulsed in time and the other is continuous wave (CW). Subsequently, they are successively interacted along the FUT during the time-of-flight of the pulsed wave. In contrast, for one-end reflecometry, called Brillouin optical time domain reflectometry (BOTDR) [53, 56], the pump wave is pulsed in time and the SpBS Stokes wave is reflected along the FUT during the pump's time-of-flight. The basic experimental configuration of BOTDR or BOTDA can be simply carried out in Fig. 5 or Fig. 6, respectively. The required modification is to insert an optical pulse generator (for example, an electro-optic intensity modulator driven by an electric pulse generator). The spatial resolution (ΔZ_{TD}) of time-domain distributed sensing is physically determined by the pulse width (τ)[43]:

$$\Delta Z_{TD} = \frac{\tau \cdot c}{2n}, \tag{33}$$

where c is the light speed in vacuum and n the group velocity of the pulse. The BGS mapping is realized by repeating the above measurement when the spectrum of the reflected Stokes in BOTDR is processed or the optical frequency offset between the pump and probe in BOTDA is tuned around the BFS ν_B.

There are two kinds of correlation-domain sensing techniques, nominated Brillouin optical correlation domain analysis (BOCDA) [59, 60] and Brillouion optical correlation domain reflectometry (BOCDR) [61, 64]. Both of them originate from the so-called synthesis of optical coherence function (SOCF) [65, 66]. Nevertheless, the SOCF in BOCDA or BOCDR is generated between the pump and probe waves or between the pump-scattered Stokes wave and the optical oscillator, respectively. In experiment, the BOCDR and BOCDA can be executed by substituting a distributed feedback laser diode (DFB-LD) driven by a function generator (such as in a sinusoidal function) for the light source in Fig. 5 and Fig. 6, respectively. Thanks to the current-frequency transferring effect of DFB-LD [67], the optical frequencies of the light

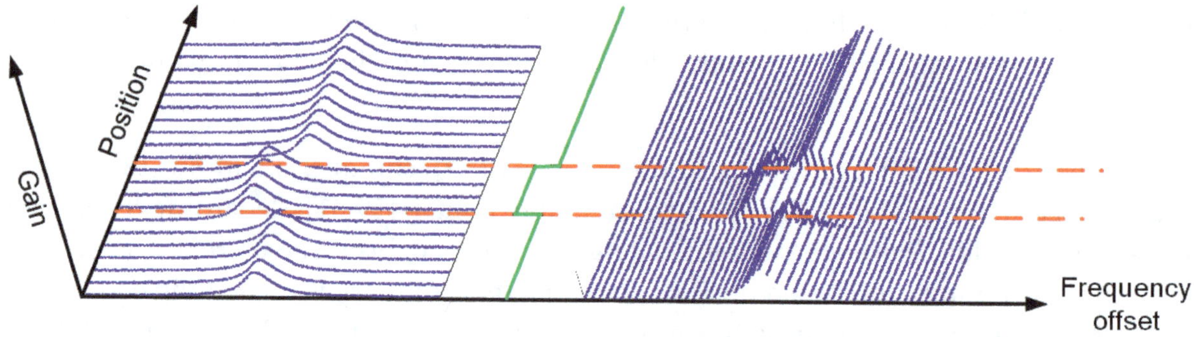

Figure 12. Schematic of sensing of location or mapping of BGS.

sources are simultaneously modulated also in a sinusoidal function. Subsequently, the optical frequency offset between pump and probe or between scattered Stokes and optical oscillator changes with time as well as position, deviating from the preset constant frequency offset around the BFS ν_B. Only at some particular locations (called correlation peaks), the frequency offset is maintained as the constant frequency offset because of the in-phase condition so that the local SBS interaction or the beating of the local Stokes and oscillator is **constructive**. At other locations rather than correlation peaks, the frequency offset is always vibrating with time, which leads to a broadened and **destructive** SBS or SpBS. The spatial resolution of BOCDA and BOCDR are both determined by[59]

$$\Delta Z_{CD} = \frac{c}{2nf_m} \cdot \frac{\Delta \nu_B}{\pi \Delta f}, \tag{34}$$

where f_m is the modulation frequency of the sinusoidal function, Δf the modulation depth, and $\Delta\nu_B$ the Brillouin linewidth defined in Eq. (10). Since the SOCF is naturally realized by an integral or summation signal processing in photonics or electronics, all SBS or SpBS along the entire FUT should be accumulated together (as an example shown in Fig. 6(a), accumulated by a LIA). Consequently, the maximum measurement length (or sensing range, L_{CD}) is decided by the distance between two neighboring correlation peaks [59]:

$$L_{CD} = \frac{c}{2nf_m}. \tag{35}$$

Because of the difference of the physical pictures between time domain and correlation domain, their sensing performance is different. For example, the spatial resolution of BOTDA/BOTDR was typically limited to be ~1 m by the lifetime of acoustic phonons (10 ns) and the nature of intrinsic Brillouin linewidth. However, BOCDA/BOCDR is of CW nature free from this limitation, and their spatial resolution can be ~cm-order [60, 68] or even ~mm-order [69]. Since BOTDA/BOTDR carries out the whole mapping of BGS along the FUT during the time-of-the-flight while BOCDA/BOCDR realizes the distributed sensing by sweeping the modulation frequency (or correlation peak), the sensing speed is different. The entire sensing speeds for both BOTDA/BOTDR and BOCDA/BOCDR are time-consuming due to the tuning of pump-probe frequency offset, averaging of mapping, and signal processing of data fitting. However, the sensing position of BOCDA/BOCDR can be random accessed [59, 61], and the dynamic sensing with high speed at the random accessed position is possible [70, 71]. The detailed difference of other performances will be described in **Section 4** and **Section 5**.

4. Challenges in Brillouin based distributed optical fiber sensors

4.1. Simultaneous measurement of strain and temperature

As explained in **Section 3**, all Brillouin based distributed optical fiber sensors interrogate the Brillouin frequency shift so as to deduce strain and temperature information based on Eq. (29). It naturally gives a physical challenge, i.e. how to distinguish the response of strain from the response of temperature based on the single parameter of BFS interrogation in a single piece of sensing fiber. In current industrial practices, two individual fibers or two fibers in a fiber cable are used to discriminate the strain and the temperature: the first one is embedded or bonded at the target material/structure to feel the total responses of strain and temperature, while the second fiber is placed beside the first one and kept in loose condition so that it feels the response of temperature only. Another way is to use two distributed sensing systems with two individual fibers [72-75]: one Raman-based or Rayleigh-based sensor is to monitor the temperature; the other Brillouin sensor to monitor the temperature and strain. After distributed sensing measurements, the strain and temperature responses can be calculated by mathematics. However, the above practices make the entire sensing system complicated and the calculated responses of strain and temperate change with service time.

Practical applications of Brillouin based distributed optical fiber sensors require a method to effectively discriminate them by use of two intrinsic parameters (denoted by y_1 and y_2) in one sensing fiber. Their changes (Δy_1 and Δy_2) depend on simultaneously the applied strain ($\Delta\varepsilon$) and temperature change (ΔT), which are governed by the following matrix:

$$\begin{pmatrix} \Delta y_1 \\ \Delta y_2 \end{pmatrix} = \begin{pmatrix} A_1 & B_1 \\ A_2 & B_2 \end{pmatrix} \begin{pmatrix} \Delta\varepsilon \\ \Delta T \end{pmatrix}, \tag{36}$$

where A_1 (A_2) and B_1 (B_2) are the strain and temperature coefficients of y_1 (y_2), respectively. Both $\Delta\varepsilon$ and ΔT can be deduced from Eq. (36), given by

$$\begin{pmatrix} \Delta\varepsilon \\ \Delta T \end{pmatrix} = \frac{1}{A_1B_2 - B_1A_2} \begin{pmatrix} B_2 & -B_1 \\ -A_2 & A_1 \end{pmatrix} \begin{pmatrix} \Delta y_1 \\ \Delta y_2 \end{pmatrix}, \tag{37}$$

It is obvious to know that the condition that the strain and temperature can be successfully distinguished is determined by

$$A_1B_2 \neq B_1A_2. \tag{38}$$

In fact, the Brillouin-based distributed sensing system always suffer a measurement uncertainty (Δy_1 and Δy_2), which is in a linear proportional relation with the discrimination errors in strain ($\Delta\varepsilon$) and temperature (ΔT), also given by Eq. (37).

A possible solution using one fiber is to monitor two acoustic resonance peaks at different orders of Brillouin gain spectrum (BGS) in a specially-designed optical fiber [13, 37, 76, 77]. So far, this method cannot ensure accurate discrimination because all the acoustic resonance frequencies exhibit similar behaviors in their dependences on strain and temperature (see Fig. 11) [49]. There is another kind of method reported for discrimination that relies on the possibility that the peak amplitude and BFS of the BGS could have quantitatively different dependences on strain and temperature [78-81]. Its accuracies is not sufficient (e.g., several degrees Celsius and hundreds of micro-strains), which is mainly due to the low signal-to-noise ratio in the BGS peak-amplitude measurement particularly for distributed sensing where troublesome noise from non-sensing locations is accumulated.

4.2. System limitation of time-domain or correlation-domain technique

There are several system limitations in time-domain BOTDA/BOTDR and correlation-domain BOCDA/BOCDR, which comes from their individual sensing techniques. For example, BOTDA/BOTDR suffers a typical limitation of spatial resolution (~1 m) mainly determined by the linewidth of BGS or the lifetime (~10 ns) of acoustic phonons. Narrower pulse width corresponding to higher spatial resolution according to Eq. (33) weakens the acoustic phonons due to the lifetime of the acoustic phonons and leads to broader BGS as well as lower frequency accuracy due to the convolution between the intrinsic BGS and broader spectrum of the pulse [82, 83]. Moreover, although the time-of-the-flight feature of BOTDA/BOTDR is suitable for long distance sensing, the nature of pump depletion and fiber transmission loss confines the maximum of measurement range within several tens of kilometers [84].

On the other hand, BOCDA/BOCDR can provide extremely high spatial resolution of cm order or mm order with a cost of system complexity. However the correlation-domain sensing nature means that there intrinsically exist periodic correlation peaks in the fiber. Besides, the nominal definitions of spatial resolution and measurement range (see Eq. (34) and Eq. (35)) show that they both depend on the modulation frequency and thus they are in a tradeoff relation with each other [59]. The accumulation of the entire BGS along the FUT corresponding to the measured BGS at the sensing location should include a high-magnitude background of the BGS at the uncorrelated positions, which makes it difficult achieve large range of strain or temperature since higher strain or temperature change shifts the measured BGS closer to the background. As introduced in **Section 3.2**, the access ability of BOCDA/BOCDR is random and the sensing speed in one location is high. However, the sensing speed along the entire FUT is still low and just comparable to BOTDA/BOTDR because the scanning of the sensing location is realized by changing the modulation frequency (see Eq. (35)), which needs quite long time to restart the communication among electronic devices (specially, function generator).

5. Advances in Brillouin based distributed optical fiber sensors

5.1. Concept of Brillouin dynamic grating

Dynamic grating can be generated by use of gain saturation effect in rare-earth-metal-doped optical fibers [85-87] or stimulated Brillouin scattering (SBS) process in optical fibers [88-93] and even in a photonic chip [94]. Dynamic grating is more advantageous for certain applications than fiber Bragg grating (FBG) [95] because it can be dynamically constructed using two coherent pump waves while FBG is static after fabrication. In comparison, the SBS-generated dynamic grating, also called Brillouin dynamic grating (BDG), is superior to the saturation gain grating due to its elasto-optic nature and lack of quantum noise [1]. In addition, the BDG is much easier to experimentally characterize [88, 90-93] while the saturation gain grating needs sophisticated double lock-in detection [86].

Up to date, there are various methods to generate BDG in optical fibers, which are schematically compared in Fig. 13 [93]. The basic principle of BDG in optical fibers is quite similar, which is shown in Fig. 13(a). Two coherent optical waves, i.e. the pump and probe (or Stokes) waves in the SBS process, are launched from the two opposite ends of optical fibers. When their optical frequency offset ($\nu_1 = f_1 - f_1'$) is equal to the BFS ν_B as well as the resonance frequency of the fundamental acoustic mode ($\nu_{ac}^{(1)}$) defined by Eq. (8):

$$\nu_1 \equiv f_1 - f_1' = \nu_B = \nu_{ac}^{(1)}, \tag{39}$$

where f_1 and f_1' are the optical frequencies of the pump and probe waves, a strong acoustic wave of the fundamental acoustic mode (the so-called BDG) is optically generated. As long as the third optical wave (i.e. the readout wave) is injected from the same end as the pump wave, there is a diffracted/reflected optical wave originating from the BDG. The diffraction or reflection efficiency (also called BDG reflectivity) is determined by the phase-matching condition, under which the pump/probe and readout wave can efficiently couple their energy via the BDG.

The method of BDG generation and detection can be classified into two different cases, which depends on the used optical fibers. In the first case of polarization maintaining fiber (PMF) [88-90] or few-mode fiber (FMF) [91] (see Fig. 13(b)), the BDG generation is separated from the BDG detection by use of orthogonal polarization states or different optical modes, respectively. The phase-matching condition means that the BFS of the BDG generation and detection should be unique as Eq. (39), which results in a frequency difference determined by the PMF's

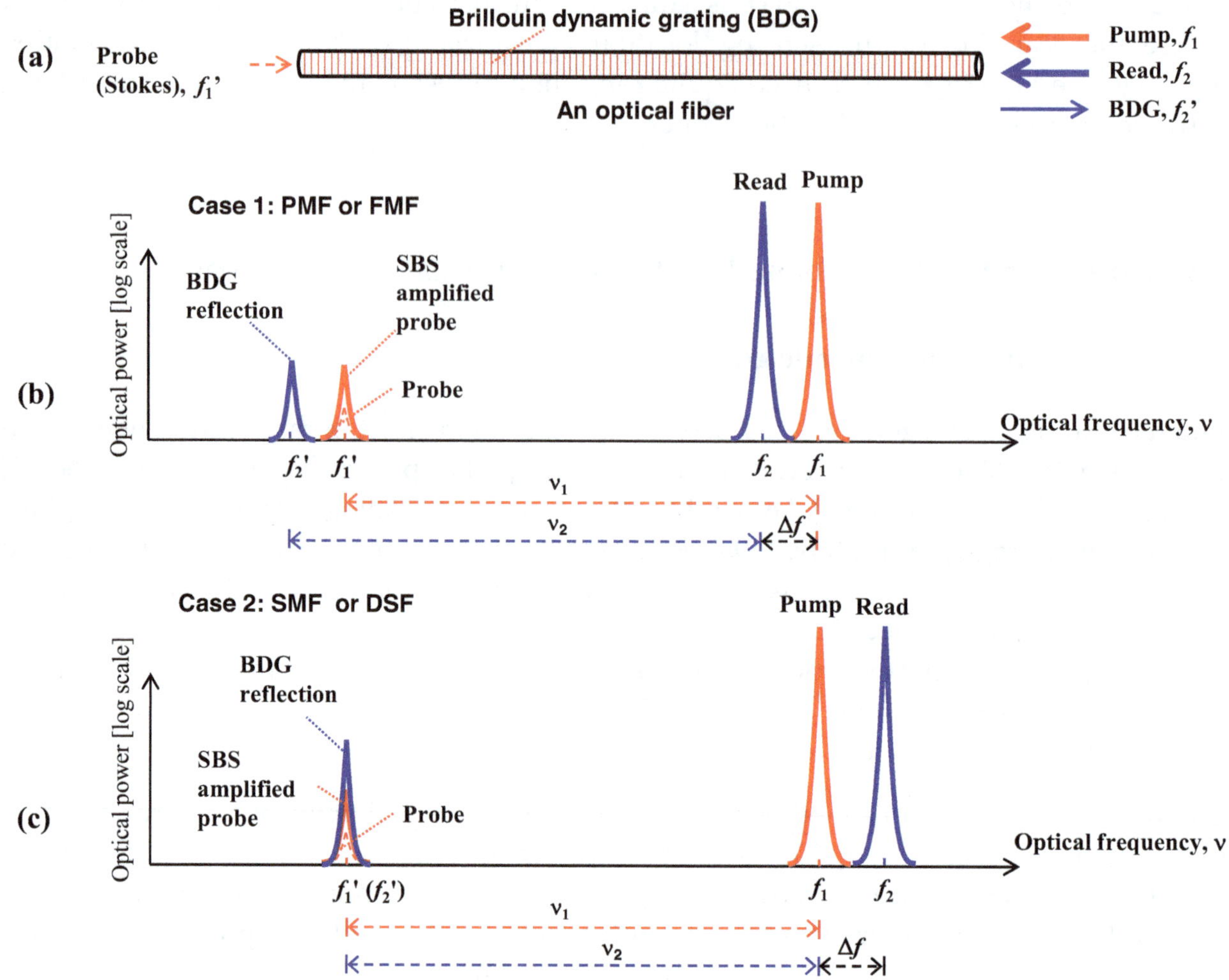

Figure 13. Principle of BDG in an optical fiber. (a): Orientation of optical injection. (b) and (c): Two different cases of the optical frequency relation among the pump, probe (Stokes), readout, and BDG reflection. (After Ref. [93]; © 2013 OSA.)

birefringence (B) or the FMF's modal refractive index difference (n_{eff}-n_{eff}' with n_{eff}' the higher-order modal refractive index):

$$\Delta f \equiv f_2 - f_1 = \frac{B}{n_{eff}} \cdot f_1, \tag{40}$$

$$\Delta f \equiv f_2 - f_1 = \frac{n_{eff} - n_{eff}^{'}}{n_{eff}} \cdot f_1, \tag{41}$$

where f_2 is the optical frequency of the readout wave.

As shown in Fig. 13 (c), the BDG in a SMF [92] or dispersion shifted fiber (DSF) [93] can be generalized into the second case. If the readout wave with the optical frequency of f_2 is launched for BDG detection, multiple-peak Stokes wave is intrinsically backscattered via SpBS. The ith-peak Stokes wave is downshifted in frequency from the readout wave by

$$\nu_2^{(i)} = \frac{2n_{eff}}{c} \cdot V_a^{(i)} \cdot f_2, \tag{42}$$

where $V_a^{(i)}$ is the acoustic velocity of the ith-order acoustic wave. The phase-matching condition of the BDG generation and detection turns to be determined by the frequency difference between the pump and readout wave:

$$\Delta f \equiv f_2 - f_1 = \frac{\nu_1^{(i)} - \nu_1^{(1)}}{1 - 2n \cdot V_a^{(i)} / c} \approx \nu_1^{(i)} - \nu_1^{(1)}, \tag{43}$$

where $\nu_1^{(i)}$ is the ith-order resonance frequency of the BGS measured by the pump-probe SBS process, also given by Eq. (42) except that f_2 is replaced by f_1. The approximation in Eq. (43) is reasonable since the acoustic velocity ($V_a^{(i)}$=~5300-5900 m/s) in silica-based fibers is far smaller than the optical velocity (c=3.0× 10^8 m/s) [11]. Note that the BDG observed in a SMF [92] can be regarded as one special example of the generalized second case with$\Delta f = 0$ or $f_2 = f_1$ in Eq. (43) because the BDG generation and detection share the same fundamental acoustic mode.

In comparison, the method based on a PMF is more attractive because the BDG generation and readout are oriented and separated in two orthogonal polarization states [88-90]. The frequency-deviation property provides an additional degree of freedom to precisely characterize the birefringence according to Eq. (28). Figure 14 depicts the optical spectra of the BDG reflection measured by an optical spectrum analyzer (OSA), including four components (leaked pump and probe waves, BDG reflected wave, and Rayleigh scattered wave from left to right). The BDG property is qualitatively confirmed by the great enhancement of the third component (BDG reflected wave), since it is transferred from weak SpBS to strong SBS process under the assistance of the BDG generated by pump and probe waves. The frequency deviation can be roughly estimated to be 44.0 GHz by a wavelength meter, which gives the birefringence value of 3.28*10^{-4}. However, the resolution is limited to about 1*10^{-6} due to 0.1 GHz-level resolution of the wavelength meter.

Most recently, a heterodyne detection was demonstrated to straightforwardly characterize the physical BDG property in a high-delta PMF [96]. Figure 15(a) summarizes a 3D distribution of the heterodyne-detected electronic spectra between the BDG reflection and the readout wave while scanning the pump–probe frequency offset (ν_1 or f_1-f_1') around the BFS of 10.510 GHz. The peak frequency and power dependence of the on f_1-f_1' are shown in Fig. 15(b) and Fig. 15(c), respectively. The frequency dependence is a linear relation because the acoustic resonance frequency of the BDG is determined by f_1-f_1' so that the diffraction wave suffers the identical frequency downshift from the readout wave. The Brillouin gain determined by the

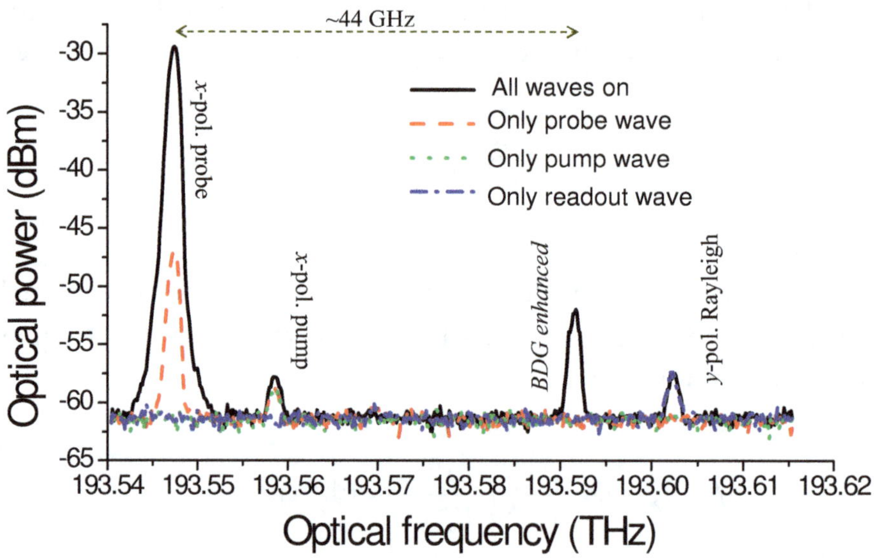

Figure 14. Qualitative characterization (optical spectra) of BDG reflection in a PMF with B=3.3*10^{-4}. Black-solid, all pump, probe and readout waves are launched; blue-dash-dotted, only readout wave; red-dashed, only probe wave; green-dotted, only pump wave.

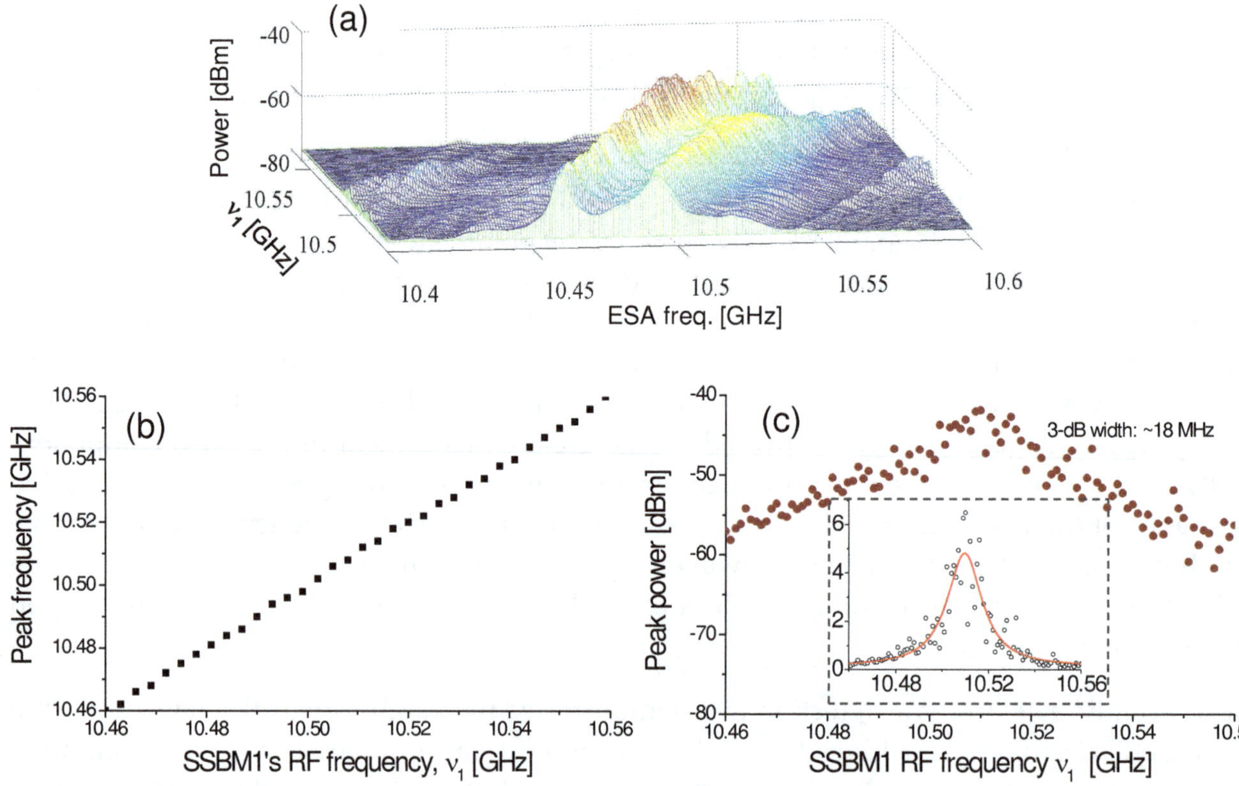

Figure 15. Characterization of the BDG reflection based on heterodyne detection. (a) 3D plot of electronic spectra of heterodyne detection when the pump-probe frequency offset (f_1-f_1') is scanned. Dependence of peak frequency (b) and power (c) of each spectrum on f_1-f_1'. The dashed inset denotes Lorentz fitting (solid curve) to the linear vertical scaled symbols in (c). (After Ref. [96]; © 2013 JJAP.)

pump-probe-based SBS process is well known to be changed when f_1-f_1' is swept [12], which is herein reflected by the measured power dependence [see Fig. 15(c)] since the diffraction/reflection wave sees the same change of the induced Brillouin gain. As shown in the inset of Fig. 15(c), Lorentz fitting provides the central frequency of ~10.510 GHz (just equal to the BFS), and Brillouin intrinsic linewidth of ~18 MHz. All experimental observation matches well the theoretical analysis of the BDG [97].

5.2. Complete discrimination of strain and temperature

As mentioned in **Section 5.1**, the BDG in a PMF can be generated by two coherent pump and probe waves in one principal polarization while readout by another wave deviated in frequency and separated spatially in the other principal polarization. This feature enables the BDG in a PMF working as a dynamic reflector for any optical wavelength of the readout wave by simply tuning the wavelengths of the pump and probe waves. The location of the BDG generation can be also dynamically assigned by changing the fiber's longitudinal structure [98] or programing the interaction position of the pump and probe waves [90, 99-103]. Up to date, the BDG in a PMF has been used for many applications in microwave photonics, all-optical signal processing, and Brillouin-based distributed sensors. For example, the BDG programmed in position or spectrum is very useful in microwave photonics of tunable optical delays [98, 104-106] or programmable microwave photonic filter [107]. Besides, it can also find significant applications in all-optical signal processing such as storing and compressing light [108], ultrawideband communications [109], and all-optical digital signal processing [110].

The first, but most successful, application of the BDG in a Panda-type PMF was demonstrated for complete discrimination of strain and temperature responses for Brillouin based distributed optical fiber sensing applications [89]. Figure 16 shows the basic experimental configuration of the high-precision BDG characterization, which can be used to precisely measure the birefringence of a PMF and to completely discriminate strain and temperature. The BDG generation is based on the pump-probe scheme, which is also the high-accuracy BGS characterization shown in Fig. 7(a). The BDG measurement is realized by the lock-in detection of the BDG reflection since the BDG is periodically chopped due to the chopping of the pump wave. The birefringence-determined frequency deviation defined in Eq. (40) is characterized within a standard error of Δf_{yx}=4 MHz, corresponding to a high-accuracy birefringence of $\Delta B = 3 \times 10^{-8}$.

The principle of the complete discrimination is based on the dependence of the BFS on strain and temperature as introduced in Eq. (29) and the orthogonal dependence of the birefringence (B) or its determined frequency deviation on strain and temperature. This is because the residual tensile stress (σ_{xy}) determining the Panda-type PMF's birefringence scales with the ambient temperature (T_i):

$$B \propto \sigma_{xy} = k \cdot (\alpha_3 - \alpha_2) \cdot (T_{fic} - T_i), \tag{44}$$

where T_{fic} denotes the fictive temperature (e.g., 850 °C) of silica glass, α_3 (α_2) the thermal coefficient of B_2O_3-doped-silica stress-applying parts (pure-silica cladding), and k a constant determined by the geometrical location of stress-applying parts in the fiber [111]. When temperature increases ($\Delta T = T_i - 25 > 0$), the residual stress is released and thus the birefringence decreases as

$$\Delta B^T = -B_0 \cdot \frac{\Delta T}{T_{fic} - 25}, \tag{45}$$

where B_0 is the intrinsic birefringence at room temperature (T_i=25 °C).

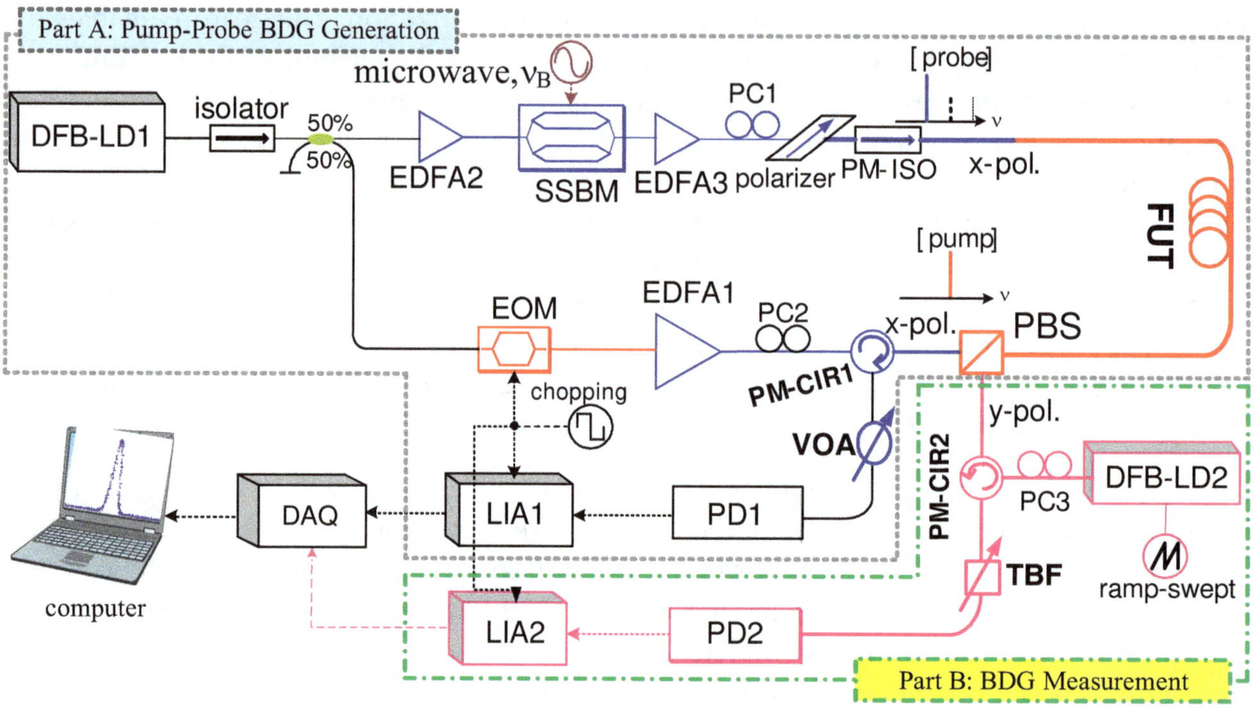

Figure 16. Configuration of the BDG characterization and strain-temperature discrimination. Part A, Pump-probe scheme to measure the BFS along *x*-axis and generate the BDG. Part B, Detection of the BDG diffraction spectrum to *y*-polarized readout wave. (After Ref. [89]; © 2009 OSA.)

In contrast, when an axial strain $\Delta\varepsilon$ is applied upon the fiber, additional stress is generated because the stress-applying parts and the cladding contract in the lateral direction differently due to their different Poisson's ratios ($\gamma_3 > \gamma_2$) [111], the birefringence is enlarged with applied strain as

$$\Delta B^\varepsilon = +B_0 \cdot \frac{(\gamma_3 - \gamma_2)}{(\alpha_3 - \alpha_2)(T_{fic} - 25)} \cdot \Delta\varepsilon. \tag{46}$$

Consequently, the birefringence-determined frequency deviation (Δf) varies linearly with respect to temperature increase and to applied strain. Suppose that $C_f\varepsilon$ and C_f^T are the strain

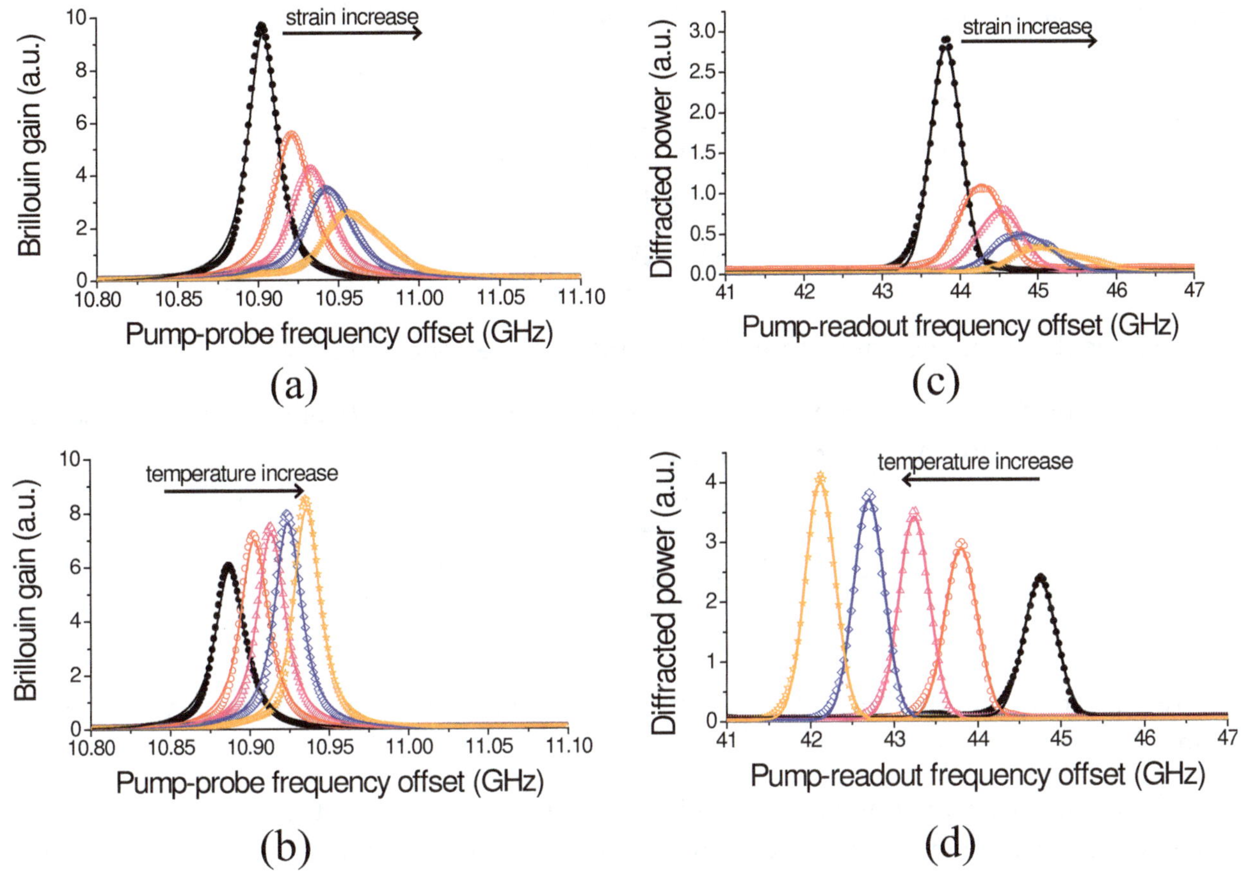

Figure 17. BGS (a, b) and BDG reflection (c, d) measured at various strain (a, c) and temperature (b, d). (After Ref. [89]; © 2009 OSA.)

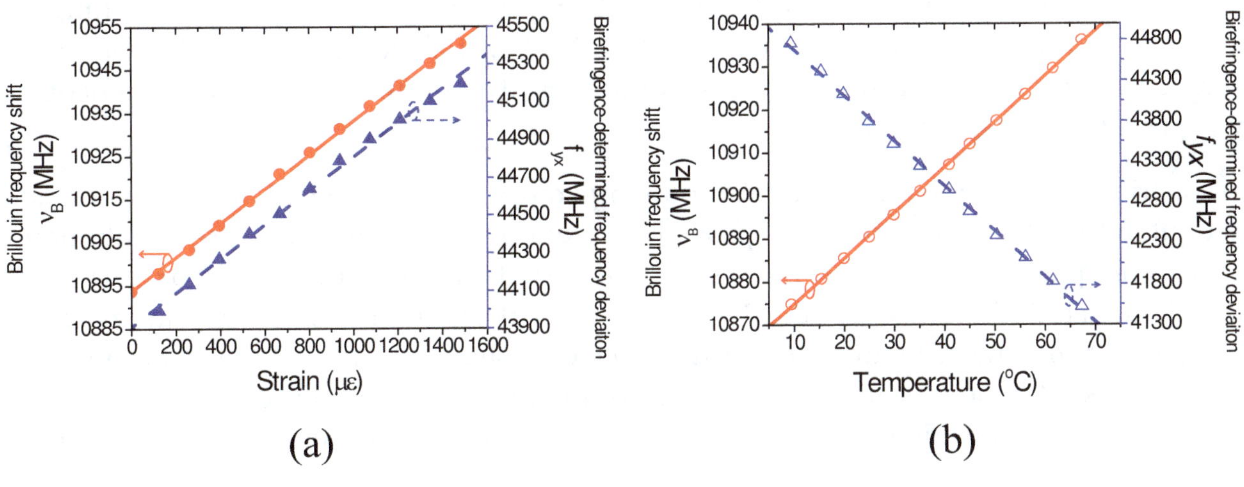

Figure 18. Brillouin frequency shift and birefringence-determined frequency deviation measured as functions of strain (a) and temperature (b). (After Ref. [89]; © 2009 OSA.)

coefficient and the temperature coefficient of the birefringence-determined frequency deviation, which can be deduced from Eqs. (40), (44), (45) and (46) as follows:

$$C_f^{\varepsilon} = +\Delta f_0 \cdot \frac{(\gamma_3 - \gamma_2)}{(\alpha_3 - \alpha_2)(T_{fic} - 25)}$$
$$C_f^{T} = -\Delta f_0 \cdot \frac{1}{(T_{fic} - 25)}, \tag{47}$$

where Δf_0 is the frequency deviation at 25°C and in loose condition.

By jointly considering the BFS (ν_B) and the frequency deviation (Δf), one can deduce the strain ($\Delta\varepsilon$) and temperature (ΔT) referred to Eq. (37):

$$\begin{pmatrix} \Delta\varepsilon \\ \Delta T \end{pmatrix} = \frac{1}{C_\nu^{\varepsilon} \cdot C_f^{T} - C_\nu^{T} \cdot C_f^{\varepsilon}} \begin{pmatrix} C_f^{T} & -C_\nu^{T} \\ -C_f^{\varepsilon} & C_\nu^{\varepsilon} \end{pmatrix} \begin{pmatrix} \nu_B\text{-}\nu_{B0} \\ \Delta f\text{-}\Delta f_0 \end{pmatrix}, \tag{48}$$

where Δf_0 and ν_{B0} are the frequency deviation and the BFS at room temperature and in loose state.

In physics, the two phenomena/quantities, i.e. the ν_B and Δf of the fiber, are inherently independent. In mathematics, C_f^T has a sign opposite to those of other three coefficients, so that the denominator ($C\nu\varepsilon\ C_f^T$-$C\nu^T\ C_f\varepsilon$) of Eq. (48) has a significant value. The experimental results are depicted in Fig. 17 and Fig. 18, which give the two groups of coefficients ($C\nu\varepsilon$=+0.03938 MHz/με and $C\nu^T$=+1.0580 MHz/°C; and $C_f\varepsilon$=+0.8995 MHz/με and C_f^T=-55.8134 MHz/ °C). Putting above strain/temperature coefficients into Eq. (48), and taking the standard errors of the measurement system ($\Delta\nu_B$=0.1 MHz and Δf_{yx}=4 MHz, respectively) into account, the accuracy of the discrimination is given as high as $\Delta\varepsilon$=± 3.1 με and ΔT=± 0.078 °C. Therefore, a complete discrimination of strain and temperature based on simultaneous measurement of the two quantities is ensured.

For distributed discrimination of strain and temperature, the localized BDG generation and readout in the PMF should be firstly proved to be effective. A correlation-based continuous-wave technique based on the BOCDA system [99] is used for random access and a pulse-based time-domain technique based on the BOTDA system [100] is employed for continuous access. It was found that the generation and readout waves based on the BOCDA system should be synchronously frequency-modulated because of the dispersion properties of all four waves (see Fig. 19) [112], including pump and probe waves, readout wave and acoustic wave (BDG as well).

The preliminary success of distributed discrimination of strain and temperature was realized by use of several lasers based on the BOCDA system [113] or the BOTDA system [114]. In [113], all pump, probe, and readout waves are synchronously modulated in frequency by sinusoidal functions to the two laser diodes. The measurement range of the distributed BGS and BDG is

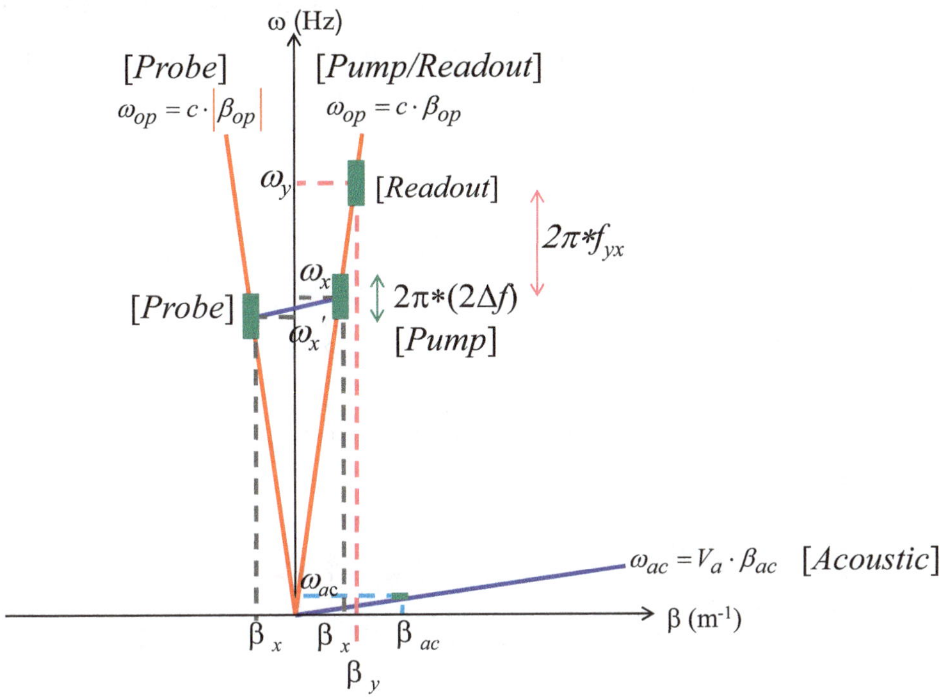

Figure 19. Dispersion properties of all optical (pump, probe and readout) and acoustic wave (Brillouin dynamic grating). (After Ref. [112]; © 2011 OSA.)

commonly given by the neighboring correlation peaks of the BOCDA system as defined in Eq. (35). Although the spatial resolution of the BGS measurement is still given by Eq. (34), that of the BDG reflection was thought to be determined by the BDG bandwidth (Δf_{yx}) as follows:

$$\Delta Z_{BDG} = \frac{c}{2nf_m} \cdot \frac{\Delta f_{yx}}{\pi \Delta f}. \tag{49}$$

The feasibility of distributed discrimination of strain and temperature was experimentally demonstrated with 10-cm spatial resolution. The f_m=12.429 MHz determines the nominal measurement range as d_m=8.35 m according to Eq. (35). For local BGS and BDG measurement, the Δf_B=1.5 GHz and Δf_D=10 GHz correspond to a nominal spatial resolution Δz_B=5 cm and Δz_D=8 cm [see Eq. (34) and Eq. (49)], respectively. As shown in Fig. 20(a), a ~8-m PMF sample is prepared, which consists of nine (A-I) cascaded fiber portions of 10-16 cm in length. The A, C, G and I portions were loosely laid at 25.1 °C for reference, while the B, D, F, and H portions were loosely inserted into a temperature-controlled water bath with 0.1-°C accuracy. The E portion was also inserted into the water bath and glued to a set of translation stages to load strain. The measured distribution of the changes of $\Delta \upsilon_B$ and Δf_{yx} are summarized in Figs. 20(b) and 20(c), respectively. Referred to the characterized coefficients in Fig. 18 and the cross-sensitivity matrix in Eq. (48), the deducted distribution of temperature and strain along the fiber is depicted in Figs. 20(d) and 20(e), which clearly shows the feasibility of distributed discrimination of strain and temperature. In [114], all pump, probe, and readout waves are pulsed in time domain; in turn, the BDG generation and readout as well as the BGS and BDG reflection are continuously localized by control of their relative delay towards the FUT and thus the local ν_B and Δf are detected for distributed discrimination of the strain and temperature responses.

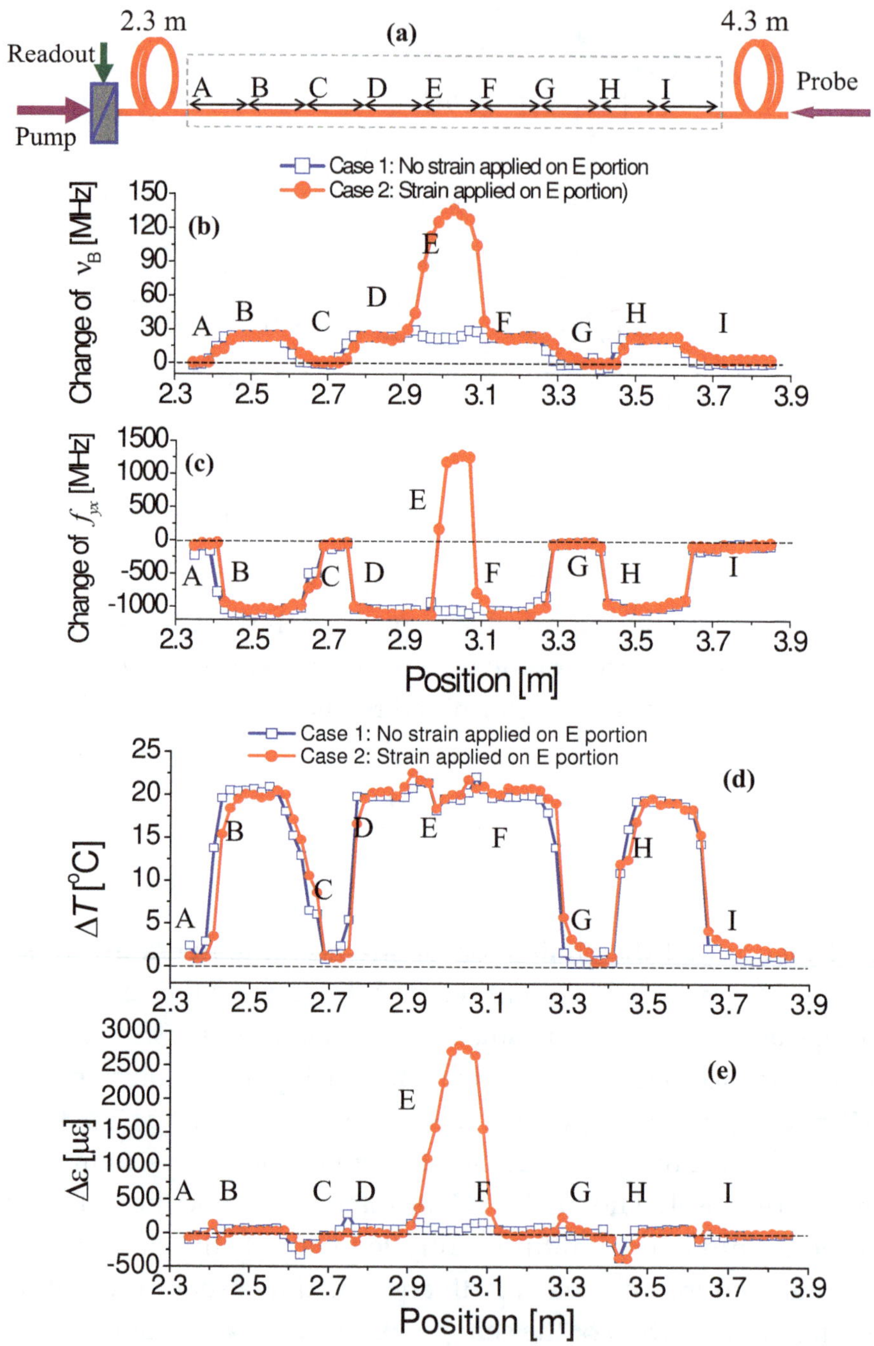

Figure 20. Preliminary experiment of distributed discrimination of strain and temperature based on two lasers modulated in frequency. (a) FUT configuration. Measured distribution of Brillouin frequency shift (b) and the birefringence-determined frequency deviation (c). Deduced distribution of temperature (d) and strain (e). (After Ref. [113]; © 2010 IEEE.)

One-laser-based Brillouin correlation-domain distributed discrimination system [112] by use of the sideband-generation technique was recently demonstrated to overcome the frequency fluctuation among the free-running lasers for pump, probe, and readout waves, and thus improve the accuracy of distributed discrimination of strain and temperature. Figure 21 represents the experimental setup. A 40-GHz intensity modulator (IM2) laid after the laser diode is driven by a radio frequency synthesizer (RF2 at ν_{RF2}) with a proper dc bias so as to generate double sidebands with suppressed carrier (DSB-SC). The optical filtering (FBG and tunable band-pass filter) is used to separate the two sidebands for the BDG generation and readout. By control of the RF1 (similar to Fig. 16), the BFS can be precisely measured and then fixed; by tuning of the RF2, the BDG reflection can be also precisely characterized; by simply change of the modulation frequency of the one laser diode, the location of the BGS and BDG can be swept for distributed measurement.

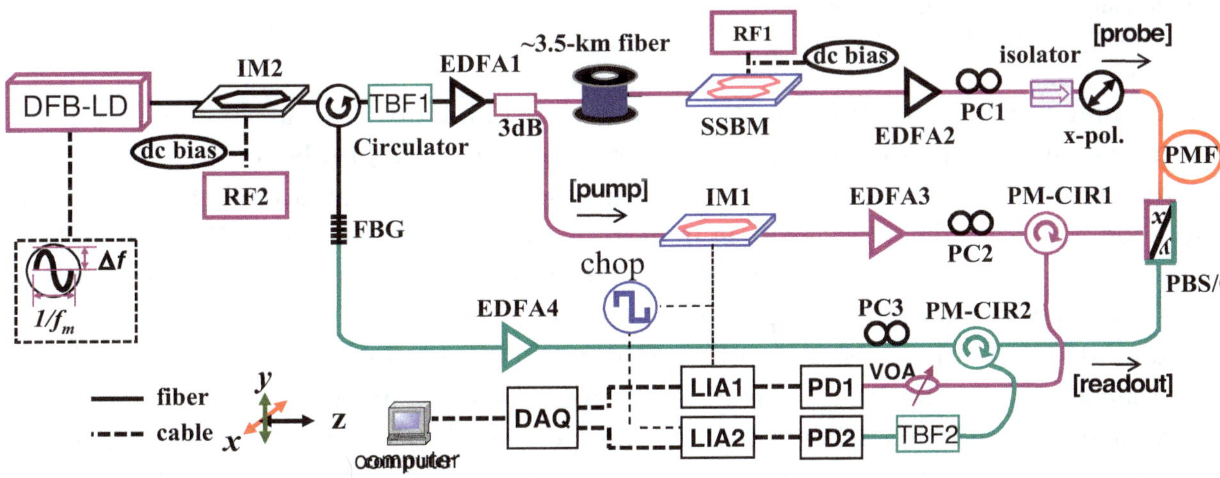

Figure 21. Experimental setup of the one-laser-based Brillouin correlation-domain distributed discrimination system. (After Ref. [112]; © 2011 OSA.)

Figure 22 shows the higher stability and accuracy (several MHz) of the one-laser scheme when compared to the two-laser scheme (several hundreds of MHz) both under no averaging process. Note that the one-laser scheme can also provide higher speed in the measurement of BGS and BDG and simpler measurement without sophisticated synchronization. Its distributed discrimination of strain and temperature was confirmed with the spatial resolution of ~10 cm and measurement range of ~5 m, which is depicted in Fig. 23 when the fiber was heated from 25 °C to 30 °C or/and the strain (ε=2000 $\mu\varepsilon$) was applied both at the location of 3.1 m. The measured results match well with the setting situation.

In order to overcome the tradeoff between the spatial resolution and measurement range always existing in the BOCDA system, a temporal gating [115] or a dual frequency modulation scheme [116] with a simple modification in Fig. 21 was used to elongate the measurement range of distributed discrimination of strain and temperature. For temporal gating scheme, the pulse modulation of RF2 makes the frequency-modulated pump, probe and readout waves optically pulsed in time and only one of the multiple correlation peaks are effectively generated in the

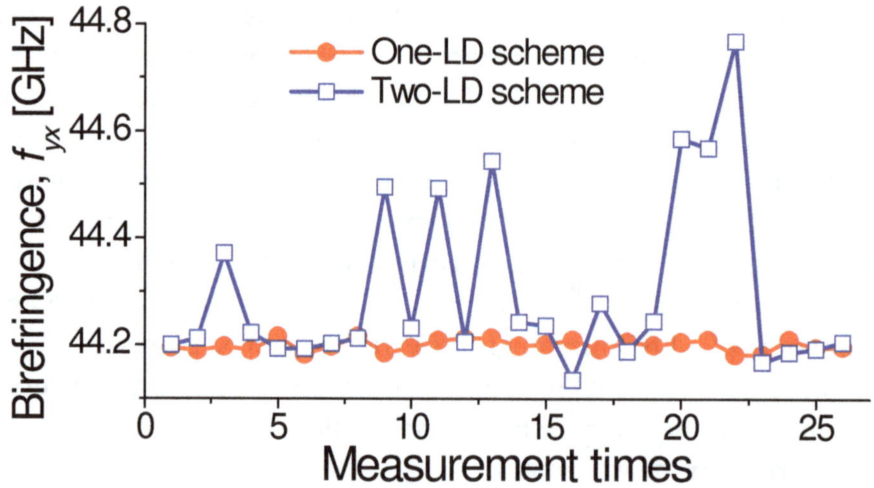

Figure 22. Comparison of the stability and accuracy of one-laser (solid dots) and two-laser (dashed squares) schemes of the BDG reflection. (After Ref. [112]; © 2011 OSA.)

FUT. For dual frequency modulation scheme, two sinusoidal functions are combined together to simultaneously modulate the optical frequencies of the pump, probe and readout waves. The greater modulation frequency ensures the higher spatial resolution while the lower modulation frequency realizes the longer measurement range. A 20 times [115] or 7 times [116] enlargement of the ratio between the measurement range and the spatial resolution was successfully demonstrated. It is expectable to achieve Brillouin optical correlation-domain distributed discrimination of strain and temperature having both a higher spatial resolution (better than 10 cm) and a longer measurement range (better than 1,000 m) by combining the dual frequency modulation scheme with the temporal gating scheme, which is now under study. Most recently, an apodization method under the assistance of intensity modulation was proposed to suppress the sidelobe of SOCF and enhance the spatial resolution of the strain-temperature discrimination by 4.5 times [117].

5.3. System improvement of sensing techniques

Many works have been involved in improving the system performance of BOTDA/BOTDR and BOCDA/BOCDR in terms of spatial resolution, measurement range, sensing speed and accuracy. In 1995, Bao *et al.* developed a Brillouin-loss-based BOTDA [118] by reversing the functions of the pulse laser and CW light. In other words, a strong CW light acts as a Brillouin pump wave and a pulse light with a scanned down-shifted frequency from that of the CW light works as a probe wave [119]. After monitoring the optical loss profile of the pump wave due to Brillouin interaction between the two light waves as a function of time or position along the fiber, a numerical signal processing of the poor SNR was used to achieve ~25 cm spatial resolution with a strain resolution of ~40 $\mu\varepsilon$ [120]. A pulse-pre-pump BOTDA (called PPP-BOTDA) was proposed to realize cm-order spatial resolution by using a wide pulse (larger than 10 ns) followed by a narrow pulse (smaller than 1 ns) [121], which is in principle similar to the BOTDA with a pulse generated by a finite extinction ratio [122]. A dark-pulse-based BOTDA was later presented to hopefully obtain 2-cm spatial resolution [123, 124], which

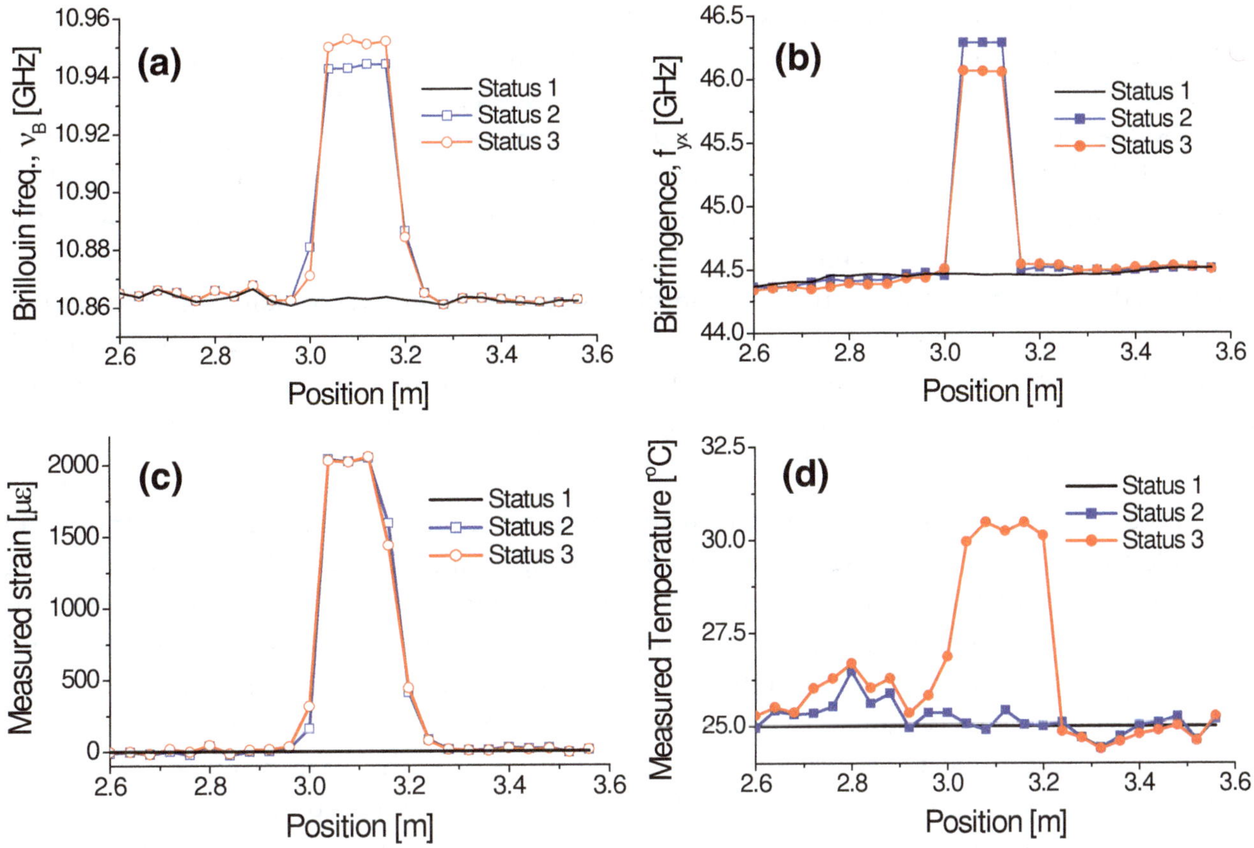

Figure 23. Experimental results of high-accuracy distributed discrimination of strain and temperature based on one-laser scheme. Distribution of Brillouin frequency shift (a), the birefringence-determined frequency deviation (b), strain (c), and temperature (d). (After Ref. [112]; © 2011 OSA.)

suffers a neglectful influence of the acoustic lifetime (~10 ns) but an experimental difficulty of high-qualify dark pulse and a scarified measurement range due to the pump depletion of the dc base. A new scheme by combination of the PPP-BOTDA and dark-pulse-based BOTDA [125] was demonstrated to achieve higher spatial resolution as well as better frequency resolution. Taking the similar principle of the PPP-BOTDA technique, a BDG-based BOTDA was proposed to obtain cm-order or sub-cm-order spatial resolution [126-128]. The only difference lies on the fact that the Brillouin interaction in the BDG-based BOTDA is generated by a long pulse along one principal polarization state (the BDG generation process) and detected by a short pulse along the orthogonal polarization state (the BDG readout process). Another type of the modified BOTDA with higher spatial resolution of less than 1 meter is based on a group of pulses, called differential pulse-width pair BOTDA (DPP-BOTDA) [129-132] or Brillouin echo BOTDA [133, 134]. In DPP-BOTDA, a pair of pulses with a small difference of the pulse widths are successively launched into the FUT; the Brillouin interaction is recorded for twice and a subtraction is performed to achieve the sensing trace with the high spatial resolution determined by the small pulse-width difference. The spatial resolution of BOTDR has been also more or less improved by an experimental optimization or signal processing process [135-137].

Although BOTDA and BOTDR are excellent for long sensing range (such as kilometers or tens of km), they still suffer the physical limitation of maximum range due to the nature of fiber loss and/or the Brillouin depletion effect. There are two typical methods, i.e. Raman-assisted BOTDR[138-140] or Raman-assisted BOTDA[141-145] and coded BOTDR[146] or coded BOTDA[147-151], to improve the poor SNR and achieve a very long sensing range. The best performance of the sensing range (longer than 120 km) [152, 153] with an acceptable spatial resolution (1 m or 2 m) has been renovated by combination of Raman assistance and coding although the system becomes extremely complicated. Most recently, specially-designed EDFA repeaters were used to extend the sensing measurements of BOTDA to more than 300 km [154]. Some efforts were also made to study the influence of Brillouin depletion on the maximum range of BOTDA [41] and to avoid it to some extent by use of Stokes together with anti-Stokes wave as Brillouin probe [155, 156].

BOCDA and BOCDR systems have natural advantages of high spatial resolution without any dependence on the acoustic lifetime and random programmable accessibility of the sensing location. Except for the great innovation of Brillouin optical correlation-domain distributed discrimination of strain and temperature introduced in **Section 5.2**, advances in BOCDA and BOCDR systems have also boosted in the past decade. The polarization disturbance along the FUT has been effectively solved by use of the polarization diversity scheme to the BOCDA system [157]. A complicated double-lock-in detection was proposed to improve the SNR of the BOCDA system [69, 158] although a modified lock-in detection based on variable chopping frequency [159] or a simplified but equivalent BOCDA system based on combination of Brillouin gain and loss [34] was later proposed. The existence of a big noise floor originated from the uncorrelated locations strongly limits the maximum strain or temperature change to be detected, which has been eliminated by use of intensity modulation for SOCF apodization [160, 161] or differential measurement scheme based on external phase modulation [162]. The measurement range of BOCDA [116, 163, 164] or BOCDR [165, 166] was extended by use of temporal gating or double frequency modulation scheme, respectively. Besides, combination of time-domain and correlation domain techniques [167, 168] has been proposed to enlarge the measurement range of the BOCDA [169] based on external phase modulation. The distributed sensing speed with cm-order spatial resolution [170-172] has been substantially increased to several Hertz along the entire FUT by optimizing the position sweeping and the BGS mapping although the local sensing speed of the BOCDA [173-175] or BOCDR [176] was well-known to be high at the random-accessed sensing location.

6. Conclusions

We have presented an essential overview of Brillouin scattering in optical fibers and Brillouin based distributed optical fiber sensors. Started from the basic principle of Brillouin scattering in optical fibers, the basic mechanism of Brillouin based distributed optical fiber sensors (linear dependence of Brillouin frequency shift on strain and temperature) and the two different groups of Brillouin based distributed optical fiber sensors (time domain: BOTDA/BOTDR; correlation domain: BOCDA/BOCDR) were described in detail. The difficulties and challenges of how to simultaneously sense strain and temperature were demonstrated and the physical limitation of the sensing abilities (spatial resolution, measurement range, accuracy etc.) were introduced, respectively. Finally, we summarized recent advances of this field towards the solutions to those difficulties and challenges.

It is valuable to address that Brillouin based distributed optical fiber sensors are nowadays in a high technical level, which have been attracting industrial companies to commercialize for structural health monitoring in civil structures, aerospace, energy (gas, oil) pipeline, and engineers (power supply).

Acknowledgements

This work was partially supported by National Natural Science Foundation of China (Grant Nos. 61007052 and 61127016), Shanghai Pujiang Program (Grant No. 12PJ1405600), and by the State Key Lab Project of Shanghai Jiao Tong University under Grant GKZD030033. Professor Kazuo Hotate at the University of Tokyo and Professor Zuyuan He at Shanghai Jiao Tong University are gratefully acknowledged for their contributions in many relevant works presented in this chapter.

Author details

Weiwen Zou*, Xin Long and Jianping Chen

*Address all correspondence to: wzou@sjtu.edu.cn

State Key Laboratory of Advanced Optical Communication Systems and Networks, Department of Electronic Engineering, Shanghai Jiao Tong University, Shanghai, China

References

[1] G. P. Agrawal, *Nonlinear Fiber Optics,* 5th ed. (Academic Press, 2012).

[2] E. Ippen and R. Stolen, "Stimulated Brillouin scattering in optical fibers," Appl. Phys. Lett. 21, 539-541 (1972).

[3] R. Shelby, M. Levenson, and P. Bayer, "Resolved forward Brillouin scattering in optical fibers," Phys. Rev. Lett. 54, 939-942 (1985).

[4] Y. Mizuno and K. Nakamura, "Brillouin Scattering in Polymer Optical Fibers: Fundamental Properties and Potential Use in Sensors," Polymers 3, 886-898 (2011).

[5] A. L. Gaeta and R. W. Boyd, "Stochastic dynamics of stimulated Brillouin scattering in an optical fiber," Phys. Rev. A 44, 3205-3209 (1991).

[6] R. B. Jenkins, R. M. Sova, and R. I. Joseph, "Steady-state noise analysis of spontaneous and stimulated Brillouin scattering in optical fibers," J. Lightwave Technol. 25, 763-770 (2007).

[7] A. S. Pine, "Brillouin Scattering Study of Acoustic Attenuation in Fused Quartz," Physical Review 185, 1187-1193 (1969).

[8] W. Zou, Z. He, and K. Hotate, "Two-dimensional finite-element modal analysis of Brillouin gain spectra in optical fibers," Photonics Technology Letters, IEEE 18, 2487-2489 (2006).

[9] A. Kobyakov, M. Sauer, and D. Chowdhury, "Stimulated Brillouin scattering in optical fibers," Advances in optics and photonics 2, 1-59 (2010).

[10] A. Kobyakov, S. Kumar, D. Q. Chowdhury, A. B. Ruffin, M. Sauer, S. R. Bickham, and R. Mishra, "Design concept for optical fibers with enhanced SBS threshold," Opt. Express 13, 5338-5346 (2005).

[11] Y. Koyamada, S. Sato, S. Nakamura, H. Sotobayashi, and W. Chujo, "Simulating and designing brillouin gain spectrum in single-mode fibers," J. Lightwave Technol. 22, 631-639 (2004).

[12] W. Zou, Z. He, A. D. Yablon, and K. Hotate, "Dependence of Brillouin frequency shift in optical fibers on draw-induced residual elastic and inelastic strains," Photonics Technology Letters, IEEE 19, 1389-1391 (2007).

[13] W. Zou, Z. Y. He, and K. Hotate, "Acoustic modal analysis and control in w-shaped triple-layer optical fibers with highly-germanium-doped core and F-doped inner cladding," Opt. Express 16, 10006-10017 (2008).

[14] L. Tartara, C. Codemard, J.-N. Maran, R. Cherif, and M. Zghal, "Full modal analysis of the Brillouin gain spectrum of an optical fiber," Opt. Commun. 282, 2431-2436 (2009).

[15] B. Ward and J. Spring, "Finite element analysis of Brillouin gain in SBS-suppressing optical fibers with non-uniform acoustic velocity profiles," Opt. Express 17, 15685-15699 (2009).

[16] Y. S. Mamdem, X. Phéron, F. Taillade, Y. Jaoüen, R. Gabet, V. Lanticq, G. Moreau, A. Boukenter, Y. Ouerdane, and S. Lesoille, "Two-dimensional FEM analysis of Brillouin Gain Spectra in acoustic guiding and antiguiding single mode optical fibers," in *COMSOL Conference*, 2010),

[17] S. Dasgupta, F. Poletti, S. Liu, P. Petropoulos, D. J. Richardson, L. Gruner-Nielsen, and S. Herstrøm, "Modeling Brillouin Gain Spectrum of solid and microstructured

optical fibers using a finite element method," Lightwave Technology, Journal of 29, 22-30 (2011).

[18] X. Qian, B. Han, and Q. Wang, "Numerical Research on Gain Spectrum of Stimulated Brillouin Scattering in Photonics Crystal Fiber," Instrumentation Science & Technology 41, 175-186 (2013).

[19] L. Dong, "Formulation of a complex mode solver for arbitrary circular acoustic waveguides," J. Lightwave Technol. 28, 3162-3175 (2010).

[20] P. D. Dragic and B. G. Ward, "Accurate modeling of the intrinsic Brillouin linewidth via finite-element analysis," Photonics Technology Letters, IEEE 22, 1698-1700 (2010).

[21] C. G. Carlson, R. B. Ross, J. M. Schafer, J. B. Spring, and B. G. Ward, "Full vectorial analysis of Brillouin gain in random acoustically microstructured photonic crystal fibers," Phys. Rev. B 83, 235110 (2011).

[22] Y. Okawachi, M. S. Bigelow, J. E. Sharping, Z. M. Zhu, A. Schweinsberg, D. J. Gauthier, R. W. Boyd, and A. L. Gaeta, "Tunable all-optical delays via Brillouin slow light in an optical fiber," Phys. Rev. Lett. 94(2005).

[23] K. Y. Song, M. G. Herraez, and L. Thevenaz, "Observation of pulse delaying and advancement in optical fibers using stimulated Brillouin scattering," Opt. Express 13, 82-88 (2005).

[24] N. Shibata, K. Okamoto, and Y. Azuma, "Longitudinal acoustic modes and Brillouin-gain spectra for GeO< sub> 2</sub>-doped-core single-mode fibers," JOSA B 6, 1167-1174 (1989).

[25] M. Nikles, L. Thevenaz, and P. A. Robert, "Brillouin gain spectrum characterization in single-mode optical fibers," Lightwave Technology, Journal of 15, 1842-1851 (1997).

[26] A. Yeniay, J.-M. Delavaux, and J. Toulouse, "Spontaneous and Stimulated Brillouin Scattering Gain Spectra in Optical Fibers," J. Lightwave Technol. 20, 1425 (2002).

[27] W. Zou, Z. He, and K. Hotate, "Experimental study of Brillouin scattering in fluorine-doped single-mode optical fibers," Opt. Express 16, 18804-18812 (2008).

[28] R. G. Smith, "Optical Power Handling Capacity of Low Loss Optical Fibers as Determined by Stimulated Raman and Brillouin Scattering," Appl. Optics 11, 2489-2494 (1972).

[29] V. I. Kovalev and R. G. Harrison, "Threshold for stimulated Brillouin scattering in optical fiber," Opt. Express 15, 17625-17630 (2007).

[30] N. Rowell, P. Thomas, H. Van Driel, and G. Stegeman, "Brillouin spectrum of single-mode optical fibers," Appl. Phys. Lett. 34, 139-141 (1979).

[31] S. M. Maughan, H. H. Kee, and T. P. Newson, "Simultaneous distributed fibre temperature and strain sensor using microwave coherent detection of spontaneous Brillouin backscatter," Measurement Science and Technology 12, 834 (2001).

[32] Y. Koyamada, "Proposal and simulation of double-pulse brillouin optical time-domain analysis for measuring distributed strain and temperature with cm spatial resolution in km-long fiber," IEICE Trans. Commun. E90B, 1810-1815 (2007).

[33] X. Bao, D. J. Webb, and D. A. Jackson, "32-km distributed temperature sensor based on Brillouin loss in an optical fiber," Opt. Lett. 18, 1561-1563 (1993).

[34] W. Zou, C. J. Jin, and J. P. Chen, "Distributed Strain Sensing Based on Combination of Brillouin Gain and Loss Effects in Brillouin Optical Correlation Domain Analysis," Appl. Phys. Express 5(2012).

[35] K. Higuma, S. Oikawa, Y. Hashimoto, H. Nagata, and M. Izutsu, "X-cut lithium niobate optical single-sideband modulator," Electron. Lett. 37, 515-516 (2001).

[36] A. Loayssa, R. Hernández, D. Benito, and S. Galech, "Characterization of stimulated Brillouin scattering spectra by use of optical single-sideband modulation," Opt. Lett. 29, 638-640 (2004).

[37] W. Zou, Z. Y. He, M. Kishi, and K. Hotate, "Stimulated Brillouin scattering and its dependences on strain and temperature in a high-delta optical fiber with F-doped depressed inner cladding," Opt. Lett. 32, 600-602 (2007).

[38] V. I. Kovalev and R. G. Harrison, "Observation of inhomogeneous spectral broadening of stimulated Brillouin scattering in an optical fiber," Phys. Rev. Lett. 85, 1879 (2000).

[39] Y. Takushima and K. Kikuchi, "Spectral gain hole burning and modulation instability in a Brillouin fiber amplifier," Opt. Lett. 20, 34-36 (1995).

[40] L. Stépien, S. Randoux, and J. Zemmouri, "Origin of spectral hole burning in Brillouin fiber amplifiers and generators," Phys. Rev. A 65, 053812 (2002).

[41] L. Thévenaz, S. F. Mafang, and J. Lin, "Effect of pulse depletion in a Brillouin optical time-domain analysis system," Opt. Express 21, 14017-14035 (2013).

[42] X. Long, W. Zou, H. Li, and J. Chen, "Critical condition for spectrum distortion of pump–probe-based stimulated Brillouin scattering in an optical fiber," Appl. Phys. Express 7, 082501 (2014).

[43] T. Horiguchi and M. Tateda, "BOTDA-nondestructive measurement of single-mode optical fiber attenuation characteristics using Brillouin interaction: Theory," Lightwave Technology, Journal of 7, 1170-1176 (1989).

[44] T. Horiguchi, T. Kurashima, and M. Tateda, "Tensile strain dependence of Brillouin frequency shift in silica optical fibers," Photonics Technology Letters, IEEE 1, 107-108 (1989).

[45] T. Kurashima, T. Horiguchi, and M. Tateda, "Thermal effects of Brillouin gain spectra in single-mode fibers," Photonics Technology Letters, IEEE 2, 718-720 (1990).

[46] http://www.itu.int/rec/T-REC-G/en.

[47] C. K. Jen, C. Neron, A. Shang, K. Abe, L. Bonnell, and J. Kushibiki, "Acoustic characterization of silica glasses," Journal of the American Ceramic Society 76, 712-716 (1993).

[48] S. P. Timoshenko and J. Goodier, *Theory of elasticity* (2011).

[49] W. Zou, Z. He, and K. Hotate, "Investigation of Strain-and Temperature-Dependences of Brillouin Frequency Shifts in GeO< sub> 2</sub>-Doped Optical Fibers," J. Lightwave Technol. 26, 1854-1861 (2008).

[50] R. W. Tkach, A. R. Chraplyvy, and R. Derosier, "Spontaneous Brillouin scattering for single-mode optical-fibre characterisation," Electron. Lett. 22, 1011-1013 (1986).

[51] N. Shibata, R. G. Waarts, and R. P. Braun, "Brillouin-gain spectra for single-mode fibers having pure-silica, GeO 2-doped, and P 2O5-doped cores," Opt. Lett. 12, 269-271 (1987).

[52] T. Horiguchi, T. Kurashima, and M. Tateda, "A technique to measure distributed strain in optical fibers," Photonics Technology Letters, IEEE 2, 352-354 (1990).

[53] T. Kurashima, T. Horiguchi, H. Izumita, S.-i. Furukawa, and Y. Koyamada, "Brillouin optical-fiber time domain reflectometry," IEICE Trans. Commun. 76, 382-390 (1993).

[54] T. Horiguchi, T. Kurashima, and Y. Koyamada, "Measurement of temperature and strain distribution by Brillouin frequency shift in silica optical fibers," in *Fibers' 92*, (International Society for Optics and Photonics, 1993), 2-13.

[55] M. Niklès, L. Thévenaz, and P. A. Robert, "Simple distributed fiber sensor based on Brillouin gain spectrum analysis," Opt. Lett. 21, 758-760 (1996).

[56] K. Shimizu, T. Horiguchi, Y. Koyamada, and T. Kurashima, "Coherent self-heterodyne Brillouin OTDR for measurement of Brillouin frequency shift distribution in optical fibers," Lightwave Technology, Journal of 12, 730-736 (1994).

[57] D. Garcus, T. Gogolla, K. Krebber, and F. Schliep, "Brillouin optical-fiber frequency-domain analysis for distributed temperature and strain measurements," Lightwave Technology, Journal of 15, 654-662 (1997).

[58] D. Garus, K. Krebber, F. Schliep, and T. Gogolla, "Distributed sensing technique based on Brillouin optical-fiber frequency-domain analysis," Opt. Lett. 21, 1402-1404 (1996).

[59] K. Hotate and T. Hasegawa, "Measurement of Brillouin Gain Spectrum Distribution along an Optical Fiber Using a Correlation-Based Technique--Proposal, Experiment and Simulation," IEICE Trans. Electron. 83, 405-412 (2000).

[60] K. Hotate and M. Tanaka, "Distributed fiber Brillouin strain sensing with 1-cm spatial resolution by correlation-based continuous-wave technique," Photonics Technology Letters, IEEE 14, 179-181 (2002).

[61] Y. Mizuno, W. Zou, Z. He, and K. Hotate, "Proposal of Brillouin optical correlation-domain reflectometry (BOCDR)," Opt. Express 16, 12148-12153 (2008).

[62] K.-Y. Song and K. Hotate, "Brillouin optical correlation domain analysis in linear configuration," Photonics Technology Letters, IEEE 20, 2150-2152 (2008).

[63] W. Zou, Z. He, and K. Hotate, "Single-End-Access Correlation-Domain Distributed Fiber-Optic Sensor Based on Stimulated Brillouin Scattering," J. Lightwave Technol. 28, 2736-2742 (2010).

[64] Y. Mizuno, W. W. Zou, Z. Y. He, and K. Hotate, "Operation of Brillouin Optical Correlation-Domain Reflectometry: Theoretical Analysis and Experimental Validation," J. Lightwave Technol. 28, 3300-3306 (2010).

[65] K. Hotate and O. Kamatani, "Optical coherence domain reflectometry by synthesis of coherence function," Lightwave Technology, Journal of 11, 1701-1710 (1993).

[66] K. Hotate and T. Okugawa, "Optical information processing by synthesis of the coherence function," Lightwave Technology, Journal of 12, 1247-1255 (1994).

[67] R. Lang and K. Kobayashi, "External optical feedback effects on semiconductor injection laser properties," Quantum Electronics, IEEE Journal of 16, 347-355 (1980).

[68] Y. Mizuno, Z. He, and K. Hotate, "One-end-access high-speed distributed strain measurement with 13-mm spatial resolution based on Brillouin optical correlation-domain reflectometry," Photonics Technology Letters, IEEE 21, 474-476 (2009).

[69] K. Y. Song, Z. Y. He, and K. Hotate, "Distributed strain measurement with millimeter-order spatial resolution based on Brillouin optical correlation domain analysis," Opt. Lett. 31, 2526-2528 (2006).

[70] K. Hotate and S. S. Ong, "Distributed dynamic strain measurement using a correlation-based Brillouin sensing system," Photonics Technology Letters, IEEE 15, 272-274 (2003).

[71] K.-Y. Song and K. Hotate, "Distributed fiber strain sensor with 1 kHz sampling rate based on Brillouin optical correlation domain analysis," in *Optics East 2007*, (International Society for Optics and Photonics, 2007), 67700J-67700J-67708.

[72] M. Alahbabi, Y. Cho, and T. Newson, "Simultaneous temperature and strain measurement with combined spontaneous Raman and Brillouin scattering," Opt. Lett. 30, 1276-1278 (2005).

[73] G. Bolognini, M. A. Soto, and F. Di Pasquale, "Simultaneous distributed strain and temperature sensing based on combined Raman–Brillouin scattering using Fabry–Perot lasers," Measurement Science and Technology 21, 094025 (2010).

[74] M. Taki, A. Signorini, C. Oton, T. Nannipieri, and F. Di Pasquale, "Hybrid Raman/ Brillouin-optical-time-domain-analysis-distributed optical fiber sensors based on cyclic pulse coding," Opt. Lett. 38, 4162-4165 (2013).

[75] D.-P. Zhou, W. Li, L. Chen, and X. Bao, "Distributed Temperature and Strain Discrimination with Stimulated Brillouin Scattering and Rayleigh Backscatter in an Optical Fiber," Sensors 13, 1836-1845 (2013).

[76] C. Lee, P. Chiang, and S. Chi, "Utilization of a dispersion-shifted fiber for simultaneous measurement of distributed strain and temperature through Brillouin frequency shift," Photonics Technology Letters, IEEE 13, 1094-1096 (2001).

[77] L. F. Zou, X. Y. Bao, V. S. Afshar, and L. Chen, "Dependence of the Brillouin frequency shift on strain and temperature in a photonic crystal fiber," Opt. Lett. 29, 1485-1487 (2004).

[78] T. Parker, M. Farhadiroushan, V. Handerek, and A. Rogers, "Temperature and strain dependence of the power level and frequency of spontaneous Brillouin scattering in optical fibers," Opt. Lett. 22, 787-789 (1997).

[79] M. Alahbabi, Y. T. Cho, and T. P. Newson, "Comparison of the methods for discriminating temperature and strain in spontaneous Brillouin-based distributed sensors," Opt. Lett. 29, 26-28 (2004).

[80] M. Belal and T. P. Newson, "Experimental Examination of the Variation of the Spontaneous Brillouin Power and Frequency Coefficients Under the Combined Influence of Temperature and Strain," J. Lightwave Technol. 30, 1250-1255 (2012).

[81] X. Y. Bao, Q. R. Yu, and L. Chen, "Simultaneous strain and temperature measurements with polarization-maintaining fibers and their error analysis by use of a distributed Brillouin loss system," Opt. Lett. 29, 1342-1344 (2004).

[82] H. Naruse and M. Tateda, "Trade-off between the spatial and the frequency resolutions in measuring the power spectrum of the Brillouin backscattered light in an optical fiber," Appl. Optics 38, 6516-6521 (1999).

[83] S.-B. Cho, Y.-G. Kim, J.-S. Heo, and J.-J. Lee, "Pulse width dependence of Brillouin frequency in single mode optical fibers," Opt. Express 13, 9472-9479 (2005).

[84] M. A. Soto and L. Thévenaz, "Modeling and evaluating the performance of Brillouin distributed optical fiber sensors," Opt. Express 21, 31347-31366 (2013).

[85] A. Minassian, G. J. Crofts, and M. J. Damzen, "Spectral filtering of gain gratings and spectral evolution of holographic laser oscillators," Quantum Electronics, IEEE Journal of 36, 802-809 (2000).

[86] X. Fan, Z. He, Y. Mizuno, and K. Hotate, "Bandwidth-adjustable dynamic grating in erbium-doped fiber by synthesis of optical coherence function," Opt. Express 13, 5756-5761 (2005).

[87] R. Elsner, R. Ullmann, A. Heuer, R. Menzel, and M. Ostermeyer, "Two-dimensional modeling of transient gain gratings in saturable gain media," Opt. Express 20, 6887-6896 (2012).

[88] K.-Y. Song and K. Hotate, "All-optical dynamic grating generation based on Brillouin scattering in polarization maintaining fiber," in *19th International Conference on Optical Fibre Sensors*, (International Society for Optics and Photonics, 2008), 70043T-70043T-70044.

[89] W. Zou, Z. He, and K. Hotate, "Complete discrimination of strain and temperature using Brillouin frequency shift and birefringence in a polarization-maintaining fiber," Opt. Express 17, 1248-1255 (2009).

[90] Y. Dong, L. Chen, and X. Bao, "Truly distributed birefringence measurement of polarization-maintaining fibers based on transient Brillouin grating," Opt. Lett. 35, 193-195 (2010).

[91] S. Li, M.-J. Li, and R. S. Vodhanel, "All-optical Brillouin dynamic grating generation in few-mode optical fiber," Opt. Lett. 37, 4660-4662 (2012).

[92] K. Y. Song, "Operation of Brillouin dynamic grating in single-mode optical fibers," Opt. Lett. 36, 4686-4688 (2011).

[93] W. Zou and J. P. Chen, "All-optical generation of Brillouin dynamic grating based on multiple acoustic modes in a single-mode dispersion-shifted fiber," Opt. Express 21, 14771-14779 (2013).

[94] R. Pant, E. Li, C. G. Poulton, D.-Y. Choi, S. Madden, B. Luther-Davies, and B. J. Eggleton, "Observation of Brillouin dynamic grating in a photonic chip," Opt. Lett. 38, 305-307 (2013).

[95] K. O. Hill and G. Meltz, "Fiber Bragg grating technology fundamentals and overview," Lightwave Technology, Journal of 15, 1263-1276 (1997).

[96] W. Zou and J. Chen, "Spectral Analysis of Brillouin Dynamic Grating Based on Heterodyne Detection," Appl. Phys. Express 6, 122503 (2013).

[97] D. P. Zhou, L. Chen, and X. Y. Bao, "Polarization-decoupled four-wave mixing based on stimulated Brillouin scattering in a polarization-maintaining fiber," J. Opt. Soc. Am. B-Opt. Phys. 30, 821-828 (2013).

[98] K. Y. Song, K. Lee, and S. B. Lee, "Tunable optical delays based on Brillouin dynamic grating in optical fibers," Opt. Express 17, 10344-10349 (2009).

[99] W. Zou, Z. He, K.-Y. Song, and K. Hotate, "Correlation-based distributed measurement of a dynamic grating spectrum generated in stimulated Brillouin scattering in a polarization-maintaining optical fiber," Opt. Lett. 34, 1126-1128 (2009).

[100] K. Y. Song, W. Zou, Z. He, and K. Hotate, "Optical time-domain measurement of Brillouin dynamic grating spectrum in a polarization-maintaining fiber," Opt. Lett. 34, 1381-1383 (2009).

[101] Y. Antman, N. Primerov, J. Sancho, L. Thévenaz, and A. Zadok, "Localized and stationary dynamic gratings via stimulated Brillouin scattering with phase modulated pumps," Opt. Express20 (7), 7807-7821 (2012).

[102] M. Santagiustina and L. Ursini, "Dynamic Brillouin gratings permanently sustained by chaotic lasers," Opt. Lett. 37, 893-895 (2012).

[103] Y. Antman, L. Yaron, T. Langer, M. Tur, N. Levanon, and A. Zadok, "Experimental demonstration of localized Brillouin gratings with low off-peak reflectivity established by perfect Golomb codes," Opt. Lett. 38, 4701-4704 (2013).

[104] Y. Antman, N. Primerov, J. Sancho, L. Thevenaz, and A. Zadok, "Long variable delay and distributed sensing using stationary and localized Brillouin dynamic gratings," in *Optical Fiber Communication Conference*, (Optical Society of America, 2012),

[105] Y. Antman, N. Primerov, J. Sancho, L. Thévenaz, and A. Zadok, "Variable delay using stationary and localized Brillouin dynamic gratings," in *SPIE OPTO*, (International Society for Optics and Photonics, 2012), 82730C-82730C-82738.

[106] S. Chin and L. Thevenaz, "Tunable photonic delay lines in optical fibers," Laser Photon. Rev. 6, 724-738 (2012).

[107] J. Sancho, N. Primerov, S. Chin, Y. Antman, A. Zadok, S. Sales, and L. Thévenaz, "Tunable and reconfigurable multi-tap microwave photonic filter based on dynamic Brillouin gratings in fibers," Opt. Express20 (6), 6157-6162 (2012).

[108] H. G. Winful, "Chirped Brillouin dynamic gratings for storing and compressing light," Opt. Express 21, 10039-10047 (2013).

[109] L. Ursini and M. Santagiustina, "Applications of the Dynamic Brillouin Gratings to Ultrawideband Communications," IEEE Photonics Technol. Lett. 25, 1347-1349 (2013).

[110] M. Santagiustina, S. Chin, N. Primerov, L. Ursini, and L. Thévenaz, "All-optical signal processing using dynamic Brillouin gratings," Scientific reports 3(2013).

[111] K. Chiang, D. Wong, and P. Chu, "Strain-induced birefringence in a highly birefringent optical fibre," Electron. Lett. 26, 1344-1346 (1990).

[112] W. Zou, Z. He, and K. Hotate, "One-laser-based generation/detection of Brillouin dynamic grating and its application to distributed discrimination of strain and temperature," Opt. Express 19, 2363-2370 (2011).

[113] W. Zou, Z. He, and K. Hotate, "Demonstration of Brillouin distributed discrimination of strain and temperature using a polarization-maintaining optical fiber," Photonics Technology Letters, IEEE 22, 526-528 (2010).

[114] Y. Dong, L. Chen, and X. Bao, "High-spatial-resolution time-domain simultaneous strain and temperature sensor using Brillouin scattering and birefringence in a polarization-maintaining fiber," Photonics Technology Letters, IEEE 22, 1364-1366 (2010).

[115] R. K. Yamashita, W. Zou, Z. He, and K. Hotate, "Measurement range elongation based on temporal gating in Brillouin optical correlation domain distributed simultaneous sensing of strain and temperature," Photonics Technology Letters, IEEE 24, 1006-1008 (2012).

[116] W. Zou, Z. He, and K. Hotate, "Range Elongation of Distributed Discrimination of Strain and Temperature in Brillouin Optical Correlation-Domain Analysis Based on Dual Frequency Modulations," Sensors Journal, IEEE 14, 244-248 (2014).

[117] R. K. Yamashita, H. Zuyuan, and K. Hotate, "Spatial Resolution Improvement in Correlation Domain Distributed Measurement of Brillouin Grating," Photonics Technology Letters, IEEE 26, 473-476 (2014).

[118] X. Bao, J. Dhliwayo, N. Heron, D. J. Webb, and D. A. Jackson, "Experimental and theoretical studies on a distributed temperature sensor based on Brillouin scattering," Lightwave Technology, Journal of 13, 1340-1348 (1995).

[119] J. Smith, A. Brown, M. DeMerchant, and X. Bao, "Pulse width dependance of the Brillouin loss spectrum," Opt. Commun. 168, 393-398 (1999).

[120] A. W. Brown, M. D. DeMerchant, X. Bao, and T. W. Bremner, "Spatial resolution enhancement of a Brillouin-distributed sensor using a novel signal processing method," J. Lightwave Technol. 17, 1179 (1999).

[121] K. Kishida, C. Li, S. Lin, and K. I. NISHIGUCHI, "Pulse pre-pump method to achieve cm-order spatial resolution in Brillouin distributed measuring technique," Technical report of IEICE, 15-20 (2004).

[122] S. Afshar V, G. A. Ferrier, X. Bao, and L. Chen, "Effect of the finite extinction ratio of an electro-optic modulator on the performance of distributed probe-pump Brillouin sensorsystems," Opt. Lett. 28, 1418-1420 (2003).

[123] A. W. Brown, B. G. Colpitts, and K. Brown, "Distributed sensor based on dark-pulse Brillouin scattering," Photonics Technology Letters, IEEE 17, 1501-1503 (2005).

[124] A. W. Brown, B. G. Colpitts, and K. Brown, "Dark-pulse Brillouin optical time-domain sensor with 20-mm spatial resolution," J. Lightwave Technol. 25, 381-386 (2007).

[125] F. Wang, X. Bao, L. Chen, Y. Li, J. Snoddy, and X. Zhang, "Using pulse with a dark base to achieve high spatial and frequency resolution for the distributed Brillouin sensor," Opt. Lett. 33, 2707-2709 (2008).

[126] K. Y. Song, S. Chin, N. Primerov, and L. Thévenaz, "Time-domain distributed fiber sensor with 1 cm spatial resolution based on Brillouin dynamic grating," J. Lightwave Technol. 28, 2062-2067 (2010).

[127] K. Y. Song and H. J. Yoon, "High-resolution Brillouin optical time domain analysis based on Brillouin dynamic grating," Opt. Lett. 35, 52-54 (2010).

[128] S. Chin, N. Primerov, and L. Thevenaz, "Sub-centimeter spatial resolution in distributed fiber sensing based on dynamic Brillouin grating in optical fibers," Sensors Journal, IEEE 12, 189-194 (2012).

[129] W. Li, X. Bao, Y. Li, and L. Chen, "Differential pulse-width pair BOTDA for high spatial resolution sensing," Opt. Express 16, 21616-21625 (2008).

[130] Y. Dong, X. Bao, and W. Li, "Differential Brillouin gain for improving the temperature accuracy and spatial resolution in a long-distance distributed fiber sensor," Appl. Optics 48, 4297-4301 (2009).

[131] Y. Dong, H. Zhang, L. Chen, and X. Bao, "2 cm spatial-resolution and 2 km range Brillouin optical fiber sensor using a transient differential pulse pair," Appl. Optics 51, 1229-1235 (2012).

[132] A. Minardo, R. Bernini, and L. Zeni, "Differential techniques for high-resolution BOTDA: an analytical approach," Photonics Technology Letters, IEEE 24, 1295-1297 (2012).

[133] S. Foaleng-Mafang, J.-C. Beugnot, and L. Thévenaz, "Optimized configuration for high resolution distributed sensing using Brillouin echoes," in *20th International Conference on Optical Fibre Sensors*, (International Society for Optics and Photonics, 2009), 75032C-75032C-75034.

[134] S. M. Foaleng, M. Tur, J.-C. Beugnot, and L. Thévenaz, "High spatial and spectral resolution long-range sensing using Brillouin echoes," J. Lightwave Technol. 28, 2993-3003 (2010).

[135] Y. Koyamada, Y. Sakairi, N. Takeuchi, and S. Adachi, "Novel technique to improve spatial resolution in Brillouin optical time-domain reflectometry," Photonics Technology Letters, IEEE 19, 1910-1912 (2007).

[136] Y. Yao, Y. Lu, X. Zhang, F. Wang, and R. Wang, "Reducing Trade-Off Between Spatial Resolution and Frequency Accuracy in BOTDR Using Cohen's Class Signal Processing Method," Photonics Technology Letters, IEEE 24, 1337-1339 (2012).

[137] F. Wang, W. Zhan, X. Zhang, and Y. Lu, "Improvement of spatial resolution for BOTDR by iterative subdivision method," J. Lightwave Technol. 31, 3663-3667 (2013).

[138] Y. T. Cho, M. N. Alahbabi, M. J. Gunning, and T. P. Newson, "Enhanced performance of long range Brillouin intensity based temperature sensors using remote Raman amplification," Meas. Sci. Technol. 15, 1548-1552 (2004).

[139] K. De Souza and T. P. Newson, "Signal to noise and range enhancement of a Brillouin intensity based temperature sensor," Opt. Express 12, 2656-2661 (2004).

[140] Y. T. Cho, M. N. Alahbabi, G. Brambilla, and T. P. Newson, "Distributed Raman amplification combined with a remotely pumped EDFA utilized to enhance the per-

formance of spontaneous Brillouin-based distributed temperature sensors," IEEE Photonics Technol. Lett. 17, 1256-1258 (2005).

[141] X. H. Jia, Y. J. Rao, L. A. Chang, C. Zhang, and Z. L. Ran, "Enhanced Sensing Performance in Long Distance Brillouin Optical Time-Domain Analyzer Based on Raman Amplification: Theoretical and Experimental Investigation," J. Lightwave Technol. 28, 1624-1630 (2010).

[142] S. Martin-Lopez, M. Alcon-Camas, F. Rodriguez, P. Corredera, J. D. Ania-Castanon, L. Thevenaz, and M. Gonzalez-Herraez, "Brillouin optical time-domain analysis assisted by second-order Raman amplification," Opt. Express 18, 18769-18778 (2010).

[143] M. A. Soto, G. Bolognini, and F. Di Pasquale, "Optimization of long-range BOTDA sensors with high resolution using first-order bi-directional Raman amplification," Opt. Express 19, 4444-4457 (2011).

[144] X. Angulo-Vinuesa, S. Martin-Lopez, P. Corredera, and M. Gonzalez-Herraez, "Raman-assisted Brillouin optical time-domain analysis with sub-meter resolution over 100 km," Opt. Express 20, 12147-12154 (2012).

[145] X. Angulo-Vinuesa, S. Martin-Lopez, J. Nuno, P. Corredera, J. D. Ania-Castanon, L. Thevenaz, and M. Gonzalez-Herraez, "Raman-Assisted Brillouin Distributed Temperature Sensor Over 100 km Featuring 2 m Resolution and 1.2 degrees C Uncertainty," J. Lightwave Technol. 30, 1060-1065 (2012).

[146] M. A. Soto, G. Bolognini, and F. Di Pasquale, "Analysis of optical pulse coding in spontaneous Brillouin-based distributed temperature sensors," Opt. Express 16, 19097-19111 (2008).

[147] M. A. Soto, G. Bolognini, F. Di Pasquale, and L. Thévenaz, "Long-range Brillouin optical time-domain analysis sensor employing pulse coding techniques," Measurement Science and Technology 21, 094024 (2010).

[148] M. A. Soto, G. Bolognini, F. Di Pasquale, and L. Thévenaz, "Simplex-coded BOTDA fiber sensor with 1 m spatial resolution over a 50 km range," Opt. Lett. 35, 259-261 (2010).

[149] M. A. Soto, P. K. Sahu, G. Bolognini, and F. Di Pasquale, "Brillouin-based distributed temperature sensor employing pulse coding," IEEE Sens. J. 8, 225-226 (2008).

[150] M. A. Soto, S. Le Floch, and L. Thevenaz, "Bipolar optical pulse coding for performance enhancement in BOTDA sensors," Opt. Express 21, 16390-16397 (2013).

[151] M. Taki, Y. Muanenda, C. J. Oton, T. Nannipieri, A. Signorini, and F. Di Pasquale, "Cyclic pulse coding for fast BOTDA fiber sensors," Opt. Lett. 38, 2877-2880 (2013).

[152] M. A. Soto, G. Bolognini, and F. D. Pasquale, "Long-range simplex-coded BOTDA sensor over 120km distance employing optical preamplification," Opt. Lett. 36, 232-234 (2011).

[153] M. A. Soto, M. Taki, G. Bolognini, and F. Di Pasquale, "Simplex-Coded BOTDA Sensor Over 120-km SMF With 1-m Spatial Resolution Assisted by Optimized Bidirectional Raman Amplification," IEEE Photonics Technol. Lett. 24, 1823-1826 (2012).

[154] F. Gyger, E. Rochat, S. Chin, M. Niklès, and L. Thévenaz, "Extending the sensing range of Brillouin optical time-domain analysis up to 325 km combining four optical repeaters," in *OFS2014 23rd International Conference on Optical Fiber Sensors*, (International Society for Optics and Photonics, 2014), 91576Q-91576Q-91574.

[155] R. Bernini, A. Minardo, and L. Zeni, "Long-range distributed Brillouin fiber sensors by use of an unbalanced double sideband probe," Opt. Express 19, 23845-23856 (2011).

[156] D. M. Nguyen, B. Stiller, M. W. Lee, J.-C. Beugnot, H. Maillotte, A. Mottet, J. Hauden, and T. Sylvestre, "Distributed Brillouin Fiber Sensor With Enhanced Sensitivity Based on Anti-Stokes Single-Sideband Suppressed-Carrier Modulation," IEEE Photonics Technol. Lett. 25, 94-96 (2013).

[157] K. Hotate, K. Abe, and K. Y. Song, "Suppression of signal fluctuation in Brillouin optical correlation domain analysis system using polarization diversity scheme," IEEE Photonics Technol. Lett. 18, 2653-2655 (2006).

[158] K. Y. Song and K. Hotate, "Enlargement of measurement range in a Brillouin optical correlation domain analysis system using double lock-in amplifiers and a single-sideband modulator," IEEE Photonics Technol. Lett. 18, 499-501 (2006).

[159] J. H. Jeong, K. Lee, K. Y. Song, J. M. Jeong, and S. B. Lee, "Variable-frequency lock-in detection for the suppression of beat noise in Brillouin optical correlation domain analysis," Opt. Express 19, 18721-18728 (2011).

[160] K. Y. Song, Z. He, and K. Hotate, "Optimization of Brillouin optical correlation domain analysis system based on intensity modulation scheme," Opt. Express 14, 4256-4263 (2006).

[161] K. Y. Song, Z. Y. He, and K. Hotate, "Effects of intensity modulation of light source on Brillouin optical correlation domain analysis," J. Lightwave Technol. 25, 1238-1246 (2007).

[162] J. H. Jeong, K. Lee, K. Y. Song, J. M. Jeong, and S. B. Lee, "Differential measurement scheme for Brillouin Optical Correlation Domain Analysis," Opt. Express 20, 27094-27101 (2012).

[163] M. Kannou, S. Adachi, and K. Hotate, "Temporal gating scheme for enlargement of measurement range of Brillouin optical correlation domain analysis for optical fiber distributed strain measurement," in *Proc. 16th Int. Conf. Optical Fiber Sensors*, 2003), 454-457.

[164] K. Hotate and H. Arai, "Enlargement of measurement range of simplified BOCDA fiber-optic distributed strain sensing system using a temporal gating scheme," in *Brug-*

es, Belgium-Deadline Past, (International Society for Optics and Photonics, 2005), 184-187.

[165] Y. Mizuno, Z. He, and K. Hotate, "Measurement range enlargement in Brillouin optical correlation-domain reflectometry based on temporal gating scheme," Opt. Express 17, 9040-9046 (2009).

[166] Y. Mizuno, Z. Y. He, and K. Hotate, "Measurement range enlargement in Brillouin optical correlation-domain reflectometry based on double-modulation scheme," Opt. Express 18, 5926-5933 (2010).

[167] D. Elooz, Y. Antman, N. Levanon, and A. Zadok, "High-resolution long-reach distributed Brillouin sensing based on combined time-domain and correlation-domain analysis," Opt. Express 22, 6453-6463 (2014).

[168] A. Denisov, M. A. Soto, and L. Thévenaz, "1'000'000 resolved points along a Brillouin distributed fibre sensor," in *OFS2014 23rd International Conference on Optical Fiber Sensors,* (International Society for Optics and Photonics, 2014), 9157D9152-9157D9152-9154.

[169] A. Zadok, Y. Antman, N. Primerov, A. Denisov, J. Sancho, and L. Thevenaz, "Random-access distributed fiber sensing," Laser Photon. Rev. 6, L1-L5 (2012).

[170] W. Zou, Z. He, and K. Hotate, "Distributed dynamic-strain sensing based on brillouin optical correlation domain analysis," in *Lasers & Electro Optics & The Pacific Rim Conference on Lasers and Electro-Optics, 2009. CLEO/PACIFIC RIM'09. Conference on,* (IEEE, 2009), 1-2.

[171] W. Zou, Z. He, and K. Hotate, "Realization of high-speed distributed sensing based on Brillouin optical correlation domain analysis," in *Conference on Lasers and Electro-Optics,* (Optical Society of America, 2009),

[172] K. Y. Song, M. Kishi, Z. Y. He, and K. Hotate, "High-repetition-rate distributed Brillouin sensor based on optical correlation-domain analysis with differential frequency modulation," Opt. Lett. 36, 2062-2064 (2011).

[173] K. Hotate and S. S. L. Ong, "Distributed dynamic strain measurement using a correlation-based brillouin sensing system," IEEE Photonics Technol. Lett. 15, 272-274 (2003).

[174] K. Y. Song and K. Hotate, "Distributed fiber strain sensor with 1-kHz sampling rate based on Brillouin optical correlation domain analysis," IEEE Photonics Technol. Lett. 19, 1928-1930 (2007).

[175] T. Yamauchi and K. Hotate, "Distributed and dynamic strain measurement by BOCDA with time-division pump-probe generation scheme," in *Conference on Lasers and Electro-Optics,* (Optical Society of America, 2004),

[176] Y. Mizuno, Z. Y. He, and K. Hotate, "One-End-Access High-Speed Distributed Strain Measurement with 13-mm. Spatial Resolution Based on Brillouin Optical Correlation-Domain Reflectometry," IEEE Photonics Technol. Lett. 21, 474-476 (2009).

7

Advanced Optical Fibers for High Power Fiber Lasers

Liang Dong

1. Introduction

The nonlinear limits of conventional single-mode fibers were well recognized in the nineties during the earlier efforts to scale the peak powers of pulsed fiber lasers. These earlier efforts focused on lowering the NA and optimizing the refractive index profile of optical fibers to increase the effective mode area [1, 2]. The emergence of photonic crystal fibers in the late nineties gave new impetus to the mode area scaling of single-mode optical fibers. The observation of the "endlessly single-mode" nature of *photonic crystal fibers* (PCF) at small hole sizes in 1996 [3] led to an early realization of the dispersive nature of a photonic crystal cladding [4], which limits the increase of normalized frequency V at short wavelengths. It was quickly realized that the scalability of Maxwell's equation allows for single-mode operation at a very short wavelengths in a small core to be directly translated into single-mode operation in a large core at longer wavelengths [5]. This led to a rapid progress in scaling of core diameters of single-mode PCF, culminating in the 100μm-core diameter demonstrated in 2006 [6].

The design and fabrication techniques developed for photonic crystal fibers also led to a realization of the potential of optical fibers with more open cladding structures than the closed concentric circles found in conventional optical fibers. It then became possible to design and fabricate leaky waveguides where differential mode losses can be used to control the number of propagation modes. As a consequence, *leakage channel fibers* (LCF) were proposed and first demonstrated in 2005 [7]. Two years later, single-mode operation in a record core diameter of ~180μm was demonstrated in a LCF [8].

Many other approaches based on conventional fibers were also pioneered during the last decade. One notable example is the *chirally coupled core* (CCC) fiber [9, 10], which relies on out-coupling of *higher-order modes* (HOM) to side cores adjacent to the main core. Phase-matching is achieved with the help of angular momentum from the helical side cores, which are formed by spinning the preform during fiber drawing. Another notable example is based on the propagation of a higher-order mode in a specially designed multimode fiber [11, 12]. It is argued that perturbations mostly have anti-symmetry in optical fibers and promote mode coupling mostly between modes of opposite symmetries. The mode spacing between a radially symmetric LP0n mode and its nearest neighbor modes with opposite symmetry is, in fact, larger for higher order modes. These higher-order modes are also more resistant to bend-induced mode compression. A special fiber design facilitates the ease of mode conversion to and from the LP0n mode using a *long period grating* (LPG).

In this chapter, we will give a brief introduction to the key approaches to effective mode area scaling which have shown great promise for future high power fiber lasers. Basic concepts are introduced and the latest developments are also discussed.

2. Photonic crystal fibers

The authors of the paper in 1996 [3] were looking for photonic bandgap guidance in the defect core of the PCF with a core size of ~4.6 μm. The reason for the absence of photonic bandgap guidance in a solid-core PCF with air holes in the cladding only became clear a few years later. To their surprise, they observed very broadband single-mode guidance from 458–1550 nm (extended to 337-1550 nm in [4]). A conventional single-mode step-index optical fiber would become multimode at shorter wavelength. Even more surprising, the diffraction angle of the output beam was smaller at shorter wavelengths, indicating a smaller fiber numerical aperture at short wavelength.

2.1. Fundamental space-filling mode of photonic crystal cladding

In a follow-on paper [4] the initial observation of the "endlessly single-mode" nature of photonic crystal fibers was explained by the dispersive nature of the cladding, which can be viewed as a composite of two materials, i.e. silica and air. The effective refractive index of the composite cladding was identified for the first time as the effective index of the fundamental mode of the composite cladding, referred to as the *fundamental space-filling mode (FSM)*. In the extreme case of conventional optical fibers with an infinite cladding, the effective index of the fundamental space-filling mode becomes the refractive index of the cladding. In PCFs with air holes, the cladding index n_{cl}, i.e. the effective index of the fundamental space-filling mode, is lower than the refractive index of the background glass n_b due to the existence of the air holes, i.e. $n_{cl}<n_b$. The core index n_{co} is the same as the index of the background glass, $n_{co}=n_b$. Fundamental optical guidance in PCFs can, therefore, in principle, be understood in a similar way to that in conventional optical fibers (see Figure 1).

This paper [4] established the principle for understanding the basic guidance properties of PCFs. In conventional optical fibers, the refractive indexes of core and cladding are only weakly dependent on wavelength due to material dispersion. This is also expected of the core refractive index of a PCF. The refractive index of the composite photonic crystal cladding, however, behaves very differently. At the long wavelength extreme, i.e. $\lambda\rightarrow\infty$, the wavelength is much larger than the air holes and the fundamental space-filling mode will occupy all areas

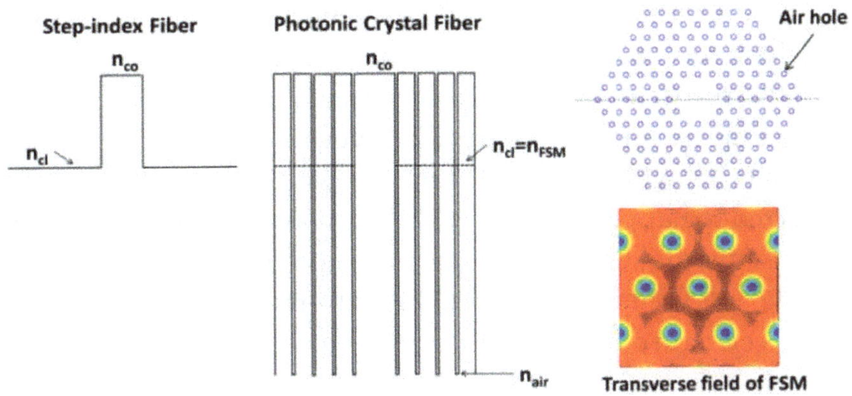

Figure 1. Refractive index of photonic crystal fiber cladding.

of the cladding equally. In this case, the refractive index of the FSM can be obtained by averaging the square of the refractive index of the composite photonic crystal cladding, i.e. $n_{cl}^2=n_b^2(1-F)+n_{air}^2F=n_b^2-F(n_b^2-n_{air}^2)$ where F is the area fraction of air; and n_{air} is the refractive index of air. The effective NA of the PCF for $\lambda\rightarrow\infty$ is, therefore, $F^{1/2}(n_b^2-n_{air}^2)^{1/2}$. At the short wavelength extreme, i.e. $\lambda\rightarrow 0$, the wavelength is much smaller than the pitch of the air holes and the fundamental space-filling mode is mostly in the region with high refractive index, i.e. the background glass. In this case, the cladding index is the refractive index of the background glass n_b. The effective NA of a PCF at $\lambda\rightarrow 0$ trends to zero. The guidance of a PCF still gets weaker at large λ, as in a conventional optical fiber, due to the inverse wavelength dependence of the normalized frequency V. In addition, its guidance also gets weaker at short wavelengths due to the diminishing NA, which also limits the growth of V and leads to the "endlessly single-mode" nature of the PCFs with small air holes. Guidance of PCF will, therefore, diminish at both long and short wavelength ends, possessing two bend-induced cut-off edges.

2.2. Single mode and multimode regime of photonic crystal fibers

In conventional fibers, a mode is guided when the effective mode index n_{eff} is between the core and cladding index, i.e. $n_{co}>n_{eff}>n_{cl}$. The mode cut-off can be identified when $n_{eff}=n_{cl}$. The second-order mode cuts off in a step-index fiber at V=2.405. In the first reported work on determining the single-mode regime of PCFs with a 1-cell core, i.e. one hole missing at the defect core, by Birks et al [4], a somewhat arbitrary equivalent step-index core radius equaling *pitch*Λ, i.e. center-to-center hole separation, was used to calculate the V value. Work by Saitoh [13] using a *finite element model* (FEM) to determine n_{FSM} and the condition $n_{eff}=n_{cl}$ to determine the second-order mode cut-off, put the effective step-index core radius to be $\Lambda/3^{1/2}$ for a one-cell core. Later work by Saitoh [14] determined the effective step-index core radius to be Λ for a 3-cell core and $2^{1/2}\Lambda$ for a 7-cell core.

The optical waveguide equation allows the scaling of all parameters measured in the length scale by the same factor. For PCF, the most convenient scaling factor is $1/\Lambda$. The second-order mode cut-off is typically plotted as a normalized wavelength λ/Λ versus normalized hole diameter d/Λ plot. This is shown for PCFs with 1-cell, 3-cell and 7-cell core in Figure 2 [14]. For each of the curves, the area above the curve is in the single-mode regime, i.e. the wavelengths above the second-order mode cut-off wavelength, and, below it, multimode regime. The "endlessly single-mode" nature of PCFs can be easily identified in Figure 2. When $d/\Lambda<0.424$, 0.165 and 0.046 respectively for 1-cell, 3-cell and 7-cell core PCFs, the PCFs will remain in the single-mode regime for all wavelengths. It is worth noting that Figure 2 is for PCFs with infinite cladding. For PCFs with finite cladding, guidance is weaker and the second-order-mode cut-off is expected to happen at slightly shorter wavelengths. The curves in Figure 2 are expected to move downwards slightly. It needs to be noted as discussed earlier, that the PCFs are too weak to guide any light at long and short extremes of wavelength. The critical bend radius at the short wavelength edge is determined to be dependent on pitch Λ and wavelength λ such that [4]

$$R_c \propto \frac{\Lambda^3}{\lambda^2} \tag{1}$$

The critical bend radius at the short length edge increases in proportion to $1/\lambda^2$ as wavelength decreases. This relation was verified experimentally for the critical bend radius at 3dB excessive bend loss [4]. It comes directly from the dispersive nature of the photonic crystal cladding.

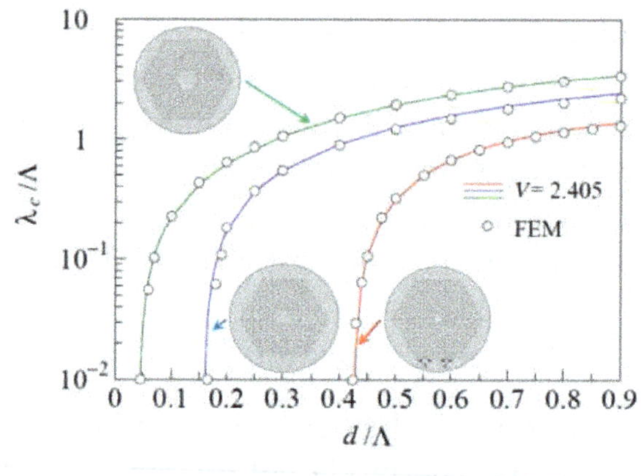

Figure 2. The second-order mode cut-off determined by $n_{eff}=n_{FSM}$ (dots) and V=2.405 (solid lines) using core radius $\varrho=\Lambda/3^{1/2}$, Λ and $2^{1/2}\Lambda$ respectively for 1-cell, 3-cell and 7-cell cores [14].

2.3. Waveguide loss of photonic crystal fibers

PCFs are intrinsically leaky, i.e. there is always a finite waveguide loss for each mode in a PCF with a finite cladding. The waveguide loss can be found by calculating the imaginary part of the effective mode index using a numerical mode solver. When plotting the waveguide loss versus wavelength, the slope is expected to change around cut-off. This can also be used to determine the second-order-mode cut-off. This was done for a 1-cell core PCF [15]. The results are consistent with those in [13].

In conventional optical fibers, the fully enclosed core and cladding boundary ensures that waveguide loss is zero for all guided modes, i.e. those that satisfy the conditions for total internal reflection at the boundary. For the open structure of PCFs with a finite cladding, all modes are leaky with finite waveguide losses. At the extreme of an infinite number of layers, the waveguide losses are zero. The waveguide losses are lower for larger air holes and can be substantially lowered by increasing the number of air-hole layers. In practice, the waveguide loss can be made below other material and process related losses. By employing appropriate polishing, etching and dehydration processes, a PCF with a loss 0.28dB/km at 1550nm has been demonstrated [16]. For applications in fiber lasers where a length of not more than a few tens of meters is used, the waveguide loss does not present any problem if appropriate designs are used.

2.4. Modeling of photonic crystal fibers

Due to the complexity of the geometrical structures of PCFs, numerical models are typically used to find mode properties including the effective mode index, waveguide loss and effective mode area. For any numerical mode solvers, electric and magnetic fields are described by an expansion series. Eigenvalue equations are then established by requiring the fields to satisfy all boundary conditions. These equations are typically expressed as a set of linear equations, which can be solved for complex effective mode index. The more complex a waveguide design is, the larger is the number of linear equations. The waveguide dispersion can be calculated from the real part of the complex effective mode index. The waveguide loss can be obtained from the imaginary part of the complex effective mode index. Electric and magnetic fields can be calculated once the effective mode index is determined. The most common and flexible numerical mode solver is a *finite element mode solver* (FEM). This is commercially available from COMSOL (*http://www.comsol.com/*). A FEM is capable of handling practically any design. It can, however, be computationally very demanding. If only circular boundaries are involved, a Multipole mode solver is a good option [17, 18]. It is based on the decomposition of fields into circular Bessel series, which are the most accurate and efficient method for modeling circular boundaries. For non-commercial research and teaching purposes, it can be downloaded from the University of Sydney website (*http://sydney.edu.au/science/physics/cudos/research/mofsoftware.shtml*). A plane wave expansion method can also be used. This, however, assumes an infinite cladding and, therefore, cannot determine the waveguide loss.

2.5. Mode area scaling with photonic crystal fibers

The first demonstration of a large-core PCF was performed by Knight el at [5]. The 1-cell PCF with relative hole diameter $d/\Lambda \approx 0.12$ and a core diameter of $2\varrho = 22.5\mu m$, provided robust single-mode guidance at 458nm. According to Figure 2, a 1-cell PCF with $d/\Lambda < 0.424$ is expected to be single-mode over its entire wavelength range. It is, therefore not surprising that the PCF was single mode at 458nm. The critical bend radius was measured to be 50cm at 458nm and 4cm at 1.55μm [5]. This fiber is expected to have a critical bend radius of ~10cm at 1μm using equation 1.

Considering the wavelength scalability of the waveguide equation, this 1-cell PCF with $2\varrho/\lambda \approx 50$ can be scaled by a factor of ~2.2 to a ~50μm core diameter to operate at 1μm. The critical bend radius for this single-mode PCF with 50μm core diameter is expected to be ~1.1m at 1μm according to equation 1! A similarly scaled 1-cell PCF with a core diameter of 30μm will have a critical bending radius of ~24cm. This is probably close to the practical limit of coiled single-mode 1-cell PCFs. Above a core diameter of ~40μm, 1-cell single-mode PCF can only be used in a straight configuration in practice. The single-mode operation of PCFs with large core diameters comes at the cost of weak guidance as a result of the diminishing effective NA. This fundamentally limits the use of single-mode PCFs with large cores in coiled configurations. If PCFs are allowed to operate in the few-moded regime, coiled PCFs with slightly larger core diameters are possible. For high average power fiber lasers with outputs exceeding kW, effective thermal management becomes increasingly critical. Long fiber lengths of many meters are required. The constraint of not being able to coil the fibers can become a major issue considering the additional space constraints.

2.6. Rare-earth-doped glass for large-core photonic crystal fibers

The diminishing NA of PCFs with large cores also has additional implications, as realized by the authors of [5]. To fabricate rare-earth-doped large-core PCFs, the core refractive index needs to be accurately controlled at levels far below the very small effective index difference between the core and cladding in order not to disturb the guidance properties of the PCFs. This requires much improved techniques for the fabrication of the active core in large-core PCFs.

The first ytterbium-doped large-core PCF with 15μm core diameter was reported in 2001 at Bath University [19]. The effective mode area at 1μm was ~100μm^2. The laser operated in single mode with poor efficiency. The key advance was the use of a repeated stack-and-draw process to achieve a uniformly doped core with a refractive index close to silica. Conventional fabrication techniques for rare-earth doped silica fibers result in significant non-uniformity in refractive index across the core as well as a raised refractive index above silica. Due to the weak guidance in large-core PCFs, any index non-uniformity across the core can lead to mode distortion. It also requires the refractive index of the core to be very close to the silica background in PCFs. The authors of [19] fabricated ytterbium-doped glass with low doping levels. The doped glass is surrounded by some undoped silica glass. The glass is stacked and drawn twice before being finally incorporated into the core of a PCF. The resulting PCF has an ytterbium-doped core which consists of 425 original doped glass sections. The dimension of each of the original doped glass sections is much smaller than the wavelength of light and the core of the PCF, therefore, appears to be homogeneous to light at the operating optical wavelength. The doped core is also heavily diluted by the addition of silica glass (90%), leading to an average refractive index close to that of silica [20]. In the same paper, it was pointed out that rare-earth-doped glass with a higher index can be stacked and drawn with undoped glass with a low index to achieve a better match to that of silica background [20].

2.7. Double-clad photonic crystal fibers with high NA air-clad for pump guidance

The concept of a composite air-glass clad with a high air-filling fraction to provide a high NA pump guide in a double-clad fiber was first proposed in 1999 at what was then then Lucent Technologies [21]. The high pump NA can enable a significant improvement in pump coupling especially from low brightness multimode diode lasers for a given pump waveguide dimension. In conventional double-clad fibers, low-index polymer coatings are typically used to form the pump cladding. The air-clad pump isolates high pump powers from the polymer coatings, leading to potentially improved reliability.

The first such fiber was demonstrated in 2000, where the pump cladding was mostly air except for a single connecting element [22]. The measured pump NA over 1m was below 0.2. The fiber was passive and there was no laser demonstration in this first attempt. The first cladding-pumping demonstration in an active fiber with an air-clad pump guide and a conventional core design was in 2001 at Southampton University [23]. The measured NA of the pump guide with air-cladding was 0.4-0.5. In a separate demonstration from Southampton University later in the same year, a conventional core with a very low NA together with a photonic crystal cladding was used in combination with a pump guide with air-cladding [24]. The core guidance came from a combination of the raised core index and the photonic crystal cladding. Cladding pumping with a low brightness laser diode at 915nm (100μm core with a NA=0.22) was demonstrated with a slope efficiency of 70% relative to the absorbed pump power. The measured pump NA over a short length (~10cm) was 0.3-0.4. NA over much longer fiber length as in a fiber laser is expected to be lower. The first true active double-clad photonic crystal fibers with a pump air-cladding was demonstrated in 2003 with a measured pump NA of 0.8 at 1μm [20].

The theoretical basis for the high NA of optical waveguides formed by air-cladding was established in [25]. It had been understood previously that the effective index of a glass and air composite can be obtained by the effective index of the fundamental space filling mode. An example of high NA air-clad used in double-clad fibers is shown in Figure 3 from [25]. In such air-glass composites, the glass webs can be considered as slab waveguides. The separations between the webs are typically far enough that the webs can be considered as isolated. In this case, the effective index of the glass-air composite cladding can be well approximated by that of the fundamental slab waveguide mode. This can be easily calculated. The resulting pump NA is only dependent on the ratio of the web width w and wavelengthλ. This is plotted in Figure 3(c), showing that a small w/λ is critical for high NA.

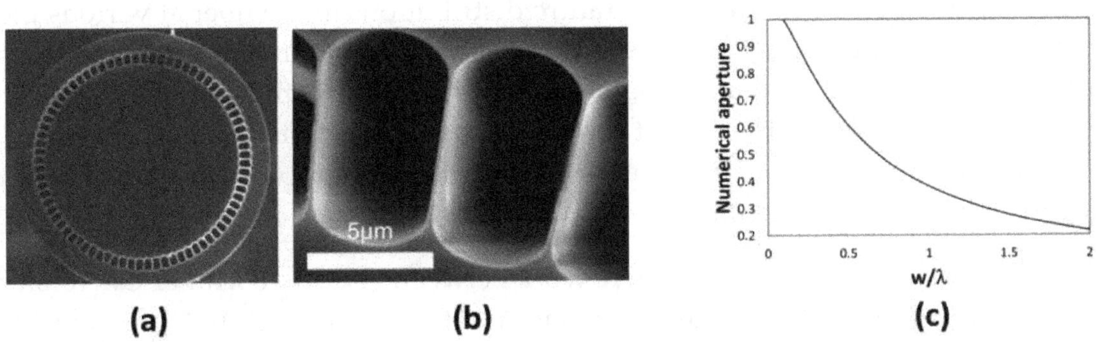

Figure 3. (a) High NA air-clad from [25], (b) close-up of the high NA air-clad, and (c) NA of air-clad waveguide based on slab waveguide model plotted against the ratio of waveguide width w and wavelength λ.

2.8. Progress of active large-core photonic crystal fibers

Since 2003, the group at Friedrich-Schiller-University Jena has been playing a significant role in the development large-mode-area photonic crystal fibers with pump air-clad. In their first work in collaboration with then Crystal Fibre A/S, now NKT Photonics, an ytterbium-doped PCF with effective mode area of ~350μm^2 and mode field diameter (MFD) of 21μm was demonstrated [26] (see Figure 4). The 3-cell PCF had a hole diameter d=2μm and pitch Λ=11.5, giving an d/Λ=0.18. The core diameter was 28μm. A relatively small circular area of 9μm was doped in the 3-cell core. The doped area had a high ytterbium-doping level of ~0.6 at%. It was further co-doped with aluminum and fluorine to provide an index merely 2×10^{-4} above the silica background. The PCF with the raised index in the doped area was simulated, showing that the fiber guides the second-order mode, which is, however, close to cut-off. The fiber had a 150μm pump guide with an air clad. The webs in the air clad were ~50μm long with a thickness of ~390nm, giving w/λ=~0.4 at 976nm. The measured pump NA was 0.55, slightly below the NA=0.68 predicated by the slab model in Figure 3. The fiber had an outer glass diameter of 450μm and was coated with standard acrylic coating. Due to the high doping level, the fiber has a pump absorption of ~9.6dB/m at 976nm. An impressive slope efficiency of 78% was demonstrated with respect to the launched pump power.

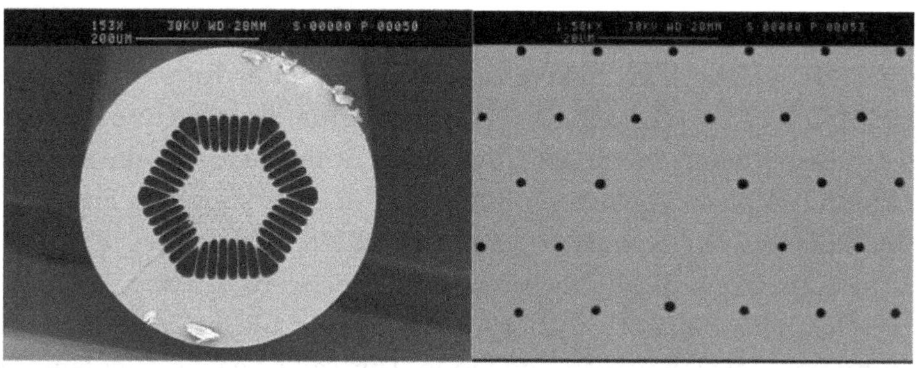

Figure 4. SEM image of the air-clad ytterbium-doped large-mode-area photonic crystal fiber in [26].

In a subsequent paper in 2003 [27], the fiber in [26] and a new fiber with narrower pump air-clad were studied with the FEM for temperature distributions in the fiber at various thermal loads in the core, considering both convective and radiative heat transportation by air at the fiber surface. The results show that, in the case of natural cooling, the thermal transportation is mainly limited by the heat transfer at the fiber surface. The air-clad impedes heat flow, especially when the width of the air clad is large. A narrower pump air-clad is advantageous, especially in actively cooled fiber lasers.

In 2004, the bar was raised in a collaborative work between Friedrich Schiller University Jena and Crystal Fibre A/S [28]. Their 7-cell fiber has an effective area of ~1000μm^2 and MFD of ~35μm (see Figure 5). Core diameter is ~40μm. Hole diameter d is 1.1μm and the pitch Λ is 12.3μm, giving a d/Λ=0.09. The fiber has a pump guide diameter of 170μm and a measured pump NA of 0.62 at 950nm. The pump absorption is ~13dB/m at 976nm. The pump air-clad is

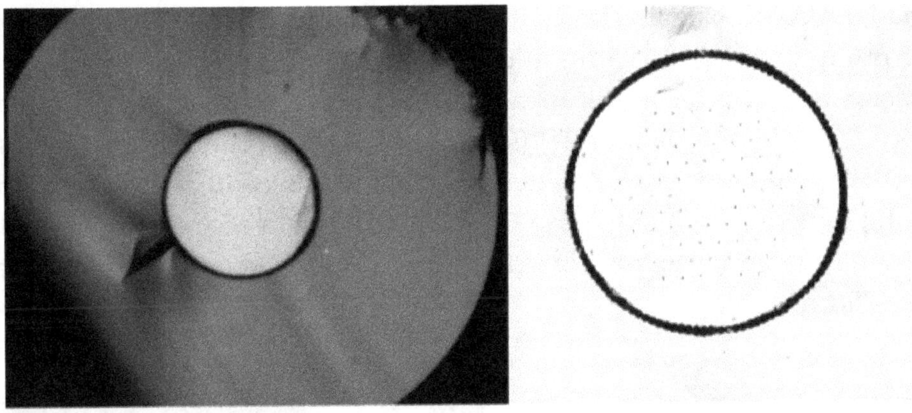

Figure 5. Microscope image of the air-clad ytterbium-doped single-mode PCF and close-up to the 40-μm core formed by seven missing air holes [28].

much narrower for more efficient heat diffusion. It is worth noting that the fiber is in the multimode regime for a 7-cell PCF (see Figure 5). This helps to ease bend loss in the weakly guided fiber.

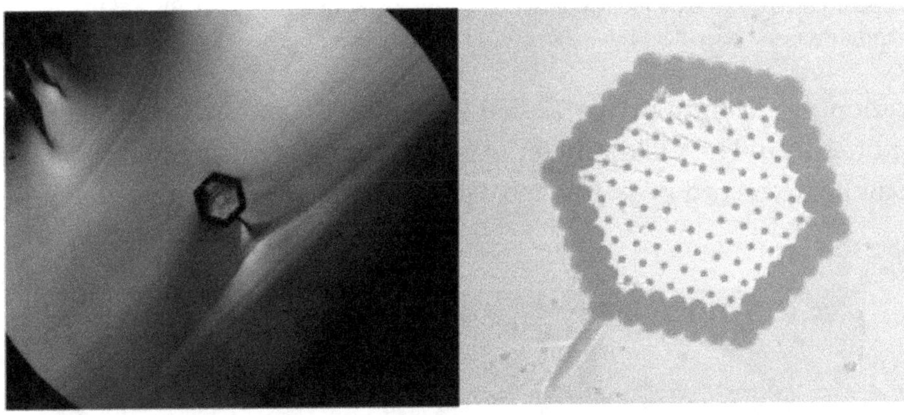

Figure 6. Microscope image of a rod-type photonic crystal fiber and close-up view of the inner cladding and core regions.

To further mitigate bend loss, a rod-type PCF was developed in 2005 [29] (see Figure 6). The rod had an outer diameter of 1.7mm so that it cannot be bent. A 48cm long length was used in the demonstration. The relative hole diameter was increased to $d/\Lambda=0.33$. This reduces bend sensitivity. This larger d/Λ in a 7-cell core PCF, however, puts this fiber firmly in the multimode regime (see Figure 2). The pump guide was reduced to a hexagon with flat-to-flat dimension of 117μm and corner-to-corner dimension of 141μm, similar to earlier fibers. The ytterbium doping level was also increased compared with earlier fibers. The increased doping level and reduced pump guide led to a high pump absorption of ~30dB/m at 976nm. The fiber was used to demonstrate power extraction of ~250W/m.

The rod-type PCF was further developed with the demonstration of a 19-cell PCF with a core diameter of 60μm, effective mode area of ~2000μm^2 and MFD of 75μm in 2006 [30] (see Figure

7). The fiber had a d/Λ=0.19, again firmly in the multimode regime (19-cell PCF not shown in Figure 2). The pump guide was 175μm in diameter. The pump air-clad had a web thickness of 400nm, which was 10μm long. The measured pump NA was 0.6 at 975nm, fairly close to the 0.67 NA predicted by the slab model (see Figure 3). The pump absorption is ~30dB/m at 976nm. The rod diameter is ~1.5mm. A 58cm long fiber was used to demonstrate 55W/m power extraction. In the same paper, a passive 100μm core fiber was also demonstrated. The 19-cell fiber had a d/Λ=0.2, again in the multimode regime. The effective mode area was ~4500μm^2 and MFD was ~75μm.

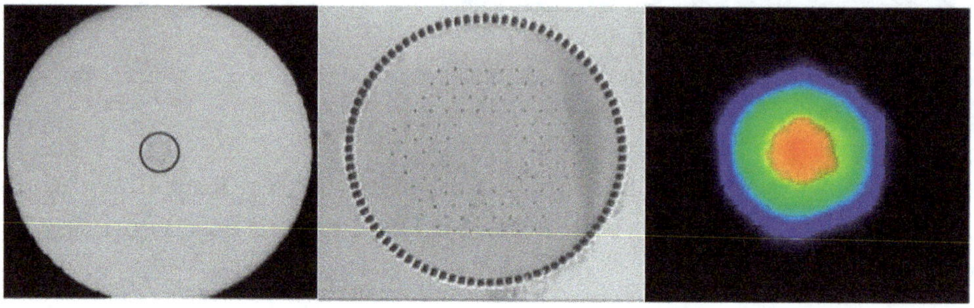

Figure 7. Microscope-image of the extended-mode-area rod-type photonic crystal fiber, SEM-picture of the micro-structured region and measured near-field intensity profile of the 60μm core fiber [30].

The demonstration of a 100μm active 19-cell PCF finally came in 2006 [6]. The pump guide had a diameter of 290μm. The rod diameter was 1.5mm. A 90cm long fiber was used to amplify 1ns pulses at 9.6kHz to record peak power of 4.5MW and pulse energy of 4.3mJ with M^2=1.3.

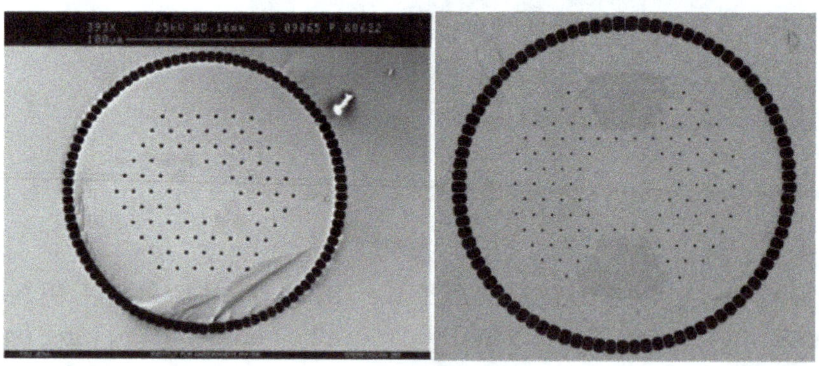

Figure 8. SEM image of the single polarization PCF with an effective area of ~700μm^2 [31] (left) and 2300μm^2 (right) [32].

Polarization-maintaining (PM) PCFs have also been developed by introducing stress elements. A PM 7-cell PCF with a mode area of ~700μm^2 was demonstrated in 2005 [31] (see left figure in Figure 8). The pitch Λ was 12.3μm with a d/Λ of 0.2. In the weakly guided PCFs, the polarization modes on slow and fast axis have different bend losses. This enables single-polarization operation where the polarization mode on the slow axis is still guided while polarization mode on the fast axis suffers high bend loss. Another single-polarization 19-cell

PCF was demonstrated in 2008 with a mode area of ~2300μm^2 (right figure in Figure 8) [32]. The corner-to-corner distance of the core was 70μm. The pitch Λ was 11μm with a d/Λ of 0.1.

3. Leakage channel fibers

A 2D micro-structured cladding, which is made possible by the stack-and-draw technique developed for photonic crystal fibers, enables new designs which do not possess the closed core-and-clad boundaries of conventional optical fibers. When a mode is guided in a conventional optical fiber, total internal reflection everywhere around the closed core-and-clad boundary, traps light entirely in the core, leading to zero waveguide loss. In designs with an open cladding, light can leak out, leading to finite waveguide loss associated with each mode. The waveguide loss is mode-dependent, providing opportunities for mode control by minimizing loss of the desired mode while maximizing loss of the unwanted modes. *Leakage channel fibers* (LCF) takes advantages of these new opportunities made possible by open cladding designs. A LCF can be precisely engineered to have high confinement loss for all higher order modes and low confinement loss for the fundamental mode and can, therefore, significantly extend the effective mode area of the fundamental mode. LCFs essentially exploit the increased ability of higher order modes to leak through small gaps in the cladding while maintaining good fundamental mode confinement.

3.1. Leakage channel fibers with air holes

The first LCF was demonstrated in 2005 [7]. It has a simple cladding design with air holes in the cladding [7]. The LCF is shown in in the left figure in Figure 9. The LCF had an outer diameter of ~270μm. The two smaller holes had a diameter of d=39.6μm and pitch Λ=51.2μm. The four larger holes had a diameter of d=46.0μm and pitch Λ=51.1μm. The passive LCF provided robust single-mode operation with a measured mode area of ~1417μm^2 (MFD=42.5μm). The most significant aspect of this work is that the LCF can be coiled down to 15cm with negligible loss penalty. This is a significant improvement over PCFs.

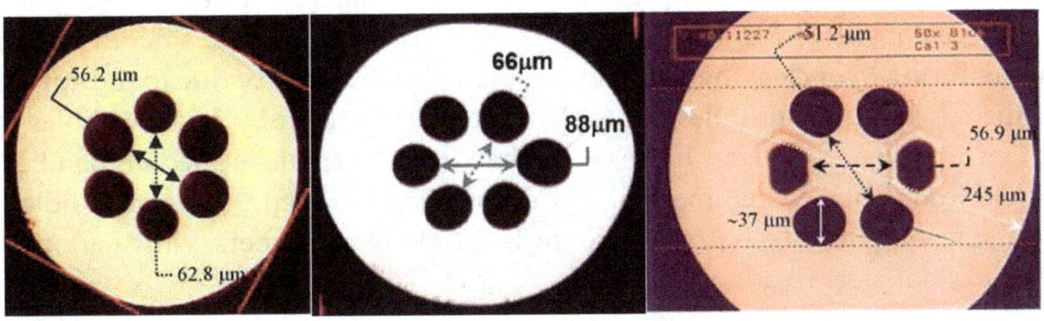

Figure 9. The LCF used in the first demonstration [7] (left), the first ytterbium-doped LCF (center) [33] and the first ytterbium-doped PM LCF (right) [34].

The first Ytterbium-doped LCF was demonstrated in 2006 [33] (see the middle figure in Figure 9). The fiber outer diameter was ~350μm and it was coated with a low index polymer to give a pump NA of 0.46. The average hole diameter was ~55μm and average pitch was ~67μm, giving an average $d/\Lambda=0.82$. The effective mode area was 3160μm^2 (MFD=63.4μm). Pump absorption was measured to be ~3.6dB/m at 976nm. Slope efficiency versus launched power was measured to be ~60% in a 5m long amplifier coiled at 40cm diameter. M^2 was measured to be ~1.3.

The first ytterbium-doped PM LCF was also demonstrated in 2006 [34] (see the right figure in Figure 9). A pair of boron-doped silica stress rods was used to replace two opposing air holes. The hole diameter was ~37μm. The effective mode area was ~1400μm^2 (MFD=42.2μm). The birefringence was measured to be ~2.1×10^{-4} over 1010-1080nm. The LCF had an outer diameter of ~245μm and was coated with a low index polymer to give a pump NA of ~0.46. The pump absorption was ~2.6dB/m at 976nm. Slope efficiency versus launched power was measured to be ~60%. M^2 was measured to be ~1.2. The PM LCF could be coiled to 12cm diameter with negligible bend loss.

3.2. All-glass leakage channel fibers

High refractive index contrast is not necessary for large core fiber designs. Low refractive index contrast is sufficient and often advantageous for further limiting higher order mode propagation. Fluorine-doped silica can be used to replace air holes in the LCFs described in the last section. The all-glass LCFs can provide much improved ease of fabrication and use, compared with fibers with air holes.

Despite the fact that air is a readily available ingredient, there are a number of drawbacks related to the use of air holes in fibers. The first one is the difficulty in precisely controlling the dimension of air holes in fiber fabrication. This is an intrinsic problem of a holey structure due to the air hole's tendency to collapse during fiber drawing. This is usually countered by a precise control of pressurization of the air holes, a process dependent on drawing conditions such as furnace temperature, feed rate, and drawing speed. When small air holes are desirable as in endless single-mode PCFs, higher pressure is required to maintain air hole dimensions due to the significantly increased tendency for the air holes to collapse by surface tension in this regime. This can make air hole dimensions to become highly sensitive to drawing conditions. This delicate balance of pressurization and collapse can lead to issues of controllability and repeatability in PCF fabrication. Air holes can also disturb smooth fracture wave propagation during fiber cleaving, leading to a poor cleaved surface due to the appearance of deep fractures behind the air holes, a problem often aggravated by large air holes and high cleaving tensions. In addition, air holes often have to be thermally sealed at the fiber ends to minimize environmental contamination. Mode distortion can occur from the air holes collapsing during splicing. This is especially true for large-mode-area fibers, which are much more susceptible to small perturbations.

A detailed analysis of all-glass LCFs was reported in [35]. For a LCF formed by one layer of features as shown in the inset of Figure 10, the core of diameter 2ρ is formed by six features with diameter d and refractive index n_f. Center-to-center feature spacing is Λ. The refractive

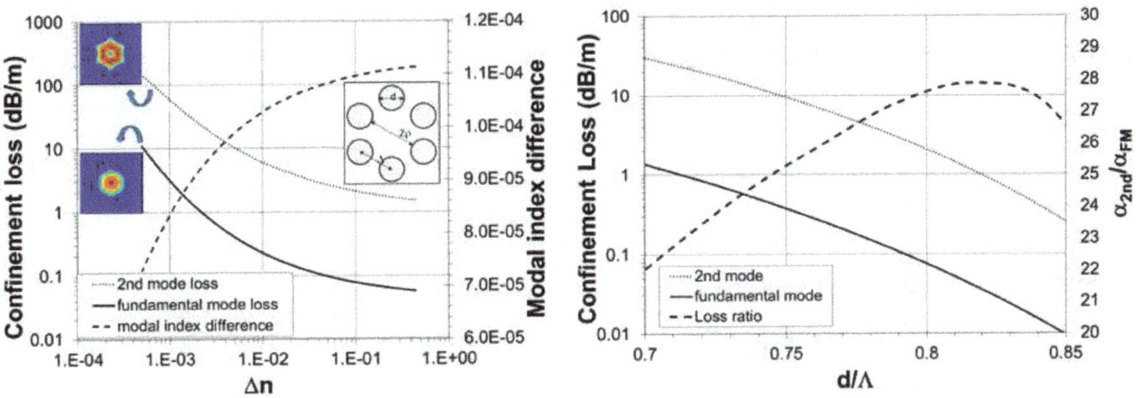

Figure 10. Effect of index contrast on confinement loss and modal index difference for one-cladding-layer LCFs with d/Λ=0.675 and 2ϱ=50μm (left) and effect of d/Λ on confinement loss and the loss ratio between the second-order mode and fundamental mode for LCFs with $\Delta n=1.2\times10^{-3}$ and 2ϱ=50μm (right) [35].

index of the background glass, usually silica, is n_b. The fiber was studied by a multipole mode solver for the effect of index contrast $\Delta n=n_b-n_f$. For the simulation in Figure 10, the following parameters were used: 2ϱ=50 μm, d/Λ=0.675, and n_b=1.444. The wavelength of the simulations was at 1.05μm. It can be seen from the left figure in Figure 10 that confinement loss for both fundamental, α_{FM}, and second-order modes, α_{2nd}, increases with a reduction of index contrast Δn, with the loss of the second-order mode, α_{2nd}, remaining over an order of magnitude higher than the loss of the fundamental mode, α_{FM}. The modal index difference, the difference between the effective mode indices of the fundamental and second modes, decreases toward low index with Δn by just ~40% over three orders of magnitude change in Δn.

The effect of normalized hole diameter d/Λ was also studied in [35] and is shown in the right figure in Figure 10 for confinement losses and the ratio of the second-mode loss to the fundamental-mode loss. The confinement loss for both the fundamental and second modes increases toward small d/Λ with the loss ratio changing very little over the entire range of d/Λ, from 22 to 28. The normalized hole diameter d/Λ is typically chosen to give an acceptable fundamental-mode loss. For LCFs with one layer of features as shown in Figure 10, the loss ratio of all-glass LCFs is very similar to that of an LCF with air holes. Slightly larger d/Λ is, however, required for achieving a similar confinement loss.

LCFs with two layers of features can be used to further improve the differential confinement loss between the fundamental and second-order modes at the expense of bending performance. Acceptable fundamental-mode loss at smaller feature sizes can be realized in LCFs with two layers of features, while leakage of higher order modes is substantially increased by a reduction of feature size despite the additional layer of features. Higher differential loss between modes can therefore be realized. Since bending loss of the fundamental mode is very strongly dependent on feature size, a reduction of feature size increases the bend loss of the fundamental mode in LCFs. An LCF with two layers of features was studied in [35] and the results are shown in Figure 11. Both the fundamental mode and second-order-mode loss shows the characteristic increase at small d/Λ, while the loss ratio α_{2nd}/α_{FM} is increased by over an order of magnitude compared to the one-layer designs in Figure 10. At d/Λ ≈ 0.548, the fundamental-mode loss $\alpha_{FM} \approx 0.1$ dB/m, while the second-order-mode loss $\alpha_{2nd} \approx 48$ dB/m, a loss ratio α_{2nd}/α_{FM} of ~480. A very high loss ratio α_{2nd}/α_{FM} of ~700 is possible at d/Λ=0.62.

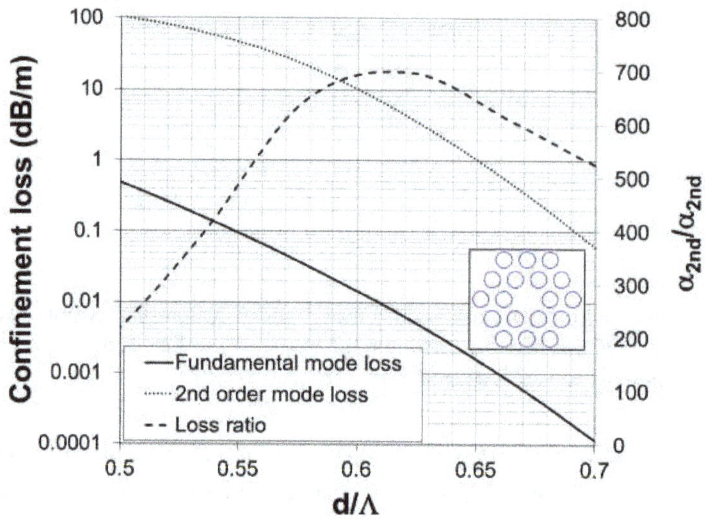

Figure 11. Effect of d/Λ on confinement loss and the loss ratio between the second-order mode and fundamental mode for an LCF with two layers of features, $\Delta n=1.2\times10^{-3}$ and $2\varrho=50\mu m$. [35].

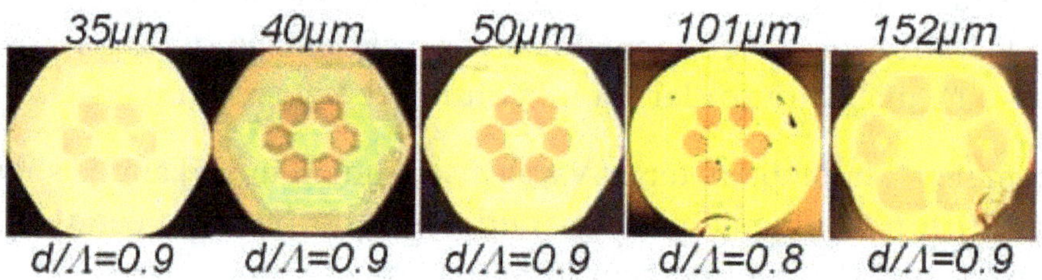

Figure 12. Some examples of fabricated all Glass LCFs. Core diameter is given above the fiber [35].

A wide range of all-glass LCFs were fabricated from core diameters from 35μm to well over 100μm [35] (see Figure 12). All fibers were made with silica glass as the background glass and slightly fluorine-doped silica glass as the cladding features and coated with standard coating with index of 1.54. The refractive index difference between the background and the low index feature wasΔ $n=1.2\times10^{-3}$. LCFs with both circular and hexagonal features were fabricated and tested. The conditions for the fabrication of LCFs with hexagonal features also created LCFs with a rounded hexagonal outline. Such a shape is known to be preferred for the pump mode mixing in a double clad fiber where the pump light propagates in a much larger cladding guide. All the fabricated LCFs in Figure 12 operated in the fundamental mode with a varying degree of bend loss performance. In general, bend loss increases rapidly with core diameter increase (see Figure 13). This effect is fundamentally related to the fact that the ability of guided modes to navigate a bend is related to how rapidly a mode can change its spatial pattern without breaking up while propagating, i.e. maintain adiabatic transition. As the mode gets larger, this ability to change diminishes very quickly due to larger Rayleigh range.

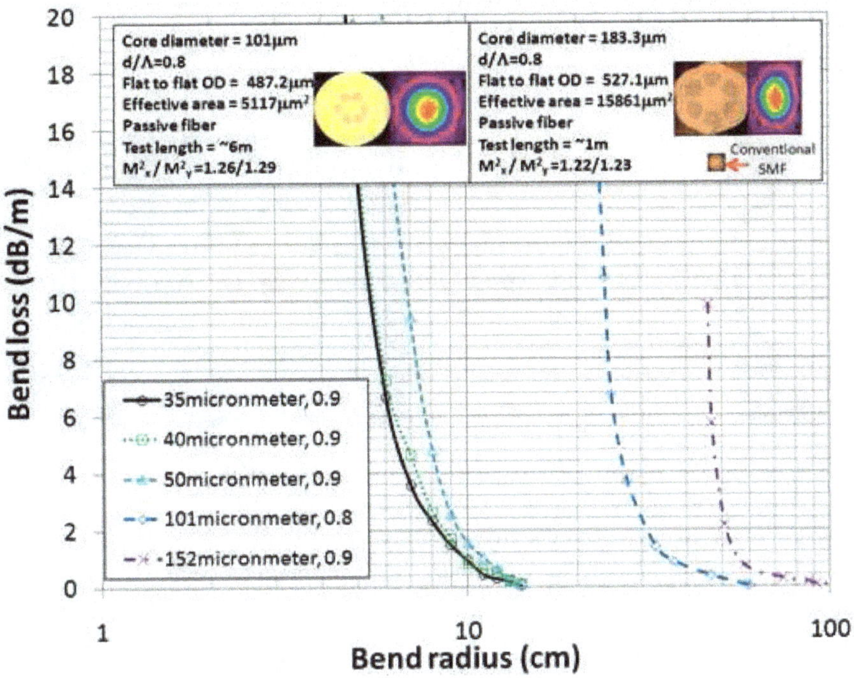

Figure 13. Cross section, measured mode and fiber details are given for the LCF with 101μm core, left inset, and the LCF with 183.3μm core, right inset. Measured bend loss for LCFs with various core diameters [36].

The LCF, shown on the top left inset in Figure 13, had 2ρ=101μm, and d/Λ=0.9. The effective mode area of the LCF was calculated to be 5117μm^2 (MFD=80.7μm). A length of the LCF ~6m long was loosely coiled in a 1m coil and the measured M^2 was M^2_x=1.26 and M^2_y=1.29. The LCF, shown on the top right inset, had 2ρ=183.3μm, and d/Λ=0.8. A conventional single mode optical fiber of the same scale is also shown for comparison. The effective mode area of this LCF was calculated to be 15861μm^2 (MFD=142.1μm), a record effective mode area for single-mode operation. The Measured M^2 of a 1m straight fiber is M^2_x=1.22 and M^2_y=1.23. Measured mode patterns at the output of the fibers are also shown in Figure 13.

3.3. Polarization maintaining all-glass leakage channel fibers

A PM all-glass LCF was first reported in [37] (see Figure 14). The passive all-glass PM LCF had a core diameter of 50μm. A high d/Λ=0.9 was used for a smaller critical bend radius. The LCF had a refractive index difference between the background and the low index feature of $\Delta n=1.2\times10^{-3}$. The low index features were made of slightly fluorine-doped silica. Two stress elements with a refractive index of ~13 × 10^{-3} below that of the background silica glass were used instead of the regular features on either sides of the core to provide birefringence. The fiber had an outer diameter of ~885μm and was coated with standard acrylic coating. The near field image measured with a single lens is shown in Figure 14(c) for the 1.8m long sample and in Figure 14(d) for the 30m sample. Due to the much higher d/Λ=0.9 used for this fiber, some bending was necessary for fundamental mode operation in a short length of this fiber. The output was robustly single mode in a 30m long sample of this passive LCF coiled in 40cm diameter coils (see Figure 14(d)). The critical bend radius for 1dB/m loss, expected to be ~11cm by FEM simulation, matched very well to the 10.5cm measured. *Polarization extinction ratio* (PER) was characterized at the output of a 1.8m long sample to be >15dB over 1010-1100nm.

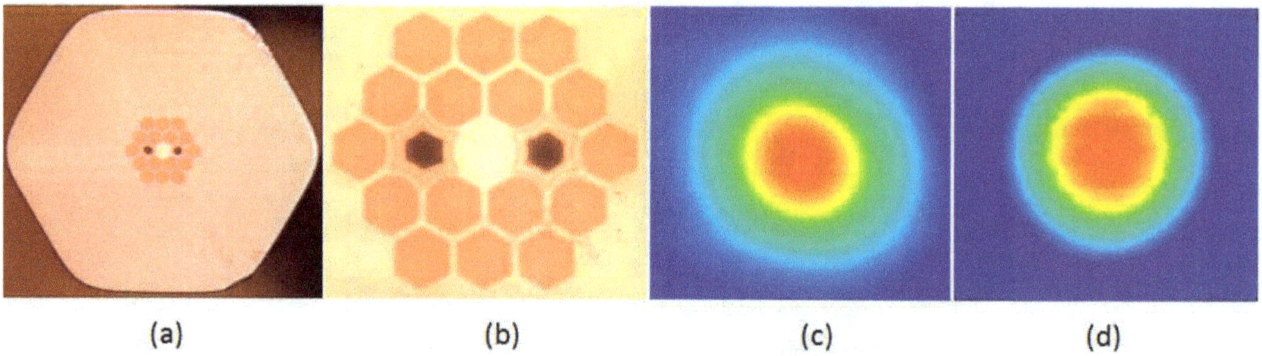

Figure 14. (a) Cross section of the passive PM LCF, (b) magnified cross section, (c) near field from the 1.8m long fiber in a 40cm-diameter coil, (d) near field from the 30m long fiber in a 40cm-diameter coil [37].

3.4. Characterization of mode losses in all-glass leaky channel fibers

Recently, fundamental and higher order mode losses have been characterized and compared to simulations based on the assumption of an infinite cladding [38]. A LCF with a ~50μm core diameter and a hexagonal cladding boundary (see Figure 15(a)) was spliced to a tunable source to ensure launch stability during the measurements. The passive LCF was coated with a low-index polymer to simulate a double-clad fiber with a pump NA of ~0.45. Power in various modes at the output was measured using the S^2 method [39]. The fiber was cut back several times to determine the propagation losses of various modes. The measurement was repeated at various coil diameters. The results are summarized in Figure 15(b) and show remarkable agreement between the measured and simulated losses of fundamental and higher-order modes. LP11 mode loss as high as ~20dB/m was measured, demonstrating the validity of the design. A similar LCF with a circular cladding boundary was also measured, showing significant less higher-order mode losses than those predicted by simulations. It is speculated that the coherent reflection at the circular cladding boundary played a significant role in enhancing the guidance of leaky higher-order modes in this case. It is, therefore, critical to have a non-circular cladding boundary to achieve the maximum possible higher-order mode losses.

3.5. Ytterbium-doped all-glass leakage channel fibers

An ytterbium-doped all-glass LCF with one layer of cladding features was also reported in [37]. The LCF was coated with a low index polymer, providing a pump NA of 0.45. This LCF had pump absorption of 11dB/m at ~976nm. The LCF also had 2ϱ=52.7μm and d/Λ=0.8. This gives a simulated effective area of 1548μm^2 at 1.05μm. The LCF has a rounded hexagonal shape and a flat-to-flat dimension of 254.2μm. The fiber was used to demonstrate amplification of 600ps pulses (with 600μJ pulse energy) to 1MW peak power.

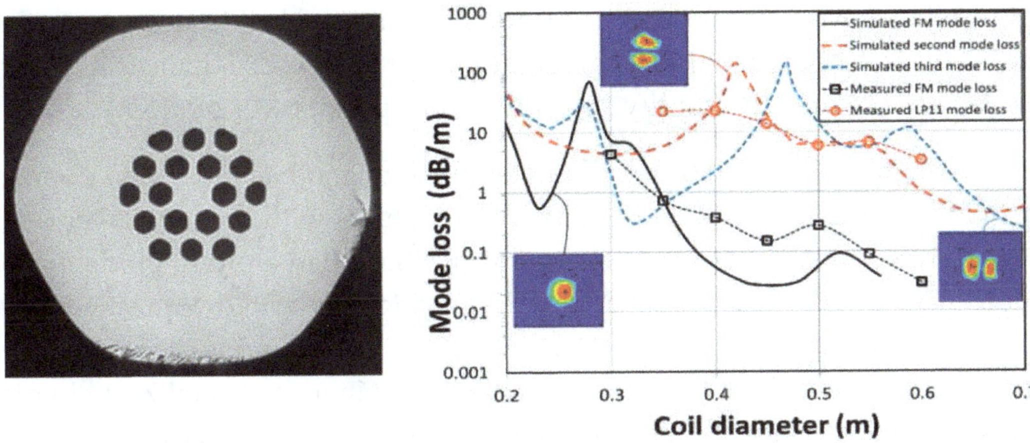

(a) (b)

Figure 15. (a) The hexagonal LCF used in the measurement and (b) simulated and measured mode loss in the hexagonal Re-LCF [38].

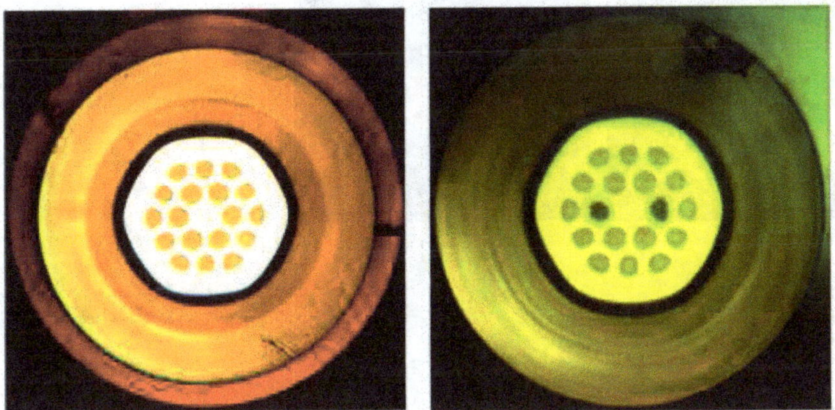

(a) (b)

Figure 16. An all glass active LCF (left) and an all glass PM active LCF with highly fluorine-doped pump cladding (right) [37].

Ytterbium double-clad all-glass LCFs with highly fluorine-doped silica as pump cladding were also demonstrated (see Figure 16) [37]. All-glass LCFs have no polymer in the pump path and have independent control of the fiber outer diameters and pump cladding dimensions. This, therefore, enables designs with smaller pump guides for higher pump absorption and, at the same time, with larger fiber diameters to minimize micro and macro bending effects, a much desired feature for large core fibers where intermodal coupling could be an issue due much increased mode density. Stress rods can also be added for PM LCFs (see Figure 16).

The LCFs had a refractive index difference between the background and the low index feature of $\Delta n=1.2\times10^{-3}$. The non-PM LCF (see the left figure in Figure 16) had an inner layer $d/\Lambda=0.8$ and outer layer $d/\Lambda=0.7$. It had a 47μm core diameter, a rounded hexagon pump guide with a dimension of 238μm by 256μm, a pump NA of 0.28, pump absorption of ~12dB/m, and an outer diameter of ~538μm coated with standard high index coating, shown as the outermost layer in Figure 16(a). A 3.5m fiber coiled in 53cm diameter was used to demonstrate a slope efficiency of 75% in an amplifier [37]. A single stage gain of 33dB was demonstrated using this fiber. It was also used to directly amplify 15ps pulses to a peak power of ~1MW.

The PM LCF (see Figure 16(b)) had a core diameter of 80μm and had a fluorine-doped pump cladding, providing a pump NA of ~0.28. Low index features with an inner layer d/Λ of 0.8 and an outer layer d/Λ of 0.7 were used. This active PM LCF had a pump guide diameter of ~400μm (flat-to-flat), a fiber outer diameter of ~835μm, and was coated with standard acrylic coating. Pump absorption was estimated to be ~12dB/m. The mode field diameter was measured to be ~62μm. The fiber was used as an amplifier in a single coil 76cm in diameter with a length of straight section at each end, demonstrating a slope efficiency of ~74% and a maximum single-path gain in excess of 30dB [37]. It demonstrated direct amplification of 14.2ps pulses to 190kW peak power with pulse energy of 2.74μJ and negligible SPM spectral broadening. M^2 was measured to be below 1.35 for the entire output power range.

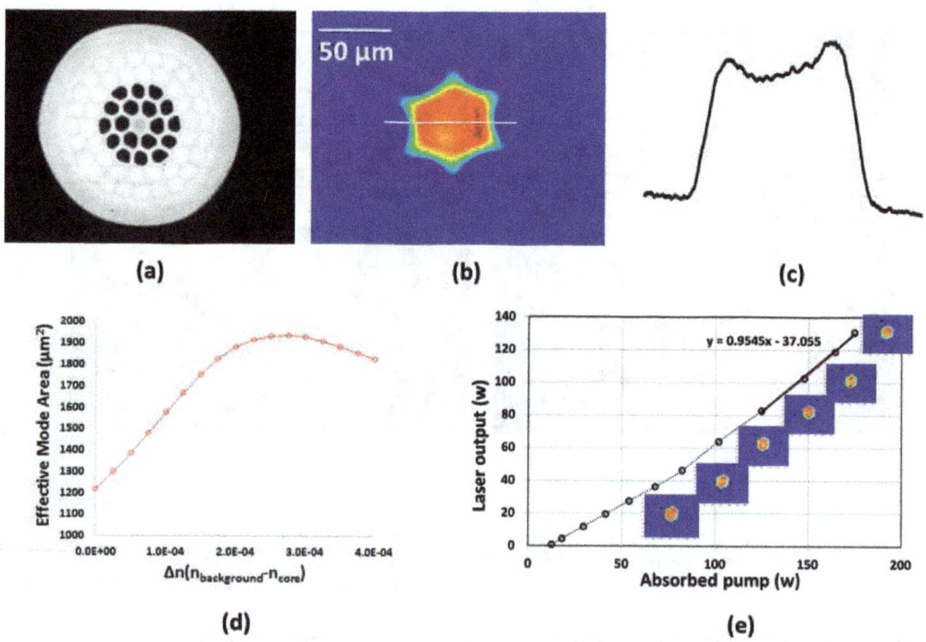

Figure 17. (a) The ytterbium-doped LCF with an index depression in the core center, (b) measured near field intensity of the guided mode, (c) measured mode intensity distribution, (d) simulated effective mode area versus the index depression, and (d) measured laser output and near field patters at various powers [40].

Recently, a flat-top mode has been demonstrated in an ytterbium-doped LCF with a ~50μm core by introducing an area ~30μm in diameter in the core center with a refractive index of ~2×10^{-4} lower than that of the background glass (Figure 17(a)) [40]. The flat-top mode (see Figure 17(b) and (c)) increased the effective mode area of the LCF from ~1200μm^2 to ~1900μm^2, a ~50% increase (see Figure 17(d)). The LCF also demonstrated near quantum-limited efficiency (see Figure 17(e)). Lasing wavelength was 1026nm and the pump was at 976nm.

4. Higher-order-mode fibers

The concept and demonstration of robust propagation of *higher-order-modes* (HOM) with large effective mode areas were first reported in 2006 [11]. It was argued that the stability of mode propagation in multimode fibers is critically dependent on the effective mode index difference between the propagating mode and its nearest neighbor anti-symmetric mode. This effective mode index difference is actually larger for the higher-order LP_{04}-LP_{07} modes than for the fundamental mode (see Figure 18). In addition, a *long period grating* (LPG), a fairly matured technology, can be used for broad band mode conversion to and from those higher-order LP0n modes.

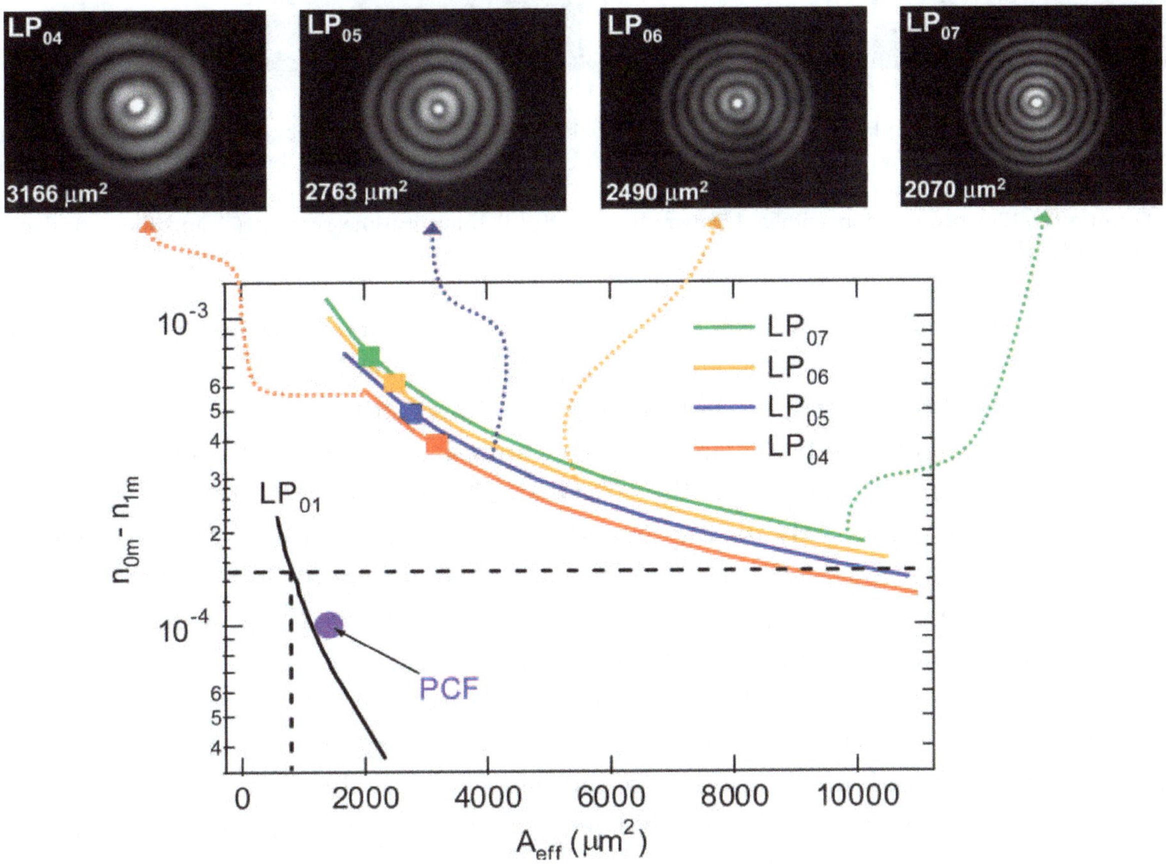

Figure 18. Effective index difference between nearest neighbor anti-symmetric mode versus A_{eff}. Top, near-field images of LPG-excited HOMs after >2m propagation with 7cm bend and with A_{eff} ranging from 2100 to 3200μm² [11].

Mode coupling first requires phase-matching. Perturbations on the fiber need to provide a vector which equals the difference in the propagation constants of the two modes involved. The larger the mode index difference between the two modes, the larger is the wave vector required for phase matching. A large wave vector implies high spatial frequency and small spatial period. For a mode index difference of 1.5×10^{-4}, the required spatial period of the perturbation at 1.55μm is ~10mm. Due to geometric constraints, the spatial frequency distribution of perturbations on fibers usually cuts off at a certain upper frequency limit. This would minimize mode coupling between modes with larger effective mode index differences. The second requirement for mode coupling is that the overlap integral among the two modes and the perturbation needs to be non-zero. This requires the perturbation to break up the mode orthogonality. Although it was not explicitly spelled out in [11], it was assumed that the perturbations on fibers are mostly anti-symmetric. In this case, mode coupling dominates between LP0n and its anti-symmetric counterpart LP1n modes in optical fibers.

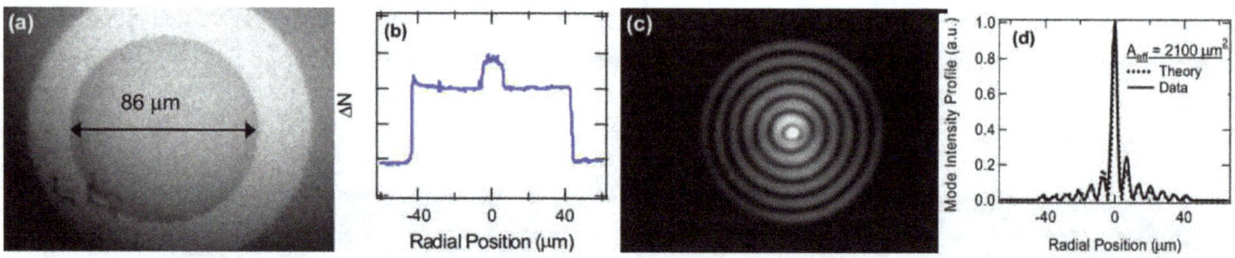

Figure 19. Characteristics of an HOM fiber. The horizontal and vertical scales of the images are identical. (a) Near-field image of a fiber facet, showing the 86μm inner cladding. (b) Refractive-index profile of the HOM fiber, with a core similar to an SMF with an 86μm inner cladding and a down-doped outer trench. (c) Near-field image of the LP_{07} mode at 1600 nm after 12m propagation with a 4.5cm radius bend. (d) Intensity line scan of (c) and theoretical profile: A_{eff}=~2100μm^2 [11].

The effective mode index difference for LP_{01}, LP_{04}, LP_{05}, LP_{06} and LP_{07} modes with their nearest neighbor anti-symmetric modes are shown versus effective mode area in Figure 18 [11]. The near-field images of the measured HOM modes are shown at the top of in Figure 18 and in Figure 19. It is clear that the effective mode index difference is significantly larger for the LP_{04}-LP_{07} modes than for the LP_{01} mode with the same effective mode area. The effective mode index difference also increases slightly for higher order LP_{0n} modes with the same effective mode area. For the same fiber, the effective mode area is actually larger for lower-mode-order LP_{0n} modes. The schematic of the proposed system is shown in Figure 20(a), where two identical LPGs are required for mode conversion from the LP_{01} mode to the LP_{07} mode and back to the LP_{01} mode. Robust propagation of the LP_{07} mode with an effective mode area of 2070μm^2 was demonstrated at 1600nm with the arrangement shown in Figure 19(c). The performance of the mode converter with high efficiency over broad bandwidth is shown in Figure 20(b). It was found that the LP_{07} mode suffered negligible bend loss at coil diameters down to 12cm. It was also found that modal stability increases with mode order. In a second paper [12], the effective mode areas of LP_{07} mode was simulated at various bend radii, showing stronger bend resistance than both LP_{03} and LP_{01} modes.

Recently, an erbium-doped HOM amplifier was demonstrated [41]. A small inner core was designed to have a LP_{01} mode MFD of 9μm, allowing for effective excitation of the LP_{01} mode when the HOM fiber was spliced to a single-mode fiber. Higher-order modes expanded to occupy the outer core. The LP_{010} mode had an effective mode area of 2700μm^2. Both the inner and the outer core were doped with erbium with absorption of ~30dB/m at 1530nm. The amplifier was both seeded at 1564nm and pumped at 1480nm in the LP_{010} mode. The pump was a Raman fiber laser.

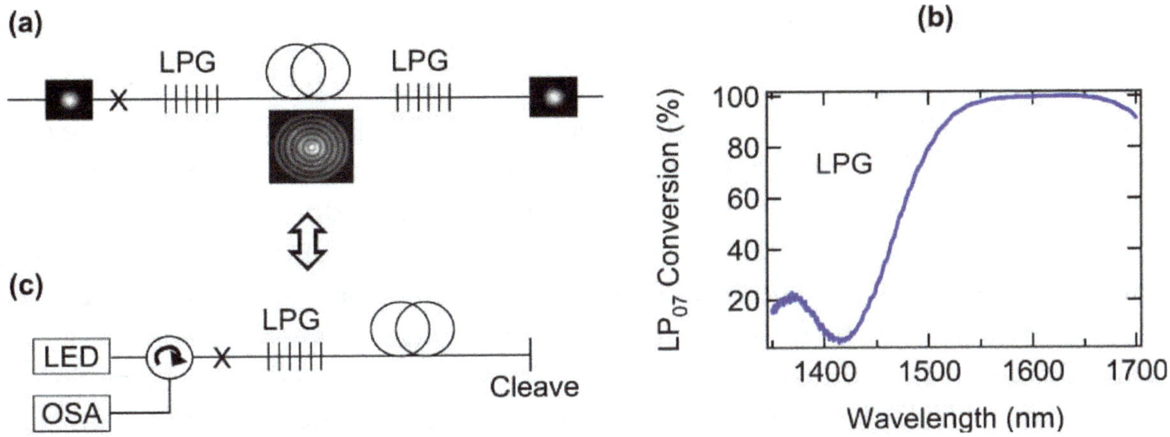

Figure 20. (a) Device schematic: light is coupled into and out of HOM with LPGs whose conversion efficiency being shown in (b). LPG with broadband coupling with efficiency >99% over 94 nm and with peak coupling efficiency as high as 99.93%. (c) Alternative schematic for characterizing HOM fibers: the cleave serves to fold the device propagation path so that the single LPG acts as both the input and the output LPG. X, splice; OSA, optical spectrum analyzer [11].

The difference in effective index between nearest neighbor anti-symmetric modes at λ=1564 nm as a function of their effective area is plotted as points in Figure 21(a) for the LP_{01} through LP_{010} modes in the fiber. The LP_{02} and LP_{03} modes have a large A_{eff}, but small mode spacing. As the mode order increases, mode spacing increases too, while A_{eff} decreases. Figure 21(b) shows the calculated intensity profiles at λ=1564nm for the LP_{01} and LP_{010} mode.

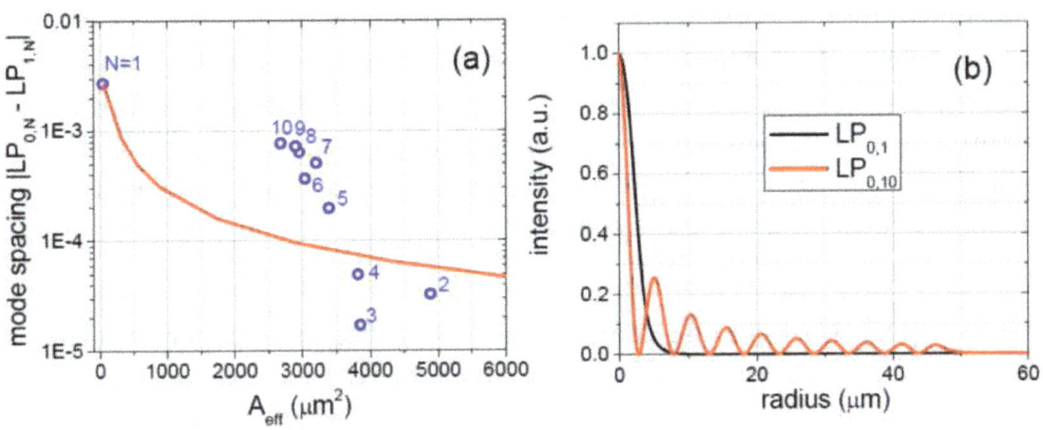

Figure 21. (a) Mode spacing as a function of effective area for the LP_{0n} modes in the HOM fiber (points) compared to a conventional LP_{01} step-index fiber with V=5 (solid curve). (b) Intensity profiles of the LP_{01} and LP_{010} modes. These calculations were done at a wavelength of 1564 nm.

A narrow line width, external cavity laser was amplified to 50mW and combined with the high-power, single mode Raman fiber laser at 1480 nm in a single-mode pump/signal combiner. The output of the pump/signal combiner was fusion-spliced to the HOM fiber. The length of the amplifier fiber after the LPG was 2.68 m. The measured slope efficiency at 1564nm was 43.2%. Over 20dB of gain was demonstrated by the amplifier.

5. Chrially-coupled core fibers

A *Chirally-coupled-core* (CCC) fiber was first reported in 2007 [9]. The fiber had a large central core and a smaller side core wound around the central core in a helical fashion (Figure 22). The preform had two parallel cores and the fiber was spun during the draw to form the helical side core. In this first report, the central core had a diameter of 35μm and a NA of 0.07. The side core had a diameter of 12μm and NA of 0.09. Edge-to-edge core separation was 2μm. The helical pitch was 6.2mm. The fiber was measured to be single-mode at 1550nm over a short length of 25cm. It was multimode below 1500nm. The simulation predicted LP_{01} mode loss to be 0.3dB/m and all HOM loss to be >130dB/m for λ>1550nm. The fiber was also confirmed to be polarization-maintaining.

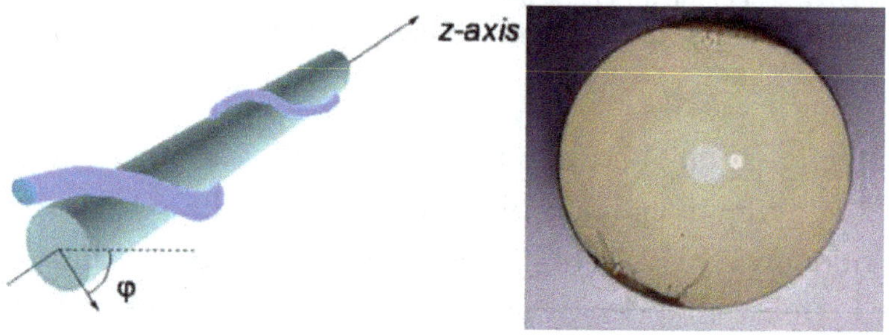

Figure 22. Structure of Chirally-coupled-core fiber [9].

The propagation of modes in the central core was affected by the coupling of modes between the central and side cores. The fiber was designed to operate where there was no fundamental mode coupling with modes in the side core. Higher-order modes in the central core were, however, coupled with the side core modes at the operating wavelength. The modes in the side core had high loss due to the tight bend from the helical arrangement. This led to high loss for the higher-order modes in the central core which were coupled to modes in the side core.

An ytterbium-doped CCC fiber was demonstrated in a subsequent paper [10]. The ytterbium-doped central core had a 33μm diameter and 0.06 NA. The undoped side core had a 16μm diameter and 0.1 NA. The helical pitch was 7.4mm and the edge-to-edge core separation was 4μm. The low index coating provided a pump NA of 0.47. The pump guide had a 250μm diameter. The measured pump absorption was 2dB/m at 915nm. The fiber demonstrated 75% slope efficiency at 1066nm in a laser configuration.

In a more recent paper [42], a more detailed theoretical analysis of quasi-phase-matching (QPM) assisted by spin and orbital angular momentum was given. For two LP modes LP_{l1m1} and LP_{l2m2}, QPM is achieved when

$$\beta_{l1m1} - \beta_{l2m2}\sqrt{1 + K^2R^2} - \Delta mK = 0 \tag{2}$$

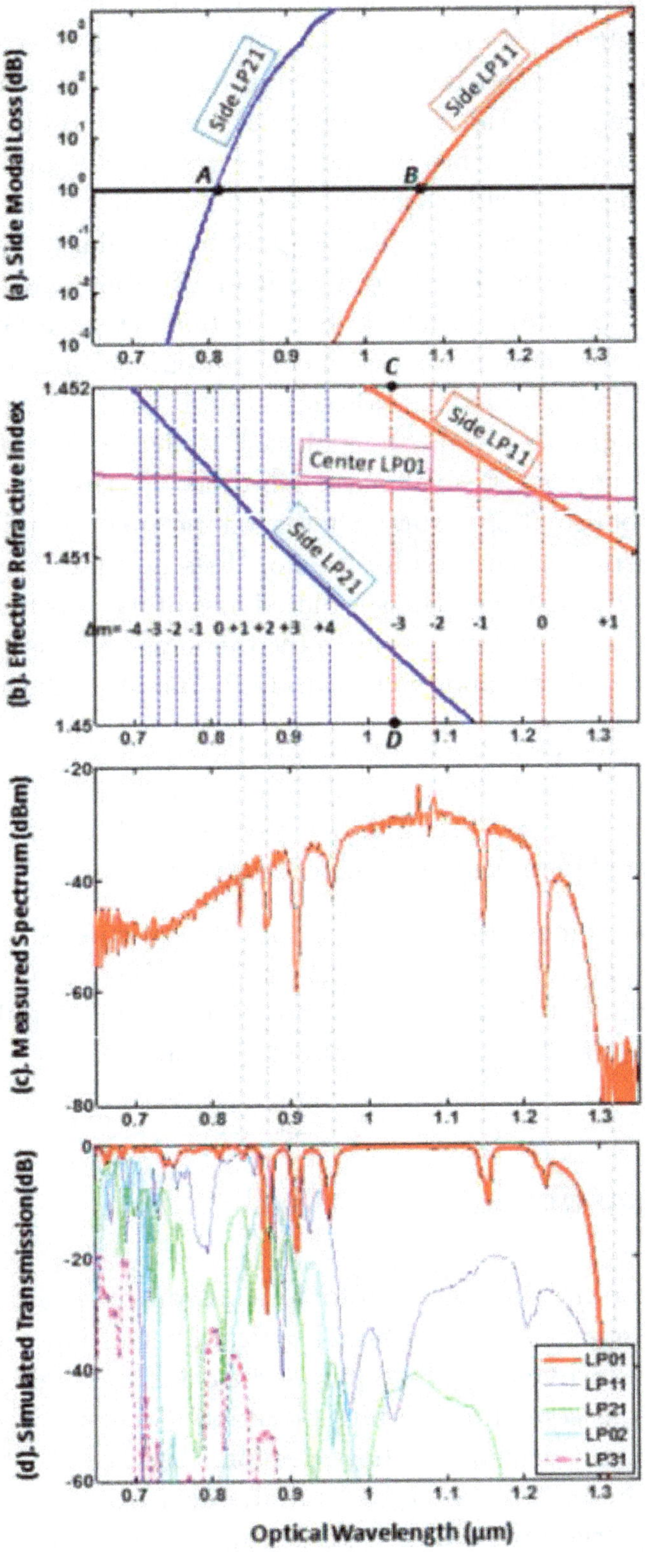

Figure 23. Calculation and measurement of quasi-phase-matching (QPM) for 1.5m-long CCC fiber sample. (a) Calculated loss for side-core LP_{11} and LP_{21} modes as a function of wavelength (top). (b) Calculated refractive indices of interacting modes and calculated QPM resonance positions. (c) Transmission of central core. (d) Simulated transmission of central core.

where β_{l1m1} and β_{l1m2} are the respective propagation constants of the two modes; $K=2\pi/\Lambda$; and Λ is helical pitch. $\Delta m=\Delta l+\Delta s$; $\Delta l=\pm l1\pm l2$; Δs=-2,-1, 0, 1, 2. The loss versus wavelength for the LP_{11} and LP_{21} modes in the side core is plotted in Figure 23(a) showing high loss for these modes in certain wavelength regimes. The mode indices of the LP_{11} and LP_{21} modes of the side core and LP_{01} mode of the central core are plotted in Figure 23(b). It can be seen clearly that the LP_{11} mode and LP_{21} modes of the side core naturally phase-match to the LP_{01} mode of the central core at ~1.22μm and ~0.81μm respectively. The QPM by angular momentum from the helical side core extends the phase matching to a number of other wavelengths determined by equation 2. These phase matching wavelengths are plotted in Figure 23(b) as red dotted vertical lines for the LP_{11} mode of the side core and the LP_{01} mode of central core coupling and blue dotted vertical lines for the LP_{21} mode of side core and LP_{01} mode of central core coupling. The measured transmission of the central core is plotted in Figure 23(c). The analysis of the CCC fiber can be simplified significantly in a helical coordination. The Maxwell equations keep the same form in the new coordination system, but the tenor form of permittivity and permeability needs to be transformed [42]. The resulting tenor does not have any z-dependence, which significantly simplifies the analysis. The simulated loss of various modes is shown in Figure 23(d), demonstrating that the narrow peaks in the transmission arise from coupling between the LP_{01} mode in the central core and the LP_{11} and LP_{21} modes in the side core.

The high loss of the LP_{01} mode in the central core at λ>1.3μm was not explained in the paper. It may be due to the angular momentum assisted coupling between LP_{01} mode in the central core and LP_{01} mode in the side core. This long wavelength cut-off has been used for the suppression of stimulated Raman scattering in fibers [43].

The higher loss for the higher-order-modes in the central core between 1-1.1μm in Figure 23(d) was not clearly explained either. For the LP_{11} mode in the central core, one possible reason for its high loss is coupling to LP_{21} mode of the side core through angular momentum-assisted QPM. Recently, a 60μm core CCC fiber was also demonstrated [44].

6. Photonic bandgap fibers

The guidance of light in a defect in a photonic bandgap fiber is based on fundamentally different principles compared to conventional optical fibers. In a conventional optical fiber, the core has a higher refractive index than that of the cladding. Light is guided through total internal reflection at the core-and-cladding boundary. In a photonic bandgap fiber, the photonic crystal cladding is designed to have photonic bandgaps, where there is an absence of modes propagating in the direction parallel to the fiber axis. Once a defect core is created within such a photonic crystal cladding, light is trapped to propagate only in modes of the defect core as it is forbidden to propagate in the cladding. This leads to significantly lower waveguide loss for the modes in the defect core within the cladding photonic bandgaps. This waveguide loss becomes zero when the photonic crystal cladding is infinite, but there is always a finite loss for practical fibers with finite cladding. This waveguide loss can, however, be significantly lowered by increasing the number of cladding layers.

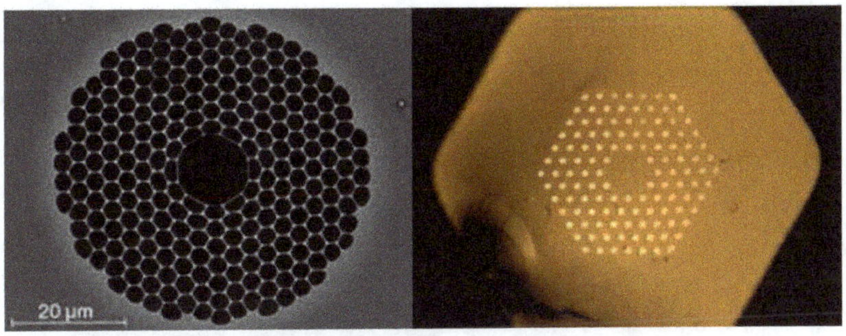

Figure 24. Images of a hollow-core photonic bandgap fiber (left) and an all-solid photonic bandgap fiber (right).

One unique feature of photonic bandgap fibers is that the defect core always has a lower refractive index than that of the higher refractive index material in the cladding. This leads to the possibility of hollow-core photonic bandgap fibers (see Figure 24, left), where light is guided mostly in the air in the hollow-core of the fiber. These fibers have extremely low nonlinearities and are well suited for high power laser delivery. The second type of photonic bandgap fibers are made entirely of glass (see Figure 24, right). These all-solid photonic bandgap fibers have a cladding which consists of a background glass and nodes of slightly higher refractive index. The much lower refractive index contrast (typically just a few percent) in the photonic crystal cladding of the all-solid photonic bandgap fibers still allows photonic bandgaps for paraxial propagation. Another important feature of photonic bandgap fibers is that their transmission is highly wavelength-dependent, i.e. low loss is possible only within the photonic bandgaps of the photonic crystal cladding. This distributive spectral filtering along a fiber can be very useful for range of potential applications including use in fiber lasers at low gain regimes to minimize gain competition, the suppression of stimulated Raman scattering, etc.

Birks et al. clearly explained the origin of the modes in the photonic crystal cladding of an all-solid photonic bandgap fiber in [45]. The density of states (DOS) of modes in the photonic crystal cladding is represented by the shades of grey in Figure 25 [45]. The refractive index of the background glass is represented by the black horizontal line. It is the cut-off line for the modes guided in the cladding nodes. The bandgap regime is represented by the red area. The fundamental mode of the defect core is illustrated by the yellow lines. The modes in the photonic crystal cladding clearly originate from the guided modes of the nodes, which are also labeled at the top of the figure. Above the background index, the modes form relatively narrow bands. Below the background index, the bands of modes significantly broaden, due to strong coupling among nodes as a result of light becoming more spread into the background glass below cut-off. The guidance property can be easily understood once it is recognized that there is a simple analogue to conventional optical fibers [46]. The core index is simply the index of the background glass and the equivalent cladding index is the upper boundary of the photonic bandgap. The effective index of the modes in the defect core (see yellow lines in Figure 25) falls between the core and cladding indices as in conventional optical fibers.

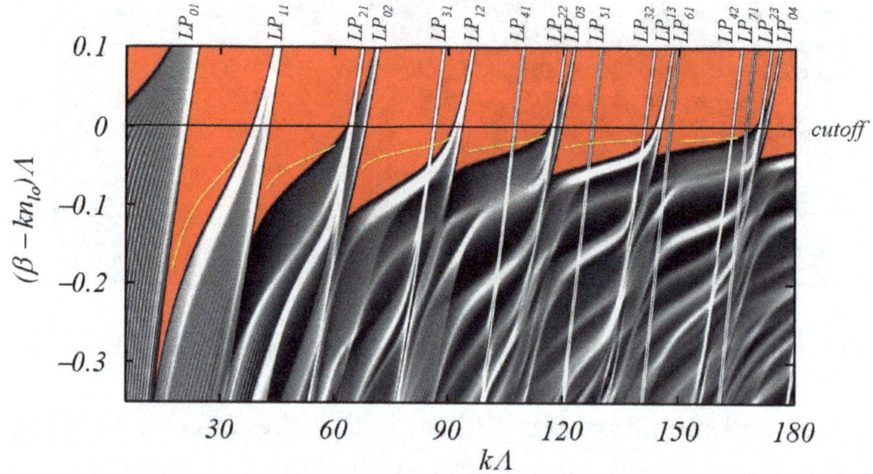

Figure 25. Plots of band structure for an example bandgap fiber. The bandgaps are shown in red. The node modes from which the bands arise are labeled along the top. Density of states (DOS) calculated using the plane-wave method, with light grey corresponding to high DOS. The yellow curve is the "fundamental" core-guided mode [45].

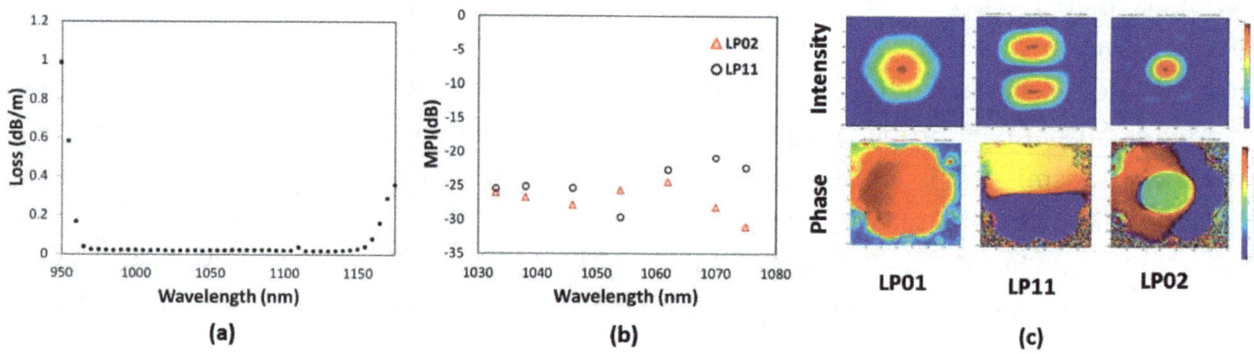

Figure 26. (a) Transmission of the 55μm core all-solid photonic bandgap fiber, (b) measured HOM contents with a S^2 method in a 5m fiber coiled at 70cm diameter, and (c) measured mode images [51].

One interesting application of ytterbium-doped all-solid photonic bandgap fibers is for the lasing of ytterbium at the long wavelength regime of 1150-1200nm [47]. It is normally difficult to lase in ytterbium-doped fibers at these extremely long wavelengths due to strong gain competition from the short wavelength regime of 1030-1070nm. The distributive spectral filtering in a photonic bandgap fiber can be used to minimize gain at short wavelengths, leading to efficient high power lasers in the extremely long wavelength regime.

In all-solid photonic bandgap fibers, a mode is guided only when it falls within the photonic bandgap of the cladding lattice. Guidance can therefore be highly mode-dependent. This provides great potential for creating designs that support only the fundamental mode, i.e. selective mode guidance. Mode area scaling to 20μm mode field diameter using all-solid photonic bandgap fibers was reported in [48]. Recently, all-solid photonic bandgap fibers with up to ~700μm^2 effective mode area have been demonstrated operating in the first bandgap [49, 50]. More recently, a fiber with a core diameter of ~55μm and an effective mode area of

920μm² at coil diameter of 50cm was demonstrated (see right figure in Figure 24). The fiber operates in the third bandgap with transmission shown in Figure 24(6). At 70cm coil diameter, a 5m fiber showed HOM contents below 25dB (see Figure 26(b) and (c)). At the design coil diameter of 50cm, HOM content was below-30dB [51].

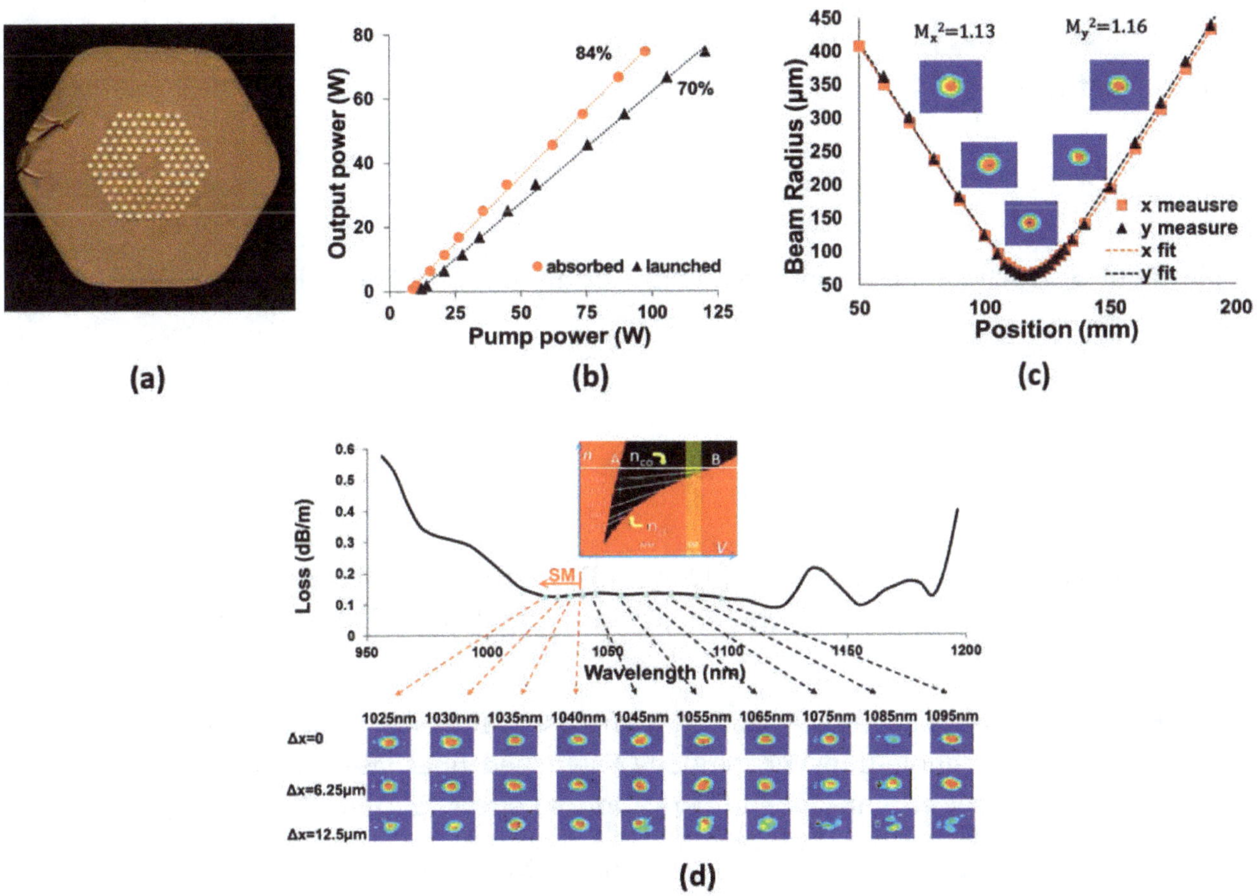

Figure 27. (a) Ytterbium-doped ~50μm core all-solid photonic bandgap fiber, (b) Laser efficiency, (c) M^2 measurement and (d) mode at various wavelengths across the bandgap [52].

Recently, an ytterbium-doped all-solid photonic bandgap fiber with ~50μm core diameter has also been demonstrated (Figure 27(a)) [52]. The fiber demonstrated high efficiency and excellent mode quality (see Figure 27(b) and (c)). The fiber also demonstrated robust single-mode behavior near the short wavelength edge of the bandgap by monitoring the output when moving away from the optimum launch condition (see Figure 27(d)). This is exactly expected from the dispersive nature of the photonic bandgap cladding.

7. Conclusions

In this chapter, we have briefly introduced a number of emerging fiber technologies for mode area scaling for fiber lasers. These new technologies are critical for the further power scaling of fiber lasers. The basic principles of these technologies were introduced and the latest developments were discussed. This is still a very active area of research. With further development, there is great potential to significantly improve throughput and expand the capabilities of fiber lasers for use in manufacturing, as well as to meet future defense needs.

Author details

Liang Dong*

Department of Electrical and Computer Engineering, University of Clemson, Clemson, SC, USA

References

[1] D. Taverner, D.J. Richardson, L. Dong, J.E. Caplen, K. Williams, R.V. Penty, "158μJ pulses from a single transverse mode, large mode-area EDFA", Optics Letters, 22, 378-380 (1997).

[2] H.L. Offerhaus, N.G. Broderick, D.J. Richardson, R. Sammut, J. Caplen and L. Dong, "High-energy single-transverse-mode Q-switched fiber laser based on a multimode large-mode-area erbium-doped fiber", Optics Letters, 23, 1683-1685 (1998).

[3] C. Knight, T. A. Birks, P. St. J. Russell, and D. M. Atkin, "All-silica single-mode optical fiber with photonic crystal cladding," Opt. Lett. 21, 1547-1549 (1996).

[4] T.A. Birks, J.C. Knight and P.St.J. Russell, "Endlessly single-mode photonic crystal fiber," Opt Lett. 22 961-963 (1997).

[5] J.C. Knight, T.A. Birks, R.F. Cregan, P.St.J. Russell and J.P. de Sandro, "Large mode area photonic crystal fibre," Elect. Lett. 34, 1347-1348 (1998).

[6] C.D. Brooks and F. Teodoro, "Multi-MW peak-power, single-transverse-mode operation of a 100μm core diameter, Yb-doped rodlike photonic crystal fiber amplifier," Appl. Phys. 89, 111119 (2006).

[7] W.S. Wong, X. Peng, J.M. McLaughlin and L. Dong, "Breaking the limit of maximum effective area for robust single-mode propagation in optical fibers", Optics Letters, 30, 2855-2857 (2005).

[8] L. Dong, X. Peng and J. Li, "Leakage channel optical fibers with large effective area", Journal of Optical Society of America B, 24, 1689-1697 (2007).

[9] C.H. Liu, G. Chang, N. Litchinitser, A. Galvanauskas, D. Guertin, N. Jacobson and K. Tankala, "Effective single-mode chirally-coupled core fiber," Advanced Solid State Photonics, paper ME2, 2007.

[10] A. Galvanauskas, M.C. Swan, and C.H. Liu, "Effectively-Single-Mode Large Core Passive and Active Fibers with Chirally-Coupled-Core structures," Conference on Lasers and Electro-Optics, paper CMB1, 2008.

[11] S. Ramachandran, J. W. Nicholson, S. Ghalmi, M. F. Yan, P. Wisk, E. Monberg, and F. V. Dimarcello, "Light propagation with ultralarge modal areas in optical fibers," Opt. Lett. 31, 1797-1799 (2006).

[12] J.M. Fini and S. Ramachandran, "Natural bend-distortion immunity of higher-order-mode large-mode-area fibers" Opt. Lett. 32, 748-750 (2007).

[13] M. Koshiba and K. Saitoh, "Applicability of classical optical fiber theories to holey fibers," Opt. Lett. 29, 1739-1740 (2004).

[14] K. Saitoh, Y. Tsuchida, and M. Koshiba, "Endlessly single-mode holey fibers: the influence of core design," Opt. Express 13, 10833-10839 (2005).

[15] B. T. Kuhlmey, R. C. Mcphedran, and C. M. de sterke, "Modal cutoff in microstructured optical fiber," Opt. Lett. 27, 1684-1686 (2002).

[16] J. Zhou, K. Tajima, K. Nakajima, K. Kurokawa, C. Fukai, T. Matsui, and I. Sankawa, "Progress on low loss photonic crystal fibers," Optical Fiber Technology 11 101–110 (2005).

[17] T. P. White, B. T. Kulmey, R. C. McPhedran, D. Maystre, G. Renversez, C. Martijn de Sterke, and L. C. Botten, "Multipole method for microstructured optical fibers. I. Formulation," J. Opt. Soc. Am. B 19, 2322-2330 (2002).

[18] B.T. Kuhlmey, T.P. White, G. Renversez, D. Maystre, L.C. Botten, C. M. de Sterke, and R. C. McPhedran, "Multipole method for microstructured optical fibers. II. Implementation and results," J. Opt. Soc. Am. B 19, 2331-2340 (2002).

[19] W.J. Wadsworth, J.C. Knight and P.St.J. Russell, "Large mode area photonic crystal fiber laser," Conference on Lasers and Electro-Optics, CWC1, 2001.

[20] W.J. Wadsworth, R.M. Percival, G. Bouwmans, J.C. Knight and P.St.J. Russell, "High power air-clad area photonic crystal fiber laser," Opt. Express 11, 48-53 (2003).

[21] D.J. DiGiovanni and R.S. Windeler: "Article comprising an airclad optical fiber," US patent 5907652 (1999).

[22] V. A. Kozlov, J. Hernández-Cordero, R. L. Shubochkin, A. L. G. Carter, and T. F. Morse, "Silica–Air Double-Clad Optical Fiber," IEEE Photonics Technology Letters 12, 1007-1009 (2000).

[23] J.K. Sahu, C.C. Renaud, K. Furusawa, R. Selvas, J.A. Alvarez-Chavez, D.J. Richardson and J. Nilsson, "Jacketed air-clad cladding pumped ytterbium-doped fibre laser with wide turning range," Elect. Lett. 37, 1116-1117 (2001).

[24] K. Furusawa, A. Malinowski, J. H. V. Price, T. M. Monro, J. K. Sahu, J. Nilsson and D. J. Richardson, "Cladding pumped Ytterbium-doped fiber laser with holey inner and outer cladding," Opt. Express 9, 714-720 (2001).

[25] W.J. Wadsworth, R.M. Percival, G. Bouwmans, J.C. Knight, T.A. Birks, T.D. Hedley, and P.St.J. Russell, "Very high numerical aperture fibers," IEEE Photonic Technology Letters 16, 843-845 (2004).

[26] J. Limpert, T. Schreiber, S. Nolte, H. Zellmer, A. Tünnermann, R. Iliew, F. Lederer, J. Broeng, G. Vienne, A. Petersson, and C. Jakobsen, "High-power air-clad large-mode-area photonic crystal fiber laser," Opt. Express 11, 818-823 (2003).

[27] J. Limpert, T. Schreiber, A. Liem, S. Nolte, H. Zellmer, T. Peschel, V. Guyenot, and A. Tünnermann, "Thermo-optical properties of air-clad photonic crystal fiber lasers in high power operation," Opt. Express 11, 2982-2990 (2003).

[28] J. Limpert, A. Liem, M. Reich, T. Schreiber, S. Nolte, H. Zellmer, A. Tünnermann, J. Broeng, A. Petersson, and C. Jakobsen, "Low-nonlinearity single-transverse-mode ytterbium-doped photonic crystal fiber amplifier," Opt. Express 12, 1313-1319 (2004).

[29] J. Limpert, N. Deguil-Robin, I. Manek-Hönninger, F. Salin, F. Röser, A. Liem, T. Schreiber, S. Nolte, H. Zellmer, A. Tünnermann, J. Broeng, A. Petersson, and C. Jakobsen, "High-power rod-type photonic crystal fiber laser," Opt. Express 13, 1055-1058 (2005).

[30] J. Limpert, O. Schmidt, J. Rothhardt, F. Röser, T. Schreiber, and A. Tünnermann, S. Ermeneux, P. Yvernault, and F. Salin, "Extended single-mode photonic crystal fiber lasers," Opt. Express 14, 2715-2720 (2006).

[31] T. Schreiber, F. Röser, O. Schmidt, J. Limpert, R. Iliew, F. Lederer, A. Petersson, C. Jacobsen, K. P. Hansen, J. Broeng, and A. Tünnermann, "Stress-induced single-polarization single-transverse mode photonic crystal fiber with low nonlinearity," Opt. Express 13, 7622-7630 (2005).

[32] O. Schmidt, J. Rothhardt, T. Eidam, F. Röser, J. Limpert, A. Tünnermann, K.P. Hansen, C. Jakobsen, and J. Broeng, "Single-polarization ultra-large-mode-area Yb-doped photonic crystal fiber," Opt. Express 16, 3918-3923 (2008).

[33] L. Dong, J. Li and X. Peng, "Bend resistant fundamental mode operation in ytterbium-doped leakage channel fibers with effective areas up to 3160μm^2", Optics Express, 14, 11512-11519 (2006).

[34] X. Peng and L. Dong, "Fundamental mode operation in polarization-maintaining ytterbium-doped fiber with an effective area of 1400μm^2", Optics Letters, 32, 358-360 (2006).

[35] L. Dong, T.W. Wu, H.A. McKay, L. Fu, J. Li and H.G. Winful, " All-glass large core leakage channel fibers, " IEEE Journal of Selected Topics in Quantum Electronics, 15, invited paper, 47-53(2009).

[36] L. Dong, H.A. Mckay, A. Marcinkevicius, L. Fu, J. Li, B.K. Thomas, and M.E. Fermann, " Extending Effective Area of Fundamental Mode in Optical Fibers, " IEEE Journal of Lightwave Technology, 27, invited paper, 1565-1570(2009).

[37] L. Dong, H.A. Mckay, L. Fu, M. Ohta, A. Marcinkevicius, S. Suzuki, and M.E. Fermann, "Ytterbium-doped all glass leakage channel fibers with highly fluorine-doped silica pump cladding, " Optical Express, 17, 8962-8969(2009).

[38] G. Gu, F. Kong, T.W. Hawkins, P. Foy, K. Wei, B. Samson, and L. Dong, "Impact of fiber outer boundaries on leaky mode losses in leakage channel fibers," Optics Express 21, 24039-24048 (2013).

[39] J. W. Nicholson, A. D. Yablon, S. Ramachandran, and S. Ghalmi, "Spatially and spectrally resolved imaging of modal contents in large-mode-area fibers," Opt. Express 16, 7233-7243(2008).

[40] F. Kong, G. Gu, T.W. Hawkins, J. Parsons, M. Jones, C. Dunn, M.T. Kalichevsky-Dong, K. Wei, B. Samson, and L. Dong, "Flat-top mode from a 50 μm-core Yb-doped leakage channel fiber," Optics Express 21, 32371-32376 (2013).

[41] J.W. Nicholson, J.M. Fini, A.M. DeSantolo, E. Monberg, F. DiMarcello, J. Fleming, C. Headley, D.J. DiGiovanni, S. Ghalmi, and S. Ramachandran, "A higher-order-mode Erbium-doped-fiber amplifier," Opt. Express 18, 17651-17657 (2010).

[42] X. Ma, C.H. Liu, G. Chang, and A. Galvanauskas, "Angular-momentum coupled optical waves in chirally-coupled-core fibers," Opt. Express 19, 26515-26528 (2011).

[43] X. Ma, I,N Hu, and A. Galvanauskas, "Propagation-length independent SRS threshold in chirally-coupled-core fibers," Opt. Express 19, 22575-22581 (2011).

[44] X. Ma, A. Kaplan, and A. Galvanauskas, "Experimental characterization of robust single-mode operation of 50μm and 60μm core chirally coupled core optical fibers," PhotonicsWest, paper 8237-59, 2012.

[45] T. A. Birks, G. J. Pearce, D. M. Bird, "Approximate band structure calculation for photonic bandgap fibers," Opt. Express 14, 9483-9490 (2006).

[46] M.J.F. Digonnet, H.K. Kim, G.S. Kino, and S. Fan, "Understanding Air-Core Photonic-Bandgap Fibers: Analogy to Conventional Fibers," Journal of Lightwave Technology 23, 4146-4177 (2005).

[47] A. Shirakawa, H. Maruyama, K. Ueda, C. B. Olausson, J. K. Lyngsø, and J. Broeng, "High-power Yb-doped photonic bandgap fiber amplifier at 1150-1200 nm," Opt. Express 17(2), 447–454 (2009).

[48] O. N. Egorova, S. L. Semjonov, A. F. Kosolapov, A. N. Denisov, A. D. Pryamikov, D. A. Gaponov, A. S. Biriukov, E. M. Dianov, M. Y. Salganskii, V. F. Khopin, M. V.

Yashkov, A. N. Gurianov, and D. V. Kuksenkov, "Single-mode all-silica photonic bandgap fiber with 20-micron mode-field diameter," Opt. Express 16, 11735–11740 (2008).

[49] M. Kashiwagi, K. Saitoh, K. Takenaga, S. Tanigawa, S. Matsuo, and M. Fujimaki, "Low bending loss and effectively single-mode all-solid photonic bandgap fiber with an effective area of 650μm^2.," Opt. Lett. 37, 1292–1294 (2012).

[50] M. Kashiwagi, K. Saitoh, K. Takenaga, S. Tanigawa, S. Matsuo, and M. Fujimaki, "Effectively single-mode allsolid photonic bandgap fiber with large effective area and low bending loss for compact high-power all-fiber lasers," Opt. Express 20, 15061–15070 (2012).

[51] F. Kong, K. Saitoh, D. Mcclane,T. Hawkins, P. Foy, G.C. Gu, and L. Dong, "Mode Area Scaling with All-solid Photonic Bandgap Fibers,", Optics Express 20, 26363-26372 (2012).

[52] G. Gu, F. Kong, T.W. Hawkins, J. Parsons, M. Jones, C. Dunn, M.T. Kalichevsky-Dong, K. Saitoh, and L. Dong, "Ytterbium-doped large-mode-area all-solid photonic bandgap fiber lasers," Optics Express 22, 13962-13968(2014).

8

Dispersion Compensating Fibres for Fibre Optic Telecommunication Systems

Michal Lucki and Tomas Zeman

1. Introduction

Chromatic dispersion (CD) of optical pulses in an optical fibre influences their width. It refers to changes in propagation of particular frequency components contained in optical pulses causing extension of optical pulses. The fibre reach is then significantly limited, unless signal regenerators are used. Because the pulse spread degrades optical systems, it is necessary to prevent its origination or to eliminate its results. CD ought to be suppressed or one shall prevent its origination by specialty optical fibres, including both conventional and microstructured optical fibres (MOF). To keep the pulse length nearly constant, it is possible to use Dispersion Compensating Fibre (DCF) that shall be installed at the regenerators that provide amplification, renewal of timing and pulse duration. DCFs have negative dispersion parameter; CD accumulating between regenerators is then suppressed. Consequently, bit error rate could be improved and the fibre reach could be extended [1].

Photonic Crystal Fibres (PCF) could be a suitable solution for the problem of CD in high-speed transmission systems, especially those using Wavelength Division Multiplexing (WDM) to increase the symbol rate [2].

We systematize design approaches for specialty dispersion-tailored optical fibres in order to offer a guideline to flexibly design optical fibres used in telecommunications whose optimized CD is a key property. Last but not least, we present selected fibre designs prepared in last few years, including CD plots and information about structural parameters.

The knowledge about how to design fibres could be useful to design optical fibres, such as for example a submicron flat CD compensating fibre. In addition, fluoride compounds, as compared with silica glass, exhibit higher effective refractive index, the wider spectrum of working wavelengths (λ) or lower insertion losses. Additives using these materials are promising for extending the application towards infrared region, where fluoride glasses are usually transparent.

2. Fibres with optimized dispersion [3]

DCFs could regenerate signals that are spread as a consequence of CD. This practically means that the bit error rate at the receiver's side could be improved or the spaces between adjacent symbols could be reduced. As a result, potential bit rate could be increased. One of the approaches how to deal with CD is to use zero CD fibres, offering near zero CD at the operating λs. (It shouldn't be exactly zero because of Four Wave Mixing problem occurring when propagation is with zero dispersion and the phases of all the frequency components are matched). Another approach is to use already mentioned DCFs at the signal regenerators. CD tailoring fibres could work at some λ, e.g. at 1.55 μm, where they could have negative value of CD being few hundred picoseconds (ps/nm/km), which expresses the delay between the slowest and fastest frequency component measured at the distance of 1 km, assuming the source of radiation emitting the spectrum of 1 nm. Sample compensating fibre was published in Refs. [4] and [5]. Very low CD parameter is observed whose properties could benefit in compensation of CD accumulating in some optical network. An unsolved problem seems to be relatively narrow window of λs. Such a fibre could then compensate CD in just one λ channel of a WDM system [3].

Resultant CD is a balance between waveguide and material component. In order to obtain flat CD, the evolution of waveguide dispersion has to be exactly opposite to the one of material dispersion, which means that both ought to be optimized at each channel step by step. The situation is simple when material dispersion is linear upon λ. Then, one could employ an algorithmic approach with iterations to calculate and precisely adjust the CD at each λ. A very important assumption is that the balance between both dispersion components could be a very useful tool in designing fibres for CD tailoring by optimizing material dispersion (through the material properties) and waveguide dispersion (as a result of optimization of the shape of the waveguide) [3] [6].

2.1. Dispersion compensating fibres [2]

When a compensating fibre works at some λ, optical symbols ought to be transmitted at small spectrum of light waves, without multiplying the bit rate in many channels. Such narrowband fibres can have very low value of CD that is possible without using additives (e.g. germanium dioxide). In Ref. [7], it could be found that CD of -18 ns/nm/km is possible. Other works concentrate rather on the optimization of the structural parameters of MOFs to obtain flat CD diagram. Microstructured fibres offer much greater flexibility because geometry could be optimized not only through the core and cladding diameter, but also by changing the air filling fraction, and the lattice pitch including its arrangements. In Ref. [8], flat CD over all the telecom bands is shown [2].

Index guiding MOFs (the one without the central inclusion, having solid core) offer high flexibility in CD compensating fibres design. For example microstructured based DCFs could have dual cores [9] or they could be doped [10]. Negative value of CD parameter in DCFs with dual core structure could also be combined with the idea of making some rings with smaller holes [7][10]. Such a DCF with low CD parameter in broad telecom band is shown in Ref. [11], in addition, a square lattice (with the defining angle of 90°) was proposed. The performance of DCFs with the concentric cores could be increased by using germanium dioxide in the core [12]. The relation between the amount of additives and resultant CD is shown in Ref. [13]. Some other substrates could be used to lower the effective index (n_{eff}), such as for example fluorides [14]. On the contrary, germanium dioxide raises n_{eff}. The key feature is not to raise or lower a certain index, but to create large index contrast between the fibre's core index and its n_{eff} of the cladding [2].

Fibres with flat CD, i.e. those having low value of CD in many λ channels concurrently are often considered as wideband fibres, but they can't be used to compensate CD in all λ channels at once, because accumulated CD in each WDM channel is different. They are rather suitable to compensate CD at one channel, which can be selected from the wide range of λs that are compatible with the used DCF.Real wideband fibres should compensate CD at every λ channel at the same time. They must have CD exactly opposite to dispersion in each channel. In Ref. [15], it is shown that larger lattice pitch in the 1st ring of holes is responsible for the CD slope that must be optimized for this purpose. In addition, the hole radius in the 1st ring should be larger to enhance dispersion. CD slope property is studied in Ref. [16] in the context of wavelength division multiplexing. Another slope compensating DCF is shown in Refs. [2] and [17].

2.2. Zero dispersion wavelengths in DCFs

Because a great part of MOFs have parabolic wavelength evolution of CD and the minimum CD at low values, those fibres can have two zero-CD wavelengths (ZDW). The 1st ZDW is the one utilized at visible wavelengths, whereas the longer ZDW is used at telecom wavelengths [18]. Such Highly Nonlinear PCF (HNPCF) could exhibit positive slope at 1st ZDW and negative slope at 2nd ZDW, as shown in Ref. [19]; the 1st ZDW was at 900 nm, the 2nd one at the λ of 1.6 μm.

The high index difference between the air-filled microstructure and pure or doped silica core enables tight mode confinement resulting in a low effective area and, thereby, a non-conventional behaviour of CD. Modifying the periodic cladding (i.e., hole-sizes, lattice pitch), the waveguide dispersion and the origination of ZDWs are influenced, keeping the fibre still in single-mode operation regime [20]. Then, the strong wavelength dependence in the characteristics of the fibre will be used to determine either huge CD with large slope or nearly-zero flat CD.

The bandwidth, at which MOFs are designed to have zero CD, could be divided into three categories: working in the region between 0.55 μm and 1 μm, where 1st zero-CD point could be determined; another region between 1 μm and 1.2 μm, where 1st or 2nd ZDW could be found. The 3rd region, starting at 1.2 μm and going up to longer λs, is used to create the 2nd zero CD and is not available for the 1st zero CD. An overview of the obtainable ZDWs as a function of MOF's structure was shown in Ref. [21].

Both core size and the radius of holes exhibit significant influence on the location of both ZDWs. Increasing core size tunes the ZDW to longer λ. The increase in air percentage in the cladding could extend the origination of the 1st ZDW at longer wavelengths and the 2nd ZDW at shorter λs (i.e. less negative CD is obtained for larger air-filling fraction).

2.3. Dispersion flattened fibres and wideband fibres [2][3]

Some compensating fibres could work at short spectrum of λs. When a compensating fibre works at a certain λ, optical pulses can be transmitted using one or – in general – low amount of WDM channels. As it has already been mentioned above, MOFs are suitable for designing the compensating fibres because they allow huge index contrast and offer many parameters to be optimized (core size, hole radius, lattice pitch, amount of rings of holes) in order to optimize waveguide dispersion [2].

Currently, flat CD fibres could for example be referred to as near zero CD transmission medium. In Ref. [22], a DCF made of pure (undoped) material is shown. Advanced DCFs with flat CD property over broad spectrum of working λs could be found in Ref. [23]. An interesting fibre with CD parameter close to zero with CD fluctuations less than 0.5 ps in S to L bands is shown in Ref. [24]. Results were obtained by careful optimization of holes in particular rings. To reduce losses, it is often required avoiding doping the MOF's core. Instead, one could consider using octagonal [25] and decagonal [26] lattice, or more sophisticated lattice arrangements [2].

Recent analyses concentrate on the low, flat CD over telecommunication λs for any telecom wavelength compensation. The value of CD and the width of the working range is a compromise. A DSF could have large negative CD value, but for medium-wide wavelength range, or, acceptable CD parameter, but designed for very broad spectrum of λ. Currently, the strongest demand is to design DCFs mainly for C and L bands, where modern WDM systems can work.

The idea how to obtain flat CD properties is to tune the 1st ZDW to shorter λs and the 2nd ZDW to longer λs, having little negative CD over the whole bandwidth at the same time. Practically, predicted CD diagram is a wide parabola. Considered properties are related to the diameter of the core and the radius of the holes in particular rings. Both were found to have different impact on the origination of each ZDW. There are requirements to locate the zero CD point at the λ of 1.55 μm or, in general, in the C-band [8]. Changing the air filling fraction and lattice pitch is not the only idea that could be used to optimize CD. For example elliptical holes instead of circular holes exhibit potential low and very stable CD properties, i.e. 0.6-1 ps/nm/km in the range from 1 μm to 1.9 μm [3].

Finally, there are a few examples that are worth to be noticed: a DCF in Ref. [27], with CD parameter being –1350 ps/nm/km at 1.55 μm; another one in Ref. [28] has CD of –440 to –480 ps/nm/km at the band of 1.5-1.62 μm [3].

3. State of the art

The employment of the considered fibres is mainly in high-speed transmission systems using wavelength division multiplexing and signal recovery, offering transmission rates of even more than 1 Tb/s, where it is proved that non-optimized CD could destroy the pulse spreading, but as well it could generate some nonlinear effects, such as four wave mixing, among others. Although the properties of the fibres significantly differ from one another, the techniques used for optimized CD could be systematized so that one could design a fibre with desired CD value and slope at a specific λ.

Understanding of the mechanisms governing CD tailoring is necessary not only permits for the fibre design, CD suppression and avoidance, but also to predict the potential manufacturing tolerances. There are a few techniques used to obtain the expected CD and we describe them in more details in the following sections.

3.1. Fibre designs [1]

One of the techniques applied is to dope the fibre's core. By inserting small doped inclusion in a MOF's core, the index contrast with result in much larger mode confinement. Because the waveguide dispersion is related to it and because the length of propagated light waves is of the range that is comparable to the core size, the inserted region will be responsible for very low waveguide dispersion component.

There is another technique that could be used to optimize dispersion, it assumes to inject some liquids into the microstructured holes. In Ref. [29], the 4th ring of holes is doped by using liquids for this purpose. But liquid's properties are strongly dependent on temperature, which means that constant temperature must be ensured. This means that such a fibre could for example be used for some experiments in laboratory conditions (e.g. sensors), but they can't be used in outdoor installations of telecommunication systems [1].

Germanium dioxide is very often used to raise the effective index. In Ref. [13] and [30], the idea how to use it to dope the microstructured fibre's core is described in details. But this solution is not perfect in terms of effective area of the mode which is rather small in the doped fibres, as well as from the perspective of increased attenuation caused by material absorption, which is high for germanium dioxide [1].

Another technique to be described and systematized is the one using air holes located close to the fibre centre. The 1st mode could be confined within the core by the index contrast, and the effective diameter of a mode is then usually small enough to control CD. Stronger confinement could be obtained in a MOF using other type of a lattice, like for example octagonal, decagonal or spiral.

An alternative way, how to control the effective index, is to have core defects. A "defect core" is usually understood as one allowing light coupling to some other regions by some imperfection that is introduced intentionally. This technique uses weak interactions between the core-guided mode and a mode localized in an intentionally introduced defect of the crystal.

Finally, the idea of concentric cores could be considered. What feature in dual core technique is responsible for low CD? It is large value of n_{eff} produced by significant asymmetry between the two cores where light is coupled. Light is propagated in both cores concurrently. Dual cores are responsible for the phase matching at a certain λ, and consequently at this wavelength large CD parameter is obtained.

Low value of CD parameter is possible in concentric cores. They could be obtained by removing [7],[10] or reducing [7],[31] the air fraction in some rings of holes [1].

There are many aspects that could influence resultant CD; it could be coupling modes and the phase matching wavelength, at which CD control is possible, could be obtained in many ways, for example by fibre bending. We found these conclusions as interesting, because many works assume that the most important consequence of bending a fibre is increased attenuation. It could be shown that large negative value of CD parameter could be the accompanying process.

3.2. Extension of transmission band towards infrared — Fluoride fibres [32]

Dealing with the structural parameters of MOFs is a solution to find optimal CD. They could be enhanced by very careful optimizing the material composition. For example fluorides could be used as additives, but they could also be used as a background material instead of silicon dioxide, which could be evaluated as significant progress in the field [32]. The fluorides have low effective index, broad range of λs and low attenuation. One could consider the use of BaF_2, CaF_2 and more advanced ZBGA (ZrF_4-BaF_2-GdF_3-AlF_3) or ZBLAN (ZrF_4-BaF_2-LaF_3-AlF_3-NaF) [32]. In addition, because those materials are composite, it is possible to change their composition in a compound in order to obtain ne_{ff} value or its slope in the investigated area of λ. Last but not least, these materials are transparent in infrared region (as well as in the telecommunication bands) which makes them interesting for some applications in sensors. The properties of fluorides allow using them in MOFs [33]. They would lower the effective index of the cladding or the index at the second, larger concentric core. It is usually combined with doping the main core by using germania. This idea could be considered as a significant progress in fibre optic technology. Then a fibre with a W-type profile or refractive index could be created. For example, the core doped by using germanium dioxide would exhibit the n_{eff} of about 1.48 and concurrently the cladding index doped by using F-SiO_2 would exhibit the index of about 1.43. Similar solution (in general modified W-type index profile fibre) could be employed in systems with WDM [32],[34].

Material and structural parameters are key features in order to obtain exact CD at each λ, but the accuracy and quality of production could become a critical issue, especially in submicron fibres whose small holes with the diameter being less than one micron is problematic for some manufacturing technologies. All this combined with the application of new materials that should be processed in different way compared to siliceous materials could result in both attenuation and CD properties far from the expectations [32].

At the telecommunication band both ZBLAN and ZBGA showed similar attenuation properties like the one of silicon dioxide, but their CD properties significantly differ [35], in addition fluorides have different mechanical properties and they have to be fused at different temperature than temperature for silica. The theoretical attenuation could even be of 0.01 dB/km, in the range from 0.3 μm to 4.3 μm. The n_{eff} in telecom band is 1.47 – 1.52 [32, 36].

4. Design approaches

To describe propagation of light in a given optical glass, we could consider the constant of propagation of a wave that is related to the phase variation [32]:

$$k = (\beta - j\alpha) \tag{1}$$

The real part β and the imaginary part ω could be specified [37]:

$$\beta = \omega\sqrt{\frac{\mu\varepsilon}{2}\left[\sqrt{1+\left(\frac{\sigma}{\omega\varepsilon}\right)^2}+1\right]} \tag{2}$$

$$\alpha = \omega\sqrt{\frac{\mu\varepsilon}{2}\left[\sqrt{1+\left(\frac{\sigma}{\omega\varepsilon}\right)^2}-1\right]} \tag{3}$$

Where ω is angular velocity, σ is conductivity, a μ is permeability and ε is permittivity. Because MOFs are dielectric it could be assumed that:

$$\left(\frac{\sigma}{\omega\varepsilon}\right) << 1 \tag{4}$$

Both real and imaginary part of propagation constant in (3) could be simplified to a form:

$$\beta = \omega\sqrt{\mu\varepsilon} \tag{5}$$

$$\alpha = 1 \tag{6}$$

MOFs could be described by an effective index n_{eff}, which expresses "the refractive index at the boundary of two entities", which in this particular case are the core and the cladding [32]:

$$n_{eff} = \frac{\beta}{k_0} \tag{7}$$

β is real part of the constant of propagation of core, k_0 is the constant of propagation in vacuum. Propagation constant depends on λ, it could be written that:

$$k = \frac{2\pi}{\lambda} \tag{8}$$

Group velocity that integrates optical waves in an "envelope" could be considered to describe its velocity. It refers to CD and is less than the speed of electromagnetic wave in vacuum [32]:

$$v_g = \frac{\partial\omega}{\partial k} = \frac{c}{\sqrt{1 + \frac{\omega^2\lambda^2}{4\pi^2c^2}}} \tag{9}$$

Then group delay τ_g could be calculated as time necessary for the propagation of wave to certain distance with velocity of v_g (9):

$$\tau_g = \frac{1}{v_g} \tag{10}$$

In general, chromatic dispersion *CD* is known as the dependence of group delay τ_g on the wavelength λ, at which the signal is transmitted [38]:

$$CD(\lambda) = \frac{\partial\tau_g(\lambda)}{\partial(\lambda)} \tag{11}$$

In simulations it is more suitable to use material dispersion equation described by using Sellmeier approximation. CD equation suitable for Sellmeier approximations is following [39]:

$$CD(\lambda) = -\frac{\lambda}{c}\frac{\partial^2 Re\left[n_{eff}\right]}{\partial\lambda^2} \tag{12}$$

Designs of MOFs could be calculated by Finite-Difference Frequency Domain (FDFD) method. The distribution of light could be calculated by discretizing electric (13) and magnetic (14) field, it is described in Refs. [32, 40], and [41]:

$$\left(\nabla_t^2 + k_0^2\varepsilon_r\right)\boldsymbol{E}_t + \nabla_t\left(\varepsilon_r^{-1}\nabla_r\varepsilon_r \cdot \boldsymbol{E}_t\right) = \beta^2\boldsymbol{E}_t \tag{13}$$

$$\left(\nabla_t^2 + k_0^2\varepsilon_r\right)\boldsymbol{H}_t + \varepsilon_r^{-1}\nabla_t\varepsilon_r \times \left(\nabla_t \times \boldsymbol{H}_t\right) = \beta^2\boldsymbol{H}_t \tag{14}$$

k_0 is wavenumber in free space, β is real part of the propagation constant, ε_r is dielectric constant as in (5). The FDFD method uses the discretization scheme shown in Ref. [42][32].

FDFD is used in many commercial simulators of mode distribution and n_{eff} calculators. In most of them, it is possible to use user friendly graphical interface, but when a very accurate result is demanded, many iterations ought to be done by creating a loop using an appropriate scripting language. We use a mode solver from the Lumerical Inc. When the fibre's cross section is proposed and the simulation parameters and monitors are set, the simulation is run [3].

Creation of realistic models of sophisticated optical glasses is an interesting feature. To introduce materials to the simulator, we used the Sellmeier approximation [32, 43],[44]:

$$n^2(\lambda) = A + \sum_i \frac{B_i\lambda^2}{\lambda^2 - C_i^2} \tag{15}$$

where A, B_i, C_i are coefficients referring to index n. As a result, the wavelength evolution of refractive index could be plot by using the Sellmeier equation. The coefficients for the expanded version of Sellmeier equation (16) could found in [32][45]:

$$n^2 - 1 = \frac{B_1\lambda^2}{\left(\lambda^2 - C_2^2\right)} + \frac{B_3\lambda^2}{\left(\lambda^2 - C_4^2\right)} + \frac{B_5\lambda^2}{\left(\lambda^2 - C_6^2\right)} \tag{16}$$

where $B_{1,3,5}$ $C_{2,4,6}$ are material constants and λ is wavelength.

5. Exemplary results

In this section, we present some selected fibre designs covering the wide area of telecom photonic fibres for CD tailoring, such as DCFs with large negative CD parameter, wideband fibres or slope compensating fibres, or fibre designs suitable for CD prevention, such as near zero CD fibres. We employ different design approaches and optimize them based on parametric investigation.

5.1. Dual core dispersion compensating fibre with large negative dispersion parameter [1]

A concentric core DCF based on microstructured optical fibre is proposed [1]. There is an assumption for the proposed structure to keep the 1st mode in a central core over the wide working spectrum of λ and in order to obtain low CD at 1.55 μm, and with low theoretical losses. The refractive index is optimized by making smaller the hole diameter in the 2nd ring. The design is specified in Table 1, the cross section is shown in Figure 1 [1].

Quantity [unit]	Value	Value	Value	Value	Value
Lattice pitch Λ [μm]	1.55	1.52	1.50	1.48	1.45
Minimum D at 1.55 μm [ps/nm/km]	-1010	-1180	-1300	-1460	-1690
Core index [-]	1.44	1.44	1.44	1.44	1.44
Theoretical loss [dB/cm]	$9\cdot10^{-10}$	$7\cdot10^{-10}$	$3.8\cdot10^{-7}$	$3.8\cdot10^{-4}$	$7\cdot10^{-2}$
Full width at half maximum [nm]	142	132	129	125	122
Hole diameter d_1 [μm]	1.33	1.34	1.314	1.27	1.23
Hole diameter d_2 [μm]	0.55	0.54	0.574	0.63	0.67
Hole diameter d_3 [μm]	1.35	1.35	1.2	1.04	0.92
Number of rings [-]	6	6	6	6	6

Table 1. Parameters of a dual core MOF for CD suppression [1].

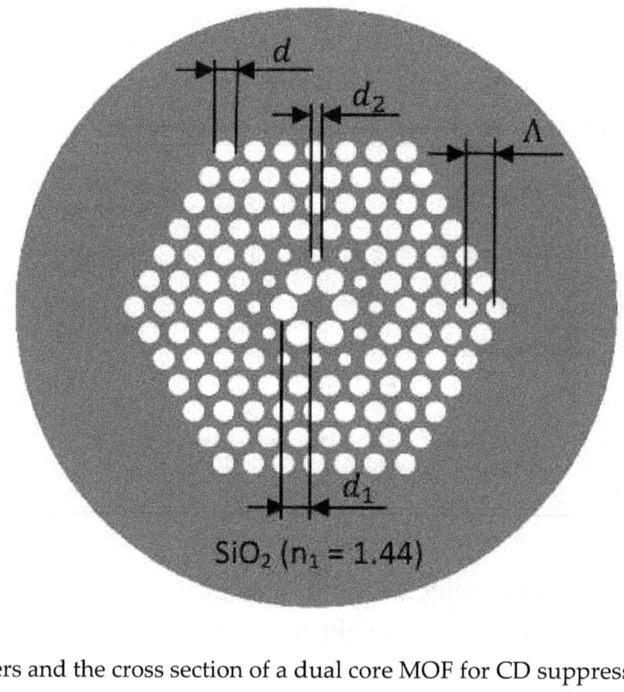

Figure 1. Structural parameters and the cross section of a dual core MOF for CD suppression purposes [1].

CD of the considered fibre is shown in Figure 2. Obtained CD parameter was -1460 ps/nm/km. The designed fibre work in the C-band, and the lowest CD is at 1.55 μm. Theoretical losses are 3.8 10^{-4} dB/cm. It could be concerned as one of the advantages of the MOF, when we have a look at losses presented for some other fibres [1].

In order to obtain very accurate results, a number of iterations were performed to find the most suitable structural parameters of a fibre. One could conclude that smaller holes in the 3rd ring, and concurrently larger radius of holes in the 2nd ring could result in improved CD properties (Figure 3). Optimized solution could be found in Table 2 [1].

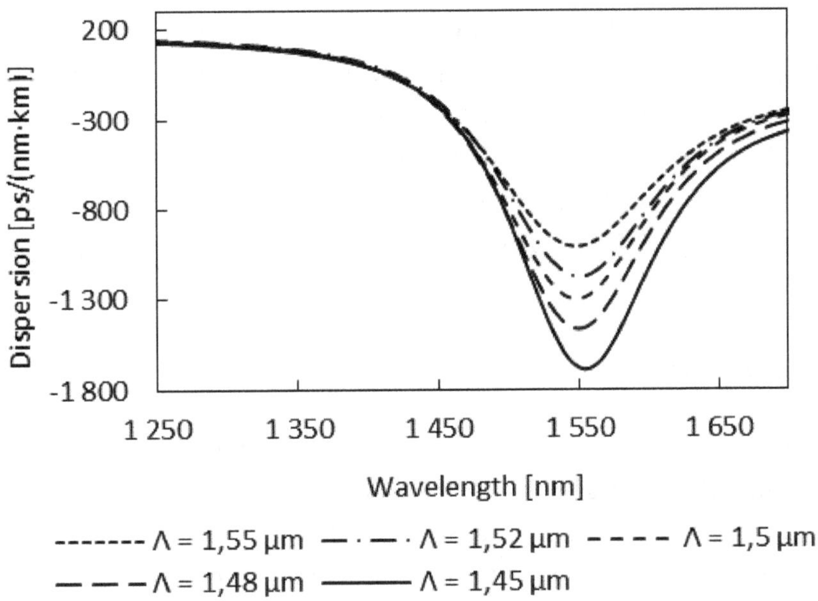

Figure 2. Chromatic dispersion in a dual core MOF for CD suppression at 1.55 µm [1].

Quantity [unit]	Value	Quantity [unit]	Value
Minimum CD [ps/nm/km]	-3290	Lattice pitch Λ [µm]	1.45
Core index [-]	1.44	Hole diam. d_1 [µm]	1.23
Theoretical losses [dB/cm]	7.50	Hole diam. d_2 [µm]	0.73
Number of rings [-]	6.00	Hole diam. d_3 [µm]	0.86

Table 2. Specification of parameters for dual core MOF for CD suppression [1].

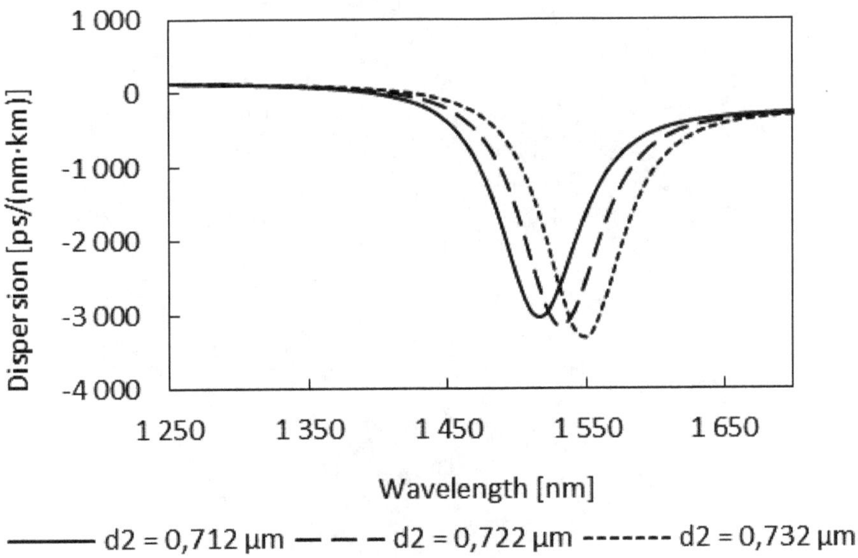

Figure 3. Dispersion in proposed dual core microstructured DCF [1].

In Figure 3, it could be observed that λ at which there is minimum CD is function of normalized hole diameter d/Λ of the holes; CD is mostly sensitive to the holes located in the 2nd ring. To move the operating λ towards 1.5 μm, one has to increase the size of the holes in this ring. Concurrently, the value of lowest CD is less. Summarizing, a one km long section of the designed DCF should be sufficient to tailor CD in a network that is created by using 75 km long conventional SMF with CD parameter being 17 ps/nm/km, (this value is in agreement with the recommendation of ITU-T for SMFs). The insertion losses of such a CD compensator are about 0.04 dB [1].

5.2. DCF with optimized dispersion slope [3]

Another design is done with the scope on fibres that could compensate CD in each channel of a system using wavelength multiplexing. It means that the fibre should be wideband, and its CD mustn't be flat, but it should have exactly opposite CD to CD of a fibre that is used in a WDM system. The proposal is shown in Figure 4 and in Table 3 [3].

n_{eff} [-]	1.47	Hole diameter [μm]	3.64
Core index [-]	1.48	Normalized hole diameter d/Λ [-]	0.58
Core size [μm]	4.46	Number of rings [-]	5

Table 3. Structural parameters of the HNPCF with dispersion evolution opposite to one in conventional MOF [3].

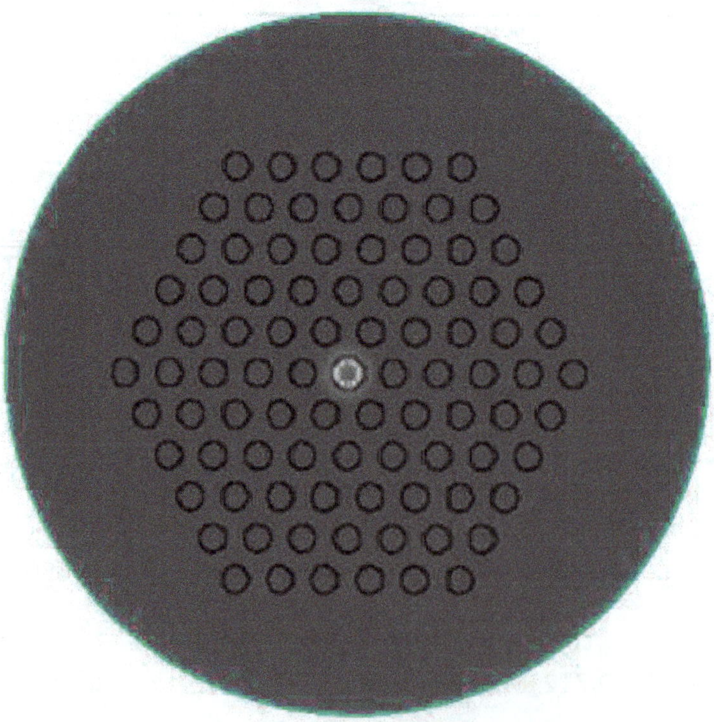

Figure 4. Cross section of a MOF with negative CD parameter [3].

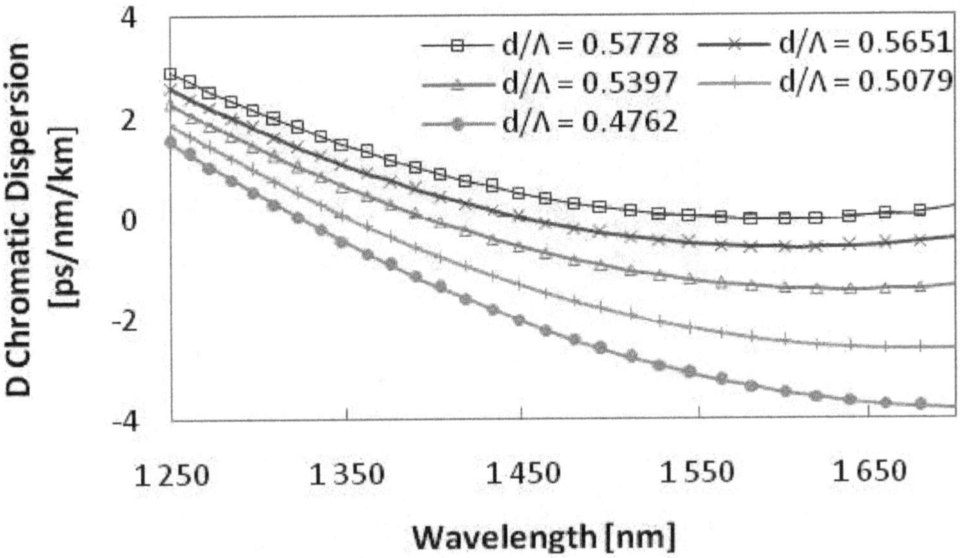

Figure 5. Reversed CD slope of the proposed DCF [3].

From Figure 5 it could be concluded that CD is reversed CD of conventional MOFs. The fibre is wideband. It could be optimized in terms of larger values of CD parameters, for example to exactly match CD evolution of conventional fibres. One shall also pay attention to the fact that larger d/Λ would result in multimode operation [3].

5.3. Dispersion flattened fibres and wideband dispersion compensating fibres [1][3]

A MOF with flat CD is demanded in design of transmission fibres (not suppression fibres) where it is d to have identical CD at each λ channel. Such a fibre could be doped in the core by using germania. The proposed core diameter could be 7.4 μm. In Figure 6, the 1st mode is kept in the core and the fibre is single-mode. Manufacturing of the MOF's core could be challenging, because of huge doping area; the fibre has large mode area. The proposed geometry is described in Table 4 [3].

Hole diameter d [μm]	1.32	Silica index [-]	1.46
Lattice pitch Λ [μm]	4.4	Propagating λ [μm]	1.55
Normalized hole diameter d/Λ [-]	0.3	Core size [μm]	7.4
Air index [-]	1	Effective cladding index at 1.55 μm	1.47
Core index [-]	1.48	Number of rings [-]	3

Table 4. Structure parameters for HNPCF flattened CD curve [3].

Obtained CD property could be evaluated as flat and oscillating around the value of -0.025 ps/nm/km. In Figure 7, comparison of CD in conventional MOF and the designed fibre is shown [3].

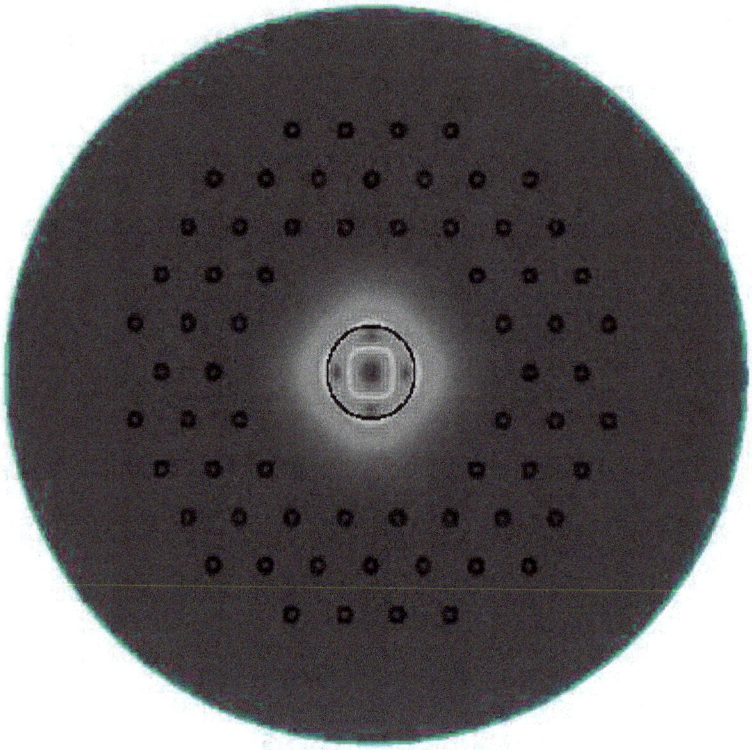

Figure 6. Wideband large mode area MOF with near zero flattened CD [3].

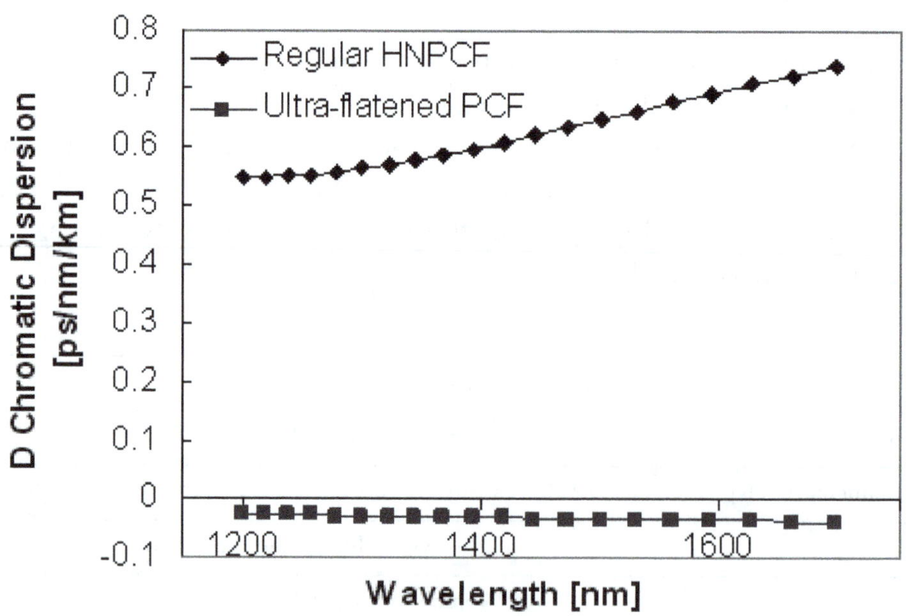

Figure 7. Dispersion in conventional microstructured fibre and proposed highly nonlinear MOF [3].

Wideband fibre could exhibit negative CD parameter, too. We propose a MOF for suppression in the band of 1.25-1.7 μm. We do not use any additives. Mode confinement is done by optimizing the geometry of the core and the normalized hole diameter d/Λ, the accepted view is that the core size should be small. In the 1st ring, there is $d_1/\Lambda=0.9$. The proposal could be found in Figure 8. Doping the core could additionally improve CD properties, but it would surely worsen attenuation. One should pay attention to the fact that optimization of CD shall be done in the context of attenuation properties. Optimizing one parameter and ignoring another is unacceptable in high-speed transmission system fibres [1].

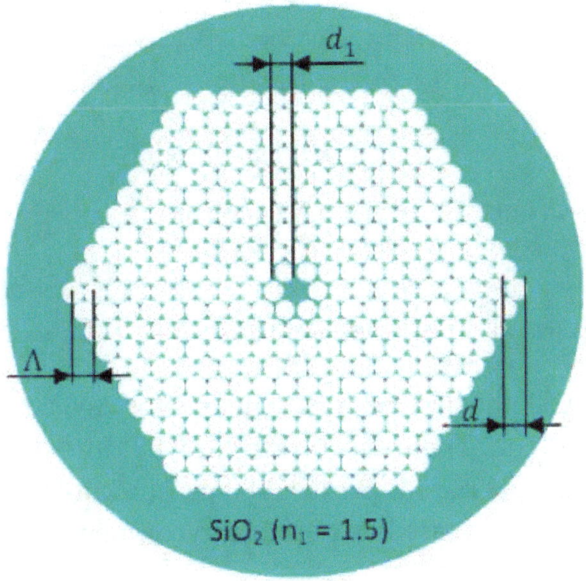

Figure 8. Cross section of designed microstructured fibre for wideband suppression of CD [1].

Having larger holes in the 1st t ring, d_1, is responsible for lower CD. On the contrary, by making smaller all the other holes (d) results in increasing dispersion. In Table 5, optimized structure is shown and its cross section is in Figure 9 [1].

Quantity [unit]	Value	Value	Value	Value	Value
CD [ps/nm/km]	-1580	-2040	-2259	-2094	-1930
Core index [-]	1.50	1.50	1.50	1.50	1.50
Theoretical losses [dB/cm]	$9.1 \cdot 10^{-6}$	$1.9 \cdot 10^{-3}$	$4.8 \cdot 10^{-2}$	$7 \cdot 10^{-2}$	$1.5 \cdot 10^{1}$
FWHM [nm]	Flat CD over 1.25-1.7 μm				
Lattice pitch [μm]	0.70	0.65	0.62	0.6	0.55
Hole diameter in 1st ring d_1 [μm]	0.70	0.65	0.62	0.6	0.55
Hole diameter d [μm]	0.70	0.65	0.62	0.6	0.55
Number of rings [-]	10	10	10	10	10

Table 5. Specification of parameters for wideband CD suppressing MOF [1].

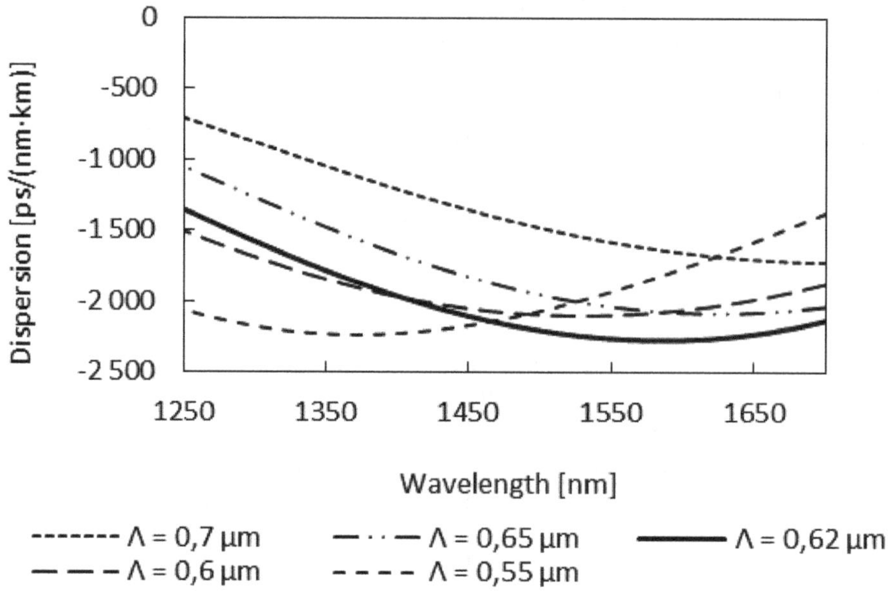

Figure 9. Resultant flattened negative value of CD over broad band of λs 1.25-1.7 μm (lattice pitch is a parameter) [1].

Concerning the values of CD, the best design is the one for Λ=0.62 μm, for which CD is -2259 ps/nm/km, obtained at 1.55 μm. Theoretical losses are 4.8 dB/cm. The optimized fibre is the one with Λ=0.70 μm, where CD parameter is -1580 ps/nm/km at 1.55 μm. Theoretical loss is lowered to 9.1 10-6 dB/cm. In this case it is possible to compensate dispersion of about 90 km of standard SMF.

5.4. Fluoride doped dispersion compensating fibres [32]

The considered fibre is a MOF employing the idea of a W-profile fibre with the core doped by using BaF_2 (n_{BaF2} = 1.468 to raise its refractive index) and containing three holes in the 1st ring doped by using $CaF_{2,}$ n_{CaF2} = 1.426 to reduce effective cladding index. Increased index contrast is responsible for enhanced CD, as in eq. (12). As for example, models optical glasses could be expressed using coefficients shown in Table 6.

Material	B_1	C_2	B_3	C_4	B_5	C_6
Silica	0.6961663	0.0684043	0.4079426	0.11624	0.897479	9.89616
BaF_2	0.6433560	0.0577890	0.5067620	0.109680	3.82610	46.3864
CaF_2	0.5675888	0.0502636	0.4710914	0.10039	3.848472	34.6490

Table 6. Sellmeier coefficients for the fluoride additives used in investigated MOF [32].

The structural and material properties are in Table 7. Large radius of CaF_2 holes and the low core size are necessary for flat CD (Figure 10). CD is -413 ~ -415 ps/nm/km at 1.4 - 1.65 μm. Theoretical loss is 1.75 10^{-4} dB/cm. CaF_2 doped holes affect the CD (Figure 11). The larger are the doped holes, the lower is resultant CD [32].

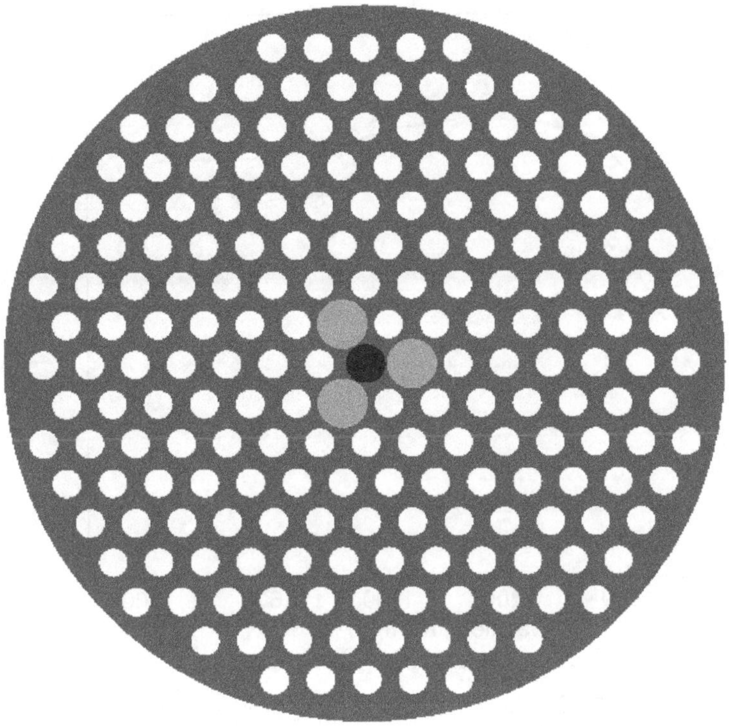

Figure 10. Fluoride-doped W-profile fibre's cross section for infrared and telecom band [32].

Proposed fibre:	
Air hole diameter d [μm]	0.4
Lattice pitch Λ [μm]	0.7
Core size [μm]	1.3
Doped hole diameter d_1 [μm]	0.76
Core dopant material	BaF_2 n = 1.468 at 1.55 μm
1st ring doping material	CaF_2 n = 1.426 at 1.55 μm
Normalized air-hole diameter d/Λ [-]	0.57

Table 7. Design specification of investigated W-profile MOF with doped holes in the 1st ring [32]

The air holes influence the dispersion slope. Larger d/Λ would limit the confinement losses. An interesting feature is that we use fluorides, ZBGA or ZBLAN, not just to dope the cladding, but we propose to use it as a background material instead of silicon dioxide (Figure 12). As a result, optimization of CD is more flexible. Last but not least, temperature and mechanical properties of such a fibre wouldn't be worse [32].

Right combination and the composition of additives is responsible for optimized properties of a compound. Let us consider zirconium (Zr), barium (Ba), gadolinium (Gd) or aluminium (Al) that are the compounds of ZBGA material: $ZrF_4 - BaF_2 - GdF_3 - AlF_3$ [32].The n_{eff} of ZBGA material could be represented by modified Sellmeier equation (17) [46], using the coefficients in Table 8 [45].

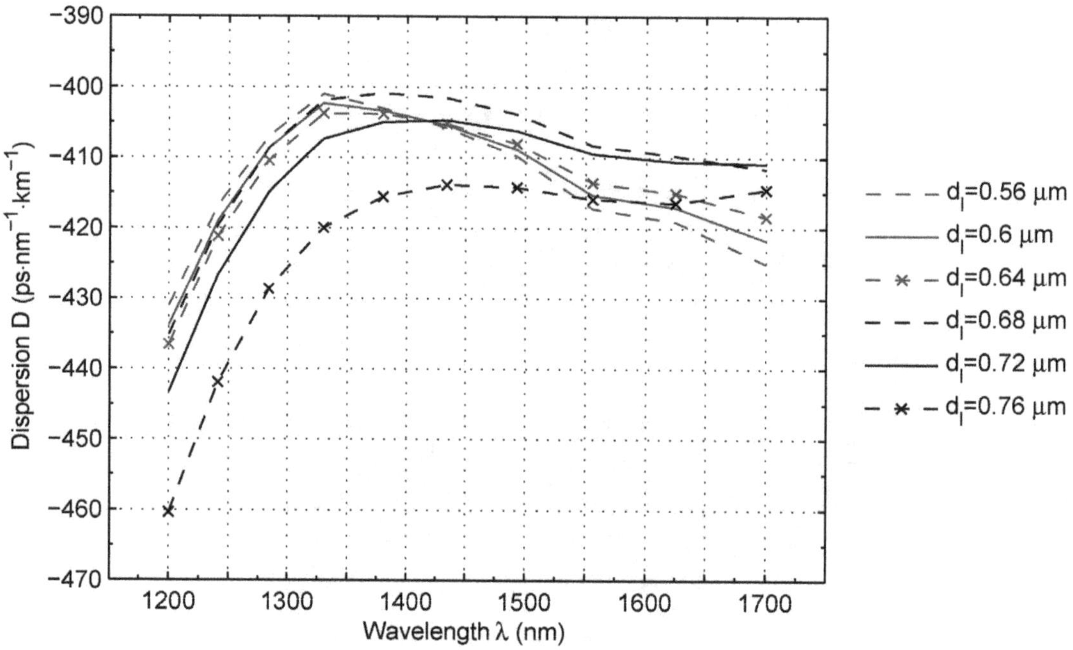

Figure 11. Wavelength evolution of CD with hole diameter of CaF_2 as a parameter [32].

$$n(\lambda)=\frac{A}{\lambda^4}+\frac{B}{\lambda^2}+C+D\lambda^2+E\lambda^4 \tag{17}$$

Material	A	B	C	D	E
ZBGA at 25°C	2.98316e-6	3.39740e-3	6.81447e-3	-1.20276e-3	-5.48085e-6

Table 8. Sellmeier coefficients for the fluoride-background ZBGA fibre [45].

ZBGA has larger n_{eff} comparing with silicon dioxide. Tailored CD is expected [32].

Parametric sweep performed for the hole diameter showed that flatten CD is possible, which is shown in Figure 6. For hole diameter being equal to 0.44 μm, the evolution of CD is similar to one obtained for silicon dioxide. Larger n_{eff} of the ZBGA fibre requires larger radius of air-holes in the cladding to tailor CD [32].

BaF_2 could be used in the core and CaF_2 in the 1st ring, ZBGA as a background material (Figure 13). CD parameter is then possible within the range of -435 ~ -438 ps/nm/km and over the entire telecommunication band. At the same time, attenuation properties are not worse than those obtainable in fibres with silica as a background material [32].

5.5. Submicron dispersion compensating fibres with modified geometry [2]

One of the trends is to use a lattice with the pitch being less than one micrometre (so-called submicron lattice). Then, by careful adjustments of the diameter of holes (d_1-d_3 in Figure 14),

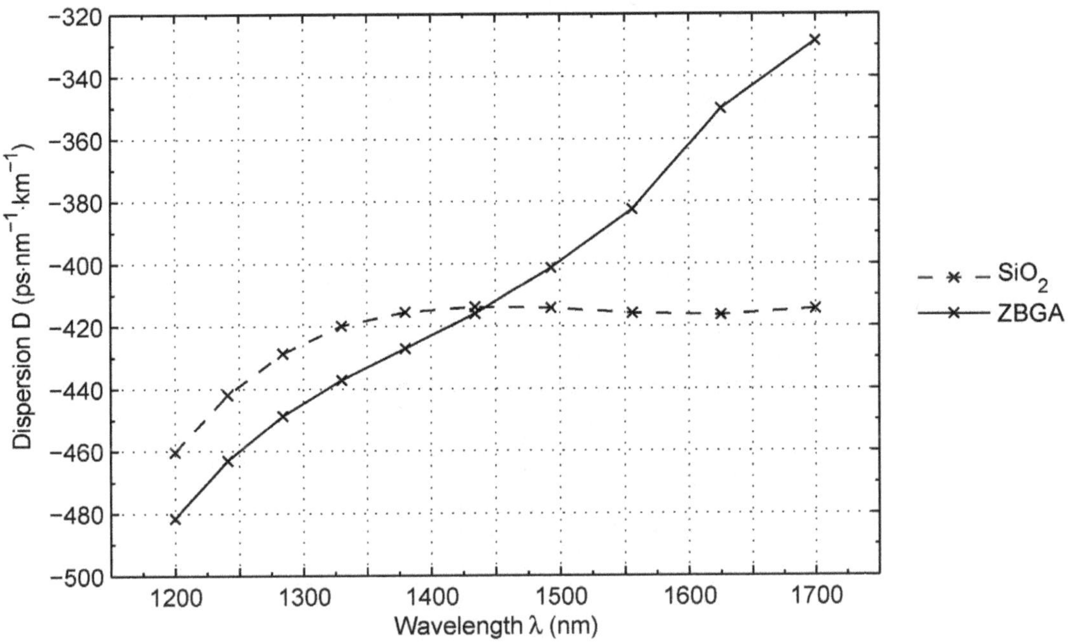

Figure 12. CD in SiO_2 and ZBGA used as a background material [32].

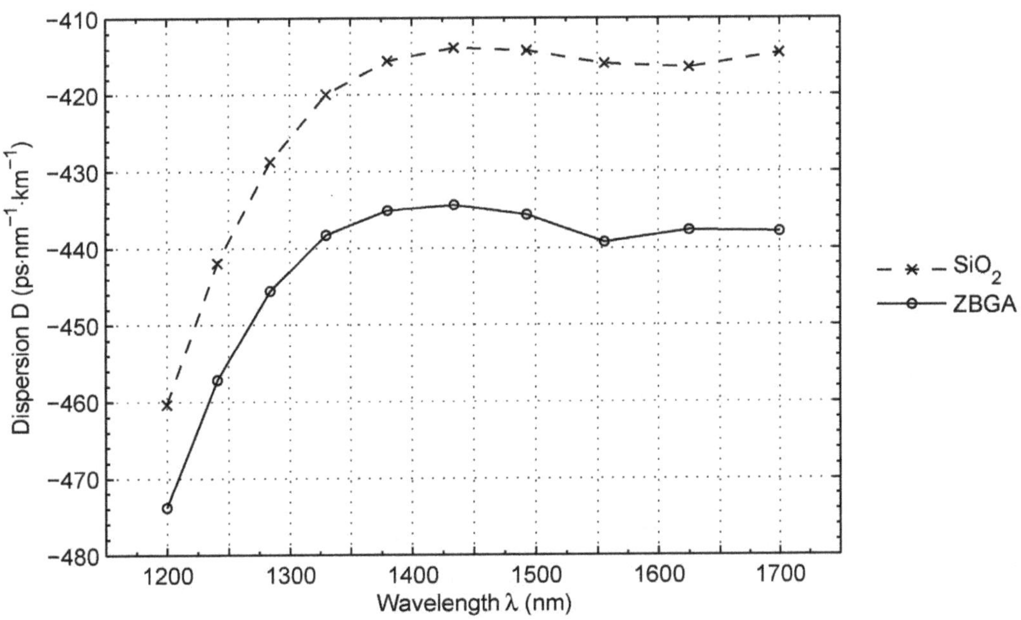

Figure 13. Proposed MOF with ZBGA material used as a background compared to a fibre with SiO_2 as a background [32].

where d_1 is hole diameter in the 1st ring of holes, one could obtain nearly zero CD for the λs from the window 1.25-1.7 μm. Hole diameter and lattice pitch are submicron. Then, high confinement of light waves and strong waveguide dispersion is possible. The geometrical parameters of the fibre are in Table 9 and in Figure 14 [2]. CD and loss are simulated for the 1st light mode. Confinement losses are 6.10^{-6} dB/km (Figure 15). Wavelength evolution of CD is shown in Figure 15 [2].

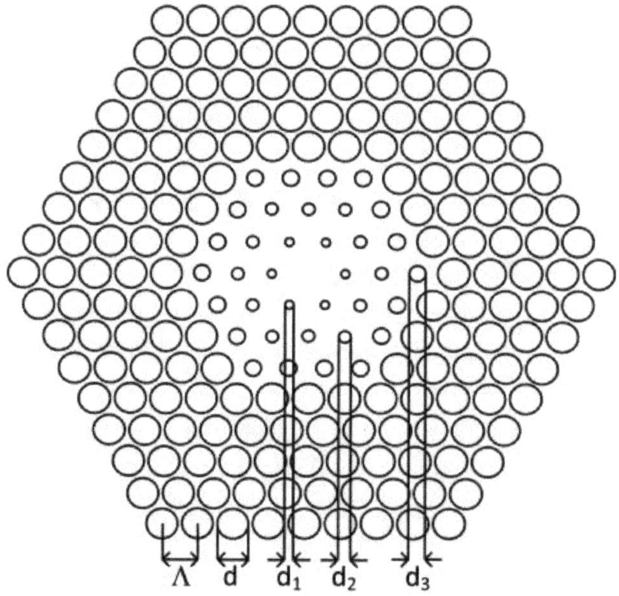

Figure 14. Cross section of a near zero CD flattened MOF with modified rings [2].

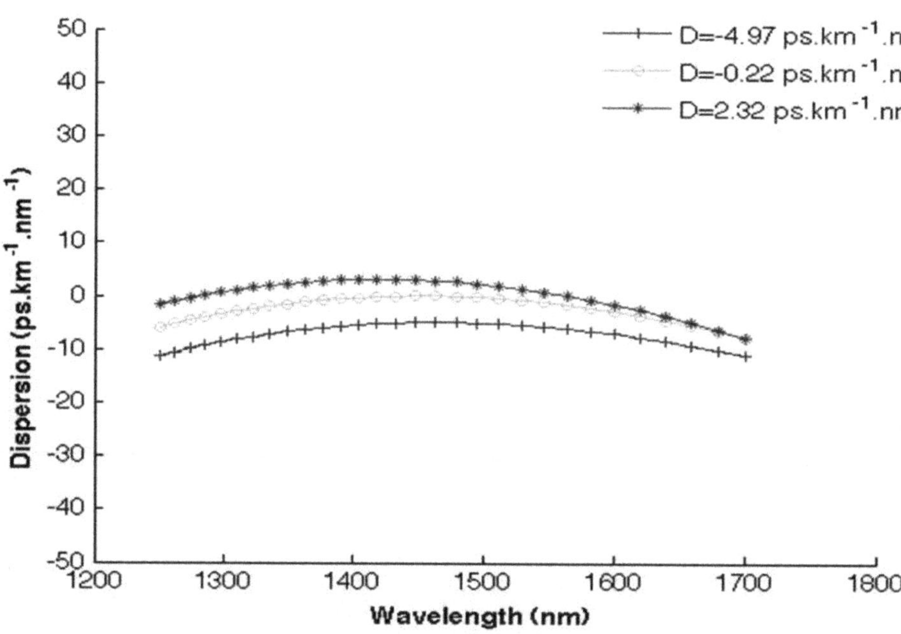

Figure 15. Near zero flattened CD for wideband utilization in telecommunications [2].

Obtained CD parameter is from -7.7 ps/nm/km to 3.1 ps/nm/km at the band of (1.25, 1.7 μm) and its average value is around 0.51 ps/nm/km. In the C-band, it is 1.35±0.46 ps/nm/km. For both C and L bands CD is 0.12±1.32ps/nm/km [2]. Results for parametric iterations for air fraction changed in three most internal rings are shown in Figures 15 and 16. From the results summarized in Table 10 it could be concluded that the fibre has CD slope of -0.09 ps.nm^{-2}.km^{-1} is obtained for little variation of d_1 from other holes and is suited to the slope of conventional ITU-T G.657 fibres. The slope is less than 0.09 ps.nm^{-2}.km^{-1} [2].

d_1	d_2	d_3	d = const.	Λ = const.	CD at 1.55 μm
μm	μm	μm	μm	μm	ps/nm/km
0.134	0.305	0.318	0.355	0.8	2.32
0.134	0.298	0.325	0.355	0.8	-0.22
0.134	0.290	0.325	0.355	0.8	-4.97

Table 9. Geometrical parameters of a fibre with nearly zero flattened CD [2].

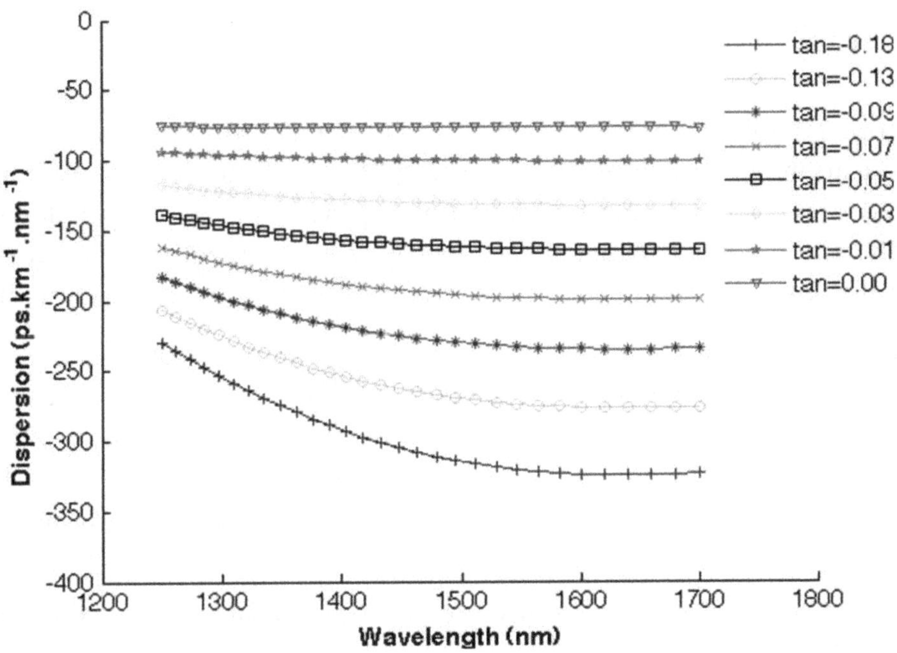

Figure 16. Optimized CD slope by varying the hole diameter in the 1st ring [2].

	d_1	d_2, constant	d_3, constant	d, constant	CD slope
	μm	μm	μm	μm	$ps.nm^{-2}.km^{-1}$
	0.18	0.31	0.355	0.355	0.01
hole diameter of	0.20	0.31	0.355	0.355	-0.03
the 1st ring	0.22	0.31	0.355	0.355	-0.07
	0.24	0.31	0.355	0.355	-0.13
	0.25	0.31	0.355	0.355	-0.18

Table 10. Optimized dispersion slope by varying diameter of holes in the 1st ring [2].

Larger CD slope is here obtained by adjusting the value of d_3 (see Table 11, Fig. 16). At d_1 being fixed, linear wavelength evolution of CD is possible. It becomes nonlinear when the slope exceeds -1 $ps.nm^{-2}.km^{-1}$. The obtained property could be applied in CD suppression performed on a fibre with the slope of 0.5 $ps.nm^{-2}.km^{-1}$ [2].

	d_1 = const.	d_2 = d, constant	d_3	d, constant	CD slope
	μm	μm	μm	μm	ps.nm^{-2}.km^{-1}
hole diameter of the 3rd ring	0.265	0.355	0.35	0.355	-0.39
	0.265	0.355	0.27	0.355	-0.63
	0.265	0.355	0.23	0.355	-0.85
	0.265	0.355	0.17	0.355	-1.32

Table 11. Large CD slope obtained by tuning the hole size in the 3rd ring [2].

6. Final conclusions

We described optical fibres from the perspective of their CD properties, which has to be solved in telecommunication. The mechanisms governing CD properties in telecom fibres are shown. The considered fibres are suitable for potential suppression of group velocity dispersion, utilized mainly in the C-band (dual core fibres with high negative CD parameter) and in wideband applications (CD flattened MOFs and exact slope suppression fibres).

A special family of fibres is so-called fluoride fibres that have CD optimized through material dispersion. This idea is combined with the proposal of a W-type fibre, where particular regions are doped by using different additives, including fluorides, among others. ZBLAN material offers broader range with low attenuation (telecom and infrared range). It makes them very attractive for getting eventually applied in spectrometry or in applications in fibre-optic sensors.

The use of fluorides assessed their flexibility in CD optimization and potential use in wideband CD suppression. Last but not least, it has been shown that optimization of fibre's CD slope is possible in submicron lattice pitch without additives, potentially for exact slope suppression of standard International Telecommunication Union fibres.

Acknowledgements

This work has been supported by the CTU foundation, SGS13-201-OHK3-3T-13.

Author details

Michal Lucki* and Tomas Zeman

*Address all correspondence to: lucki@fel.cvut.cz

Department of Telecommunication Engineering, Faculty of Electrical Engineering, Czech Technical University in Prague, Prague, Czech Republic

References

[1] Lucki, M., Zeleny, R. Broadband Dispersion Compensating Photonic Crystal Fibre: conference proceedings, SPIE Vol. 8306. Bellingham (Washington): SPIE, 83060Z-1-83060Z-6, 2011.

[2] Lucki, M., Jiruse, D., Kraus, S. Single-Material Submicron Microstructured Fibres for Broadband Applications in Exact Slope Compensation or Zero- Dispersion Propagation: conference proceedings, 14th International Conference on Transparent Optical Networks (ICTON). Piscataway: IEEE, Tu.P.28., 2012.

[3] Lucki, M. Optimization of Microstructured Fibre for Dispersion Compensation Purposes: conference proceedings, 2011 13th International Conference on Transparent Optical Networks (ICTON). Piscataway: IEEE, p. Tu.P.12, 2011.

[4] Veng M., et al. Dispersion compensating fibres. Opt. Fibre Technol. 2000; 6 164-80.

[5] Antos A., Smith D. Design and characterization of dispersion compensating fibre based on the LP_{01} mode. J. Lightwave Technol. 1994; 12(10) 1739-1745.

[6] Ferrando A., et al. Nearly zero ultraflattened dispersion in photonic crystal fibres. Opt. Lett.2000; 25 790-792.

[7] Ni Y. et al. Dual-Core Photonic Crystal Fibre for Dispersion Compensation. IEEE Photonics Technology Letters 2004; 16(6) 1516-1518.

[8] Liu Z. et al. A broadband ultra flattened CD microstructured fibre for optical communications. Optics Communications 2007; 272(1) 92-96.

[9] Fujisawa T. et al. Chromatic dispersion profile optimization of dual-concentric-core photonic crystal fibres for broadband dispersion compensation. Opt. Express 2006; 14(2) 893-900.

[10] Yang S. et al. Theoretical study and experimental production of high negative dispersion photonic crystal fibre with large area mode field. Opt. Express 2006; 14(7) 3015-3023.

[11] Nejad S. M., Ehtehsami N. A Novel Design to Compensate Dispersion for Square-lattice Photonic Crystal Fibre over E to L Wavelength Bands: conference proceedings, Communication Systems, Networks and Digital Signal Processing Symposium, 654-658, 2010.

[12] Hosaka T. et al. Dispersion of pure GeO2 glass core and F-doped GeO2 glass cladding single-mode opticalfibre. Eelectron. Letters 1987; 23(1) 24-26.

[13] Hoo Y. et al. Design of photonic crystal fibres with ultra-low, ultra-flattened chromatic dispersion. Optics Communications 2004; 242(4-6) 327-332.

[14] Ono-Kuwahara M. et al. Fluorine-doped silica fibre with high transparency and resistivity to deep ultra violet light, Lasers and Electro-Optics Lasers and Electro-Op-

tics: conference proceedings, Conference on Quantum Electronics and Laser Science. CLEO/QELS 1-2, 2008.

[15] Olyaee S., Taghipour F. Ultra-Flattened Dispersion Photonic Crystal Fibre with Low Confinement Loss: conference proceedings, Proceedings of the 11th International Conference on Telecommunications, pp. 531-534, 2011.

[16] Peckham D. W. et al. Reduced Dispersion Slope, Non-Zero Dispersion Fibre: conference proceedings, European Conferencce and Exhibition on Optical Communication, 139-140, 1998.

[17] Aikawa K. et al. High-performance Dispersion-slope and Dispersion Compensation Modules. Fujikura Technical Review 2002; 31.

[18] Andersen P. A., Paulsen H. N., and Larsen J. J. A photonic crystal fibre with zero dispersion at 1064 nm. Optical Communication 2002; 2 1-2.

[19] Hansen K. P. Introduction to nonlinear photonic crystal fibres. J. Opt. Fibre Commun. 2005; 2 226-254.

[20] Ferrando A., Silvestre E., and Andrés P. Designing the properties of dispersion flattened photonic crystal fibres. Opt. Express 2001; 9(13) 687-697.

[21] Lægsgaard J., Asger Mortensen N., and Bjarklev A. Mode areas and field energy distribution in honeycomb photonic bandgap fibres. J. Opt. Soc. Am. B 2003; 20(10) 2037-2045.

[22] Birks T. et al. Dispersion compensation using single-material fibres. IEEE Photonics Technology Letters 1999; 11 674-676.

[23] Haxha S., Ademgil H. Novel design of photonic crystal fibres with low confinement losses, nearly zero ultra-flatted chromatic dispersion, negative chromatic dispersion and improved effective mode area. Optics Communications 2008; 15(2) 278-286.

[24] Saitoh K. et al. Chromatic dispersion control in photonic crystal fibres: application to ultra-flattened dispersion. Opt. Express 2003; 11(8).

[25] Razzak S. M. A. Proposal for Highly Nonlinear Dispersion Flattened Octagonal Photonic Crystal Fibres. IEEE Photonics Technology Letters 2010; 20(4).

[26] Razzak S. M. A. et al. Chromatic Dispersion Properties of a Decagonal Photonic Crystal Fibre: conference proceedings, International Conference on Information and Communications Technology, 159-162, 2007.

[27] Ming W., et al. Broadband dispersion compensating fibre using index-guiding photonic crystal fibre with defected core. Chin. Opt. Lett 2008; 6 22-24.

[28] Zsigri B., Lægsgaard J., Bjarklev A. A novel photonic crystal fibre design for dispersion compensation. J. Opt. A: Pure Appl. Opt. 2004; 6 717-720.

[29] Yu Ch. et al. Tunable dual-core liquid-filled photonic crystal fibres for dispersion compensation. Optics Express 4443 2008; 16(17) 4443-4451.

[30] Zhong, Q., Inniss, D. Characterisation of lightguiding structure of optical fibres by atomic force microscopy. J. Lightwave. Technol. 1994; 12 1517-1523.

[31] Huttunen, A., Torma, P. Optimization of dual-core and microstructure fibre geometries for dispersion compensation and large mode area. Optics Express 2005; 13(2) 627-635.

[32] Lucki, M., Zeleny, R.: Broadband submicron flattened dispersion compensating fibre with asymmetrical fluoride doped core: conference proceedings, Micro-structured and Specialty Optical Fibres II. Bellingham: SPIE, 87750M-1-87750M-8, 2013.

[33] Tefelska, M. et al. Propagation Effects in Photonic Liquid Crystal Fibres with a Complex Structure. Optical and Acoustical Methods in Science and Technology, 2010; 118(6) 1259-1261.

[34] Petruzzi, P., Lowry, C., Goldhar, J., Sivanesan, P., Dispersion compensation using only Fibre Bragg Gratings: conference proceedings, Optical Fibre Communication Conference 4, 14-16, 1999.

[35] Saad, M. Fluoride glass fibres. Photonics Society Summer Topical Meeting Series 81-82; 2011.

[36] Tran, D., Sigel, G., Bendow, B. Heavy metal fluoride glasses and fibres: A review. Journal of Lighwave Technology 1984; 2 566-586.

[37] Poli, F., Cucinotta, A., Selleri, S. Photonic Crystal Fibres: Properties and Applications. Springer Series in Materials Science 102. The Nederlands; 2007.

[38] Hájek, M., Holomeček, P. Chromatická disperze jednovidových optických vláken a její měření Mikrokom; 2006.

[39] Saitoh, K., Koshiba, M. Chromatic dispersion control in photonic crystal fibres: application to ultra-flattened dispersion. Optics Express 2003; 11(8) 843-852.

[40] Stern, M. S. Semivectorial polarised finite difference method for optical waveguides with arbitary index profiles. Optoelectronics, IEE Proceedings 1988; 135 56-63.

[41] Huang, W. P., Xu, C. L. Simulation of three-dimensional optical waveguides by a full-vector beam propagation method. IEEE Journal of Quantum Electronics 1993; 29 2639-2649.

[42] Yee, K. Numerical solution of initial boundary value problems involving maxwell's equations in isotropic media. IEEE Transactions on Antennas and Propagation 1966; 14 302-307.

[43] Ghosh, G., Yajima, H. Pressure-dependent Sellmeier coefficients and material dispersions for silica fibre glass. Journal of Lightwave Technology 1998; 16 2002-2005.

[44] Iwasaki, T., Endo, M., Ghosh, G. Temperature-dependent Sellmeier coefficients and chromatic dispersions for some optical fibre glasses. Journal of Lightwave Technology 1994; 12 1338-1342.

[45] Weber. J. M. et al. Handbook of Optical Materials. The CRC Press Laser and Optical Science and Technology Series; 2003.

[46] Mitachi, S. Dispersion measurement on fluoride glasses and fibres. Lightwave Technology Journal 1989; 7(8) 1256-1263.

Permissions

All chapters in this book were first published in AOFTFOPA, by InTech Open; hereby published with permission under the Creative Commons Attribution License or equivalent. Every chapter published in this book has been scrutinized by our experts. Their significance has been extensively debated. The topics covered herein carry significant findings which will fuel the growth of the discipline. They may even be implemented as practical applications or may be referred to as a beginning point for another development.

The contributors of this book come from diverse backgrounds, making this book a truly international effort. This book will bring forth new frontiers with its revolutionizing research information and detailed analysis of the nascent developments around the world.

We would like to thank all the contributing authors for lending their expertise to make the book truly unique. They have played a crucial role in the development of this book. Without their invaluable contributions this book wouldn't have been possible. They have made vital efforts to compile up to date information on the varied aspects of this subject to make this book a valuable addition to the collection of many professionals and students.

This book was conceptualized with the vision of imparting up-to-date information and advanced data in this field. To ensure the same, a matchless editorial board was set up. Every individual on the board went through rigorous rounds of assessment to prove their worth. After which they invested a large part of their time researching and compiling the most relevant data for our readers.

The editorial board has been involved in producing this book since its inception. They have spent rigorous hours researching and exploring the diverse topics which have resulted in the successful publishing of this book. They have passed on their knowledge of decades through this book. To expedite this challenging task, the publisher supported the team at every step. A small team of assistant editors was also appointed to further simplify the editing procedure and attain best results for the readers.

Apart from the editorial board, the designing team has also invested a significant amount of their time in understanding the subject and creating the most relevant covers. They scrutinized every image to scout for the most suitable representation of the subject and create an appropriate cover for the book.

The publishing team has been an ardent support to the editorial, designing and production team. Their endless efforts to recruit the best for this project, has resulted in the accomplishment of this book. They are a veteran in the field of academics and their pool of knowledge is as vast as their experience in printing. Their expertise and guidance has proved useful at every step. Their uncompromising quality standards have made this book an exceptional effort. Their encouragement from time to time has been an inspiration for everyone.

The publisher and the editorial board hope that this book will prove to be a valuable piece of knowledge for researchers, students, practitioners and scholars across the globe.

List of Contributors

Renxian Li, Lixin Guo, Bing Wei, Chunying Ding and Zhensen Wu
School of Physics and Optoelectronic Engineering, Xidian University, China

Mohammad Syuhaimi Ab-Rahman, Hadi Guna and Norhana Arsad
Universiti Kebangsaan Malaysia, Selangor Darul Ehsan, Malaysia

Y.O. Barmenkov and A.V. Kir'yanov
Centro de Investigaciones en Optica, Leon, Guanajuato, Mexico

S. Straullu and S. Abrate
Istituto Superiore Mario Boella, Torino, Italy

R. Gaudino
Politecnico di Torino, Torino, Italy

João M.P. Coelho
Laboratório de Óptica, Lasers e Sistemas, Faculdade de Ciências, Universidade de Lisboa, Pólo do Lumiar, Estrada do Paço do Lumiar, Lisboa, Portugal
Instituto de Biofísica e Engenharia Biomédica, Faculdade de Ciências, Universidade de Lisboa, Campo Grande, Lisboa, Portugal

Catarina Silva, Marta Nespereira, Manuel Abreu and José Rebordão
Laboratório de Óptica, Lasers e Sistemas, Faculdade de Ciências, Universidade de Lisboa, Pólo do Lumiar, Estrada do Paço do Lumiar, Lisboa, Portugal

Weiwen Zou, Xin Long and Jianping Chen
State Key Laboratory of Advanced Optical Communication Systems and Networks, Department of Electronic Engineering, Shanghai Jiao Tong University, Shanghai, China

Liang Dong
Department of Electrical and Computer Engineering, University of Clemson, Clemson, SC, USA

Michal Lucki and Tomas Zeman
Department of Telecommunication Engineering, Faculty of Electrical Engineering, Czech Technical University in Prague, Prague, Czech Republic

Index

www.ingramcontent.com/pod-product-compliance
Lightning Source LLC
LaVergne TN
LVHW080719210926
848485LV00003B/492